MAGNESIUM ALLOYS

Edited by **Mahmood Aliofkhazraei**

Magnesium Alloys
http://dx.doi.org/10.5772/61904
Edited by Mahmood Aliofkhazraei

Contributors

Radovan Kovacevci, Masoud Harooni, Zuzanka Trojanova, Hamouda M Mousa, Chan Hee Park, Cheol Sang Kim, Yu Jae-In, Jerome Frayret, Sebastian Feliu Batlle, Takahiro Ishizaki, Ai Serizawa, Mika Tsunakawa, Tobias Schilling, Michael Bauer, Thomas Hassel, Axel Haverich, Hans Jürgen Maier, Leslie La Londe

Published by InTech

Janeza Trdine 9, 51000 Rijeka, Croatia

Publishing Process Manager Edi Lipovic
Technical Editor SPi Global
Cover InTech Design team

First published February, 2017
Printed in Croatia

Additional hard copies can be obtained from orders@intechopen.com

Magnesium Alloys, Edited by Mahmood Aliofkhazraei
p. cm.
Print ISBN 978-953-51-2897-7
Online ISBN 978-953-51-2898-4

For EU product safety concerns:
IN TECH d.o.o., Prolaz Marije Krucifikse Kozulić 3, 51000 Rijeka, Croatia,
info@intechopen.com or visit our website at intechopen.com.

Meet the editor

Dr. Mahmood Aliofkhazraei works in the corrosion and surface engineering group at the Tarbiat Modares University. He is the head of Aliofkhazraei research group. Dr. Aliofkhazraei has received several honors, including the Khwarizmi International Award and the Best Young Nanotechnologist Award of Iran. He is a member of the National Association of Surface Sciences, Iranian Corrosion Association, and National Elites Foundation of Iran. His research focuses on materials science, nanotechnology and its use in surface, and corrosion science.

Contents

Preface

Magnesium alloys usually have desirable properties including high chemical stability, easy processing and manufacturing, and also lightweight. Magnesium alloys weigh about 70% of aluminum alloy weight and 30% of iron and steel weight. Most of these alloys are used for fabrication of structures in aerospace industries. Magnesium belongs to the second main group of the periodic table of elements (alkaline earth metal) and therefore can't be found in pure state in nature and only exists as a chemical composition. Pure magnesium has an irregular hexagonal crystal structure, and hence it is not suitable for cold forming. However, most of the magnesium alloys show good processing ability, and complex shapes can be easily produced. Magnesium is highly reactive and high shrinkage occurs during the casting of magnesium. High shrinkage and thermal expansion coefficient (10% more than aluminum) limited the applications of magnesium alloys. The mentioned negative properties lead to a rejection of magnesium alloys as a competitive alternative to aluminum or steel. Therefore, attempts to improve features of magnesium have been made by adding different alloying elements.

This book collects new developments about magnesium alloys and their use in different industries. I would like to thank all contributors to this book for their high-quality manuscripts. I wish open-access publishing of this book helps all researchers to benefit from this collection.

Dr. Mahmood Aliofkhazraei
Tarbiat Modares University, Iran

Section 1

Processing

Chapter 1

Laser Welding of Magnesium Alloys: Issues and Remedies

Masoud Harooni and Radovan Kovacevic

Additional information is available at the end of the chapter

http://dx.doi.org/10.5772/66608

Abstract

Automotive industry tends to use lightweight alloys to save on mass in order to have more economic and environment friendly automobiles. A variety of alloys have been used in automotive industry such as magnesium, aluminum, and galvanized steel. Magnesium is the lightest structural metal that can significantly help decrease the body structure weight. Laser welding is one of the main joining processes used in automotive industry due to its superior joint properties. In the current study, the main issue that was pore formation during the laser welding of magnesium alloy is investigated. First, the process was performed using different process parameters to study their effect on the weld quality. Then a variety of approaches were used to mitigate pores in the weld. The results showed that these approaches could effectively mitigate pore formation in the weld bead. In addition, the pore formation issue was nondestructively detected using real-time methods such as spectrometer and high speed charge-coupled device (CCD) camera. The results showed that there was a good correlation between pore formation and the real-time-monitoring detected data.

Keywords: laser welding, magnesium alloy, porosity, spectrometer, online monitoring, numerical modeling

1. Introduction

One of the main concerns for the automotive industry is the CO_2 emission targets requested by governments globally. One reaction to this demand is a renewed focus on mass savings. Lightweight components improve human life quality by two aspects: providing more economic means of transportation and a cleaner environment due to less greenhouse gas generation [1, 2]. A mass reduction of 10% in an automobile will reduce 7% of the fuel consumption [3]. Magnesium is one of the most favorite selections for mass savings since it is the lightest structural metal and has the best strength-to-weight ratio among commercial metals. The density of magnesium is 36% less than aluminum and 78% less than steel. The production of magnesium increased by 390% from 1995 to 2007. This increase demonstrated the growing demand for magnesium in different industries [3]. Repots reveal that automotive industry achieved an average of 25% reduction in fuel consumption between 1990 and 2005 [4].

Magnesium alloy consumption takes different shapes and forms for different industries [1]. To satisfy the production of different shapes to fit industry needs, it is necessary to study the joining processes of this alloy [5]. Different joining processes such as laser welding [1, 6], arc welding [1], hybrid laser-arc welding [1, 7], friction stir welding (FSW) [1, 8], resistance spot welding [1, 9], and electromagnetic pulse welding [1, 10] are used for welding magnesium alloys. Among these processes, laser welding is paid more attention due to its advantages. High power density, low heat input, and consequently a narrow fusion zone and heat-affected zone (HAZ), a high depth-to-width ratio, and suitability for joining complex shapes are some of the advantages of laser welding [1]. Previous authors' studies on laser welding of magnesium alloys are summarized in **Table 1**.

Title	Ref.
Mitigation of pore generation in laser welding of magnesium alloy AZ31B in lap joint configuration	[11]
Studying the effect of laser welding parameters on the quality of Zek100 magnesium alloy sheets in lap joint configuration	[12]
Effect of process parameters on the weld quality in laser welding of AZ31B magnesium alloy in lap joint configuration	[13]
Dual-beam laser welding of AZ31B magnesium alloy in zero-gap lap joint configuration	[14]
Detection of defects in laser welding of AZ31B magnesium alloy in zero-gap lap joint configuration by a real-time spectroscopic analysis	[15]
Pore formation mechanism and its mitigation in laser welding of AZ31B magnesium alloy in lap joint configuration	[16]
Two-pass laser welding of AZ31B magnesium alloy	[17]
Experimental and numerical studies on the issues in laser welding of lightweight alloys in a zero-gap lap joint configuration	[18]

Table 1. Authors' previous studies on laser welding of magnesium alloys.

2. Joining of magnesium alloys

There are a variety of methods that has been used for joining of magnesium alloys. In the following section, a brief description of each method along with its advantages and disadvantages is discussed.

2.1. Adhesive bonding

This method employs adhesive to make a bonding between two parts. It usually can be used for any material and it is suitable for complex shape and size. However, the surface of the alloys should be treated before applying the adhesive. This is one of the disadvantages of this method. The treatment includes degrease, abrasion, and chemical methods [1]. Again, this will take time and needs an extra operation which is not desirable for production industry. After treatment, the adhesive should be applied evenly and sometime small plastic spheres are added to adhesive to assure the consistent gap between two parts.

After applying the adhesive on between two parts, it should be cured. The curing process could be performed in room temperature or by using heat [1]. This might take few minutes to several hours based on the type of the applied adhesive. This could be a time-consuming step that is not advantageous for industry. Overall, this method is desirable because it saves on mass for joining parts and also it does not melt the base metal. On the other hand, it is relatively long process and it is not suitable for mass production joining processes.

2.2. Frictions stir welding

This is a solid-state welding process that does not use heat source to melt the base metal. Instead, joining process is performed using a plastic deformation in the base metals. Generally, this process can be compared with forming manufacturing processes. This process was mainly applied for aluminum alloys at the beginning, but it was used for other alloys later. This method is a solid phase process; therefore, issues that occur during the fusion welding such as solidification cracking, distortion, spatter, and radiation are not a problem during the friction stir welding (FSW) [1]. Due to the same property of this method, the microstructure of the metal after FSW process have fine grains which results in a superior mechanical property. However, this method is not suitable for complex shapes and thin metal sheets that are commonly used in industry. This method also uses a tool which has a limited life and in comparison with laser welding which is noncontact is considered a more expensive process [1].

2.3. Resistance spot welding

This method is one of the main processes used in automotive industry to join metals, but it is not performed for magnesium alloys. This process is at the stage of research mainly but needs more work and efforts to get to production levels for this alloy. This method is considered one of the most economic processes in automotive industry to join sheet metal in body structure. The generated heat is based on the electrical resistance of the metal, and the process should be performed very fast to avoid heat dissipation in the whole part. This rapid local heat generation results in high solidification rate that finally produces dendritic microstructure [1]. Due to low corrosion resistance property in magnesium alloy, mainly a coating layer is used on the surface of the metal sheet. This layer is not conductive and should be removed prior to resistance spot welding [1]. In addition, magnesium is very reactive metal and can form oxide layer on its surface rapidly that will affect its weldability in this method. It should be noted that this method is suitable for complex shapes and geometries. It seems that this method is well developed for aluminum but requires more research on magnesium alloys.

2.4. Laser

Laser is a high intensity heat source that can be used for different processes such as cladding, welding, and heat treatment. Laser welding is one of the most efficient joining processes having high power density which is used to join similar and dissimilar materials. One of the main advantages of laser welding is the relatively high welding speed compared to other fusion welding processes. This helps industries to reduce their mass production cycle time. Heat input is one of the main causes of metallurgical issues in the welding of metals. This parameter shows that to achieve a nominal weld depth, how much heat should be inserted to the workpiece. Since laser has a high intensity (power/beam area at focus), a higher depth could be achieved using a lower heat input compared to other welding methods. As a result, the high depth-to-width ratio of the weld bead shape in laser welding introduces a better weld quality due to the limited size of the HAZ [18].

3. Issues in welding of magnesium and the suggested remedies

3.1. Porosity

Among a variety of issues in welding of magnesium alloys, porosity has a major effect upon the weld joint mechanical properties. There are several reasons for pore formation in the welding of magnesium alloys. Therefore, it is important to understand the pore formation mechanisms and their mitigation procedures during the welding process. Different mechanisms cause pore formation in the laser welding process, including hydrogen pores [1, 19], unstable keyhole [20], preexisting pores, surface coating [1, 15, 16], gas entrapment [1, 21], and alloy elements with a low vaporization point [22].

3.1.1. Hydrogen gas

The only gas that can be dissolved in molten magnesium is hydrogen [6]. Hydrogen pores are formed by the difference in solubility of hydrogen in solid and liquid magnesium. As molten magnesium solidifies, hydrogen solubility sharply decreases and rejects the hydrogen gas resulting in pore formation in the weld bead [1]. Mikucki and Shearouse [19] concluded that the hydrogen content should be kept as low as possible in magnesium in order to prevent pore formation during the solidification process [19].

3.1.2. Preexisting pores

Preexisting pores are another cause of pore formation in the weld bead of magnesium alloys [1, 6]. The initial high gas content in magnesium alloys is one of the reasons for pore formation in the weld bead, especially in alloys that are die-casted [6]. The welding procedure provides the condition for the pores to coalesce and form larger pores [1, 6, 23]. It is noted that a higher welding speed reduces the available time to form and grow pores, resulting in fewer pores in the weld bead [23].

3.1.3. *Oxidation and surface coatings*

Surface coatings could generate pores during the laser welding process. The results showed that the existing oxide layer on the surface of magnesium alloy is the main cause of porosity in laser-welded samples [11, 16]. It was concluded that the pore formation mechanism was due to magnesium hydroxide decomposition during the laser welding process.

3.1.4. *Unstable keyhole*

An unstable keyhole is one of the causes of pore formation in the laser welding process [6]. Keyhole stability depends on properties of the molten metal such as surface tension and vapor pressure [23]. The vapor pressure inside the keyhole forces the keyhole to be open, whereas surface tension tends to close the keyhole [23]. In laser welding of aluminum alloys, the keyhole oscillates and contributes to the pore formation [24]. However, magnesium has a much higher vapor pressure and lower surface tension than aluminum alloys [23] resulting in higher stability of keyhole during the laser welding process.

3.1.5. *Gas entrapment*

Gas entrapment in the molten pool is another reason for pore formation in the weld bead [6, 21]. Pores created by this mechanism are a result of the oscillation of the surface of the molten pool that entraps gases from the air [21]. The oscillation of the surface happens in the laser welding of magnesium because magnesium has an extremely low surface tension and viscosity [6]. In order to prevent gas entrapment in the laser welding of magnesium, it is suggested to use a shielding gas with the appropriate parameters [21].

3.2. Cracking

Cracking is another defect that is common in the welding of a magnesium alloy. Magnesium alloys are susceptible to two types of cracking during the welding process. Solidification cracking is one of them and the crack occurs during the solidification. The other one is liquation cracking which mainly occurs in heat affected zone (HAZ). Hot cracking (solidification cracking) is reported to be the main crack type in magnesium alloys [6]. The coefficient of thermal expansion for magnesium is high (twice that of steel). Therefore, during the welding process, thermal stress results in the deformation of magnesium sheets that finally cause cracking in the weld bead. Alloying elements in the magnesium alloy make the cracking even worse. Magnesium forms intermetallics with alloying constituents such as Mg-Cu, Mg-Al, and Mg-Ni. These intermetallics have a low melting point in comparison to the magnesium melting point. Therefore, these intermetallics remain liquid among the grain boundaries of solidified magnesium resulting in brittle boundaries that are susceptible to cracking [1]. In order to decrease cracking in the weld bead, different solutions such as back plating, adjusting the process parameters and the addition of alloying elements are suggested.

3.3. Oxidation

Magnesium is a very active material that can easily react with oxygen and be detrimental in the welding process. The magnesium oxide forms on the molten pool resulting in its irregular shape and low mechanical properties. Therefore, providing sufficient shielding gas during welding process is essential. In addition, magnesium has a low vaporization point that causes some of its loss during the welding process. The melting point and boiling temperature of magnesium are 630 and 1090°C resulting in a limited temperature range around 460°C. Therefore, it is necessary to employ a low heat input source that is easily controllable. One of the best choices is a laser beam heat source that has low heat input and also can be controlled. Another solution is to substitute the loss of material with the addition of alloying elements or fluxes in order to increase the vaporization point [1].

4. A variety of methods to mitigate porosity

In this section, a variety of methods are performed in order to mitigate pore formation caused by oxide layer existing on the surface of magnesium alloy AZ31B sheets. Magnesium is a very reactive alloy and oxidation is inevitable [1]. The oxide layer may form during the manufacturing process or even when the alloy is exposed to atmosphere. Magnesium has a low corrosion resistance and recently oxide-based coating is used to improve the corrosion resistance properties of this metal. AHC Surface Coating [25], Keronite [26], and Tagnite [27] are examples of coating processes that apply oxide-based coating on the surface of magnesium alloys. To satisfy the wide application of magnesium alloy in different industries, it has been necessary to develop a reliable joining process such as welding. Also, based on what was discussed about the existence of oxide layer on the surface of magnesium alloy, it seems necessary to investigate the effect of oxide layer on the weld quality. So far, there is no report on a cost-effective, efficient, and easy to-apply welding procedure that is capable of joining magnesium alloys sheets without removing the oxide layer prior to welding. In this chapter, a variety of methods are introduced to mitigate pore formation caused by oxide layer. In addition, prior to using these methods to mitigate pore formation, authors studied the effect of laser parameters on the pore formation in another work which is not presented at this work [13].

4.1. Mitigation of pore formation using a plasma arc preheating

For evaluating the influence of the oxide layer on the weld quality, three different surface conditions are introduced: as-received (AR) with the oxide layer remaining on the surface, mechanical removal (MR) of oxides by sand paper, and a heat-treated surface by plasma arc

(PA). In addition, in a separate experiment, a furnace is employed to compare results with the plasma arc preheating process. The results are discussed in detail in the following section.

4.1.1. Experimental setup

The laser welding experiments are performed using a 4-kW fiber laser. The experimental setup including a six-axis robot and a welding laser head is shown by the schematic view in **Figure 1**. As it is shown, a preheating source is used 100 mm prior to laser head. The nominal chemical composition of this alloy is shown in **Table 2**. Pure argon with a flow rate of 60 standard cubic feet per hour (SCFH) is employed to shield the molten pool. In all experiments, the focused laser beam with a beam diameter of 0.6 mm is set up on the top surface of the overlapped sheets to perform the laser welding process.

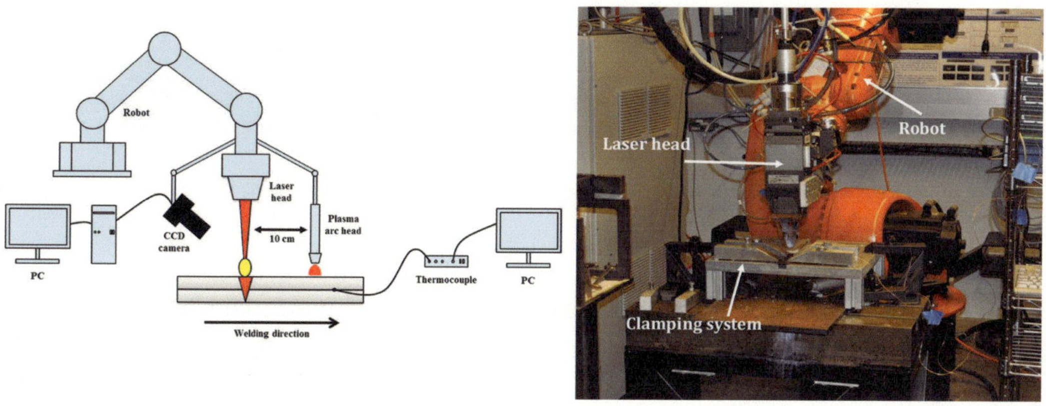

Figure 1. The schematic view of the experimental setup.

Element	Al	Zn	Mn	Si	Fe	Cu	Ni	Mg
Wt.%	2.46	1.70	0.58	<0.1	<0.005	<0.05	<0.005	Bal.

Table 2. The nominal chemical composition of the AZ31B magnesium alloy.

4.1.2. Experimental results

Figure 2 shows the AR samples cross-sectional views. It should be noted that the generated pores in the AR samples are mainly emanating from the faying surface of the two overlapped metal sheets. The presence of pores and also the arrangement of pores in the AR samples imply only that the source of pore generation might be the existing oxide layer on the surface of the AZ31B magnesium alloy sheet. That the oxide layer causes pore formation in the weld is also reported in literature [1, 6, 28, 29], however, without any discussion on how to mitigate its effect.

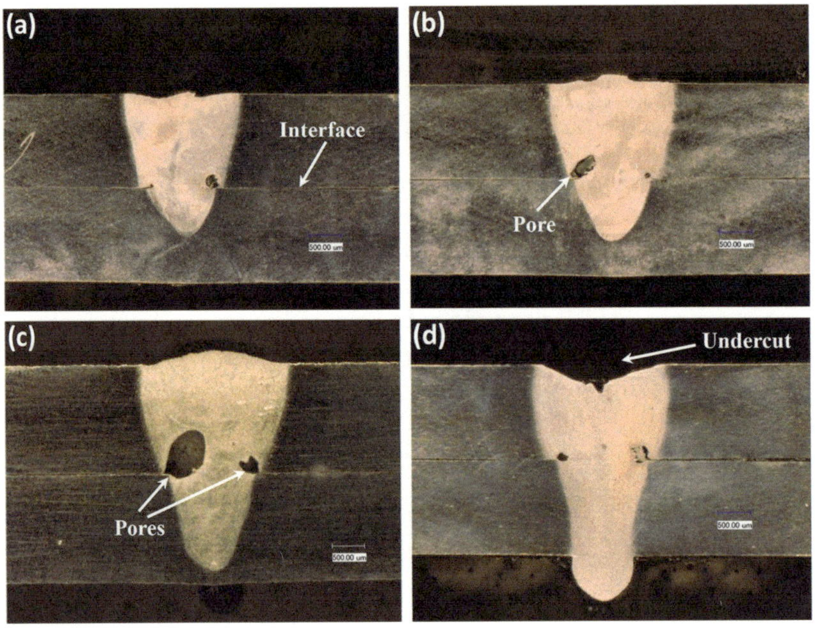

Figure 2. Weld cross-sectional views of AR samples with different laser powers (a) 800 W, (b) 900 W, (c) 1000 W, and (d) 1100 W.

In order to identify the oxide layer on the surface of the AZ31B magnesium alloy, energy-dispersive X-ray spectroscopy (EDS) analysis is applied for the AR and MR samples. The results of EDS are shown in **Figure 3(a)** and **(b)**. Considering these figures, the decrease in oxygen level is clearly shown on the surface of the MR-treated sample. The result denotes that the oxide layer on the AR sample is a dense oxide layer. Magnesium begins to oxidize around 450°C [6], and the oxide layer on AZ31B-H24 is formed by hot rolling process during the manufacturing process [30].

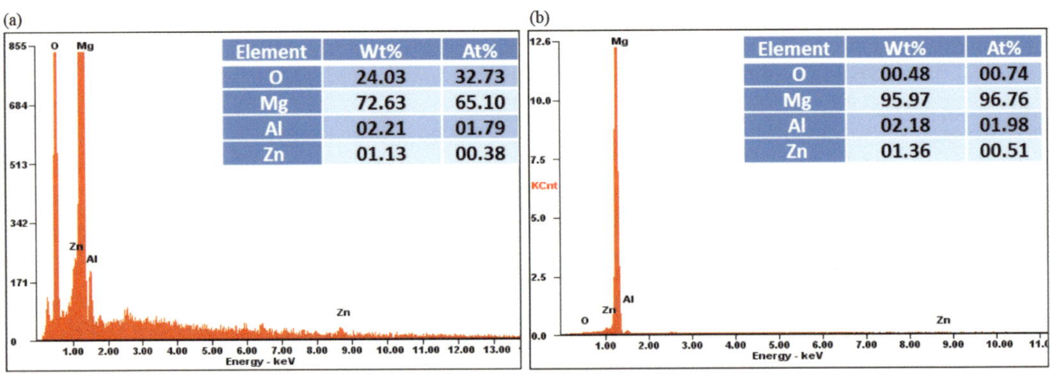

Figure 3. EDS results of (a) as-received and (b) mechanically removed samples.

http://dx.doi.org/10.5772/66608

The properties of oxide formed on metals are divided in two groups including protective and non-protective coatings [1, 31]. This category is based on the Pilling-Bedworth ratio (P-B ratio) R that is defined by the ($R = Wd/wD$) formula. Where W is the molecular weight of oxide, w is the molecular weight of metal, D is the density of oxide, and d is the density of metal [1, 31]. Pilling and Bedworth perceived that for R less than 1, the oxide layer on the metal is porous and non-protective [1]. As an example, aluminum and magnesium oxides are compared. The P-B ratio of Al_2O_3/Al is 1.280 and it is well accepted that aluminum oxide is a protective layer on the surface. On the other hand, the P-B ratio for MgO/Mg is 0.806, less than 1 [1]. Therefore, magnesium oxide is categorized under the porous and a non-protective group that easily forms the non-protective, porous, and thick oxide layer [1]. In order to clarify the P-B ratio effect on the oxide layer formation, a schematic view is presented in **Figure 4** [1].

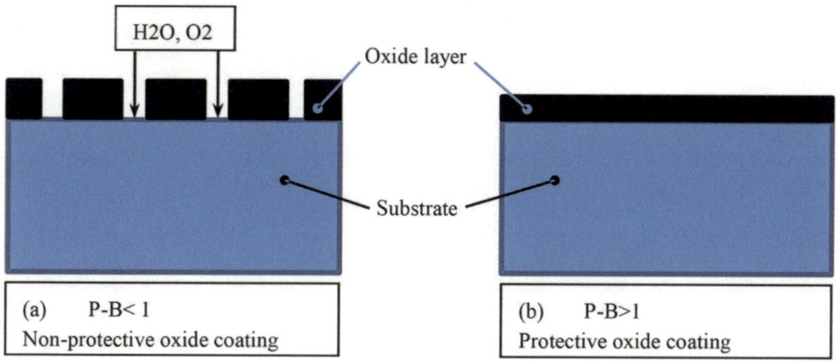

Figure 4. Schematic presentation of surface oxide formation on (a) magnesium alloy and (b) aluminum alloy [1].

The cross-sectional views of MR-treated samples are shown in **Figure 5**. The results show that by removing the oxide layer from the surface of the AZ31B magnesium alloy, pore formation is mitigated at the interface of the two overlapped metal sheets. This result implies that the source of pore formation is related to the oxide layer existing on the surface of the AZ31B magnesium alloy, as assumed.

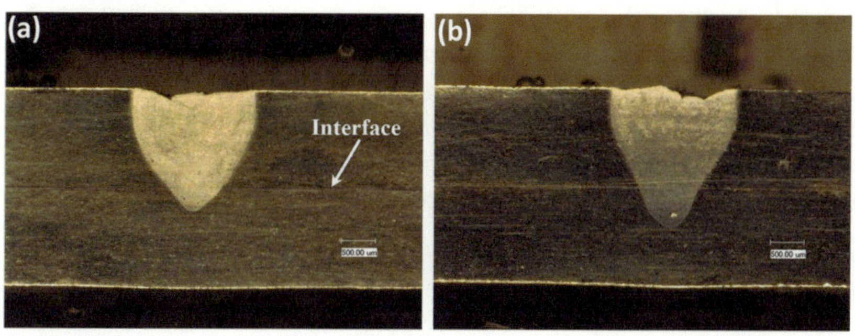

Figure 5. Welds cross-sectional views of MR-treated samples: (a) 1000 W and (b) 1100 W.

PA-treated samples cross-sectional views are illustrated in **Figure 6**. The results show that pores are effectively mitigated at the interface of the two overlapped metal sheets. Preheating the samples results in a change in the oxide layer and indirectly mitigates the pore formation in the weld bead.

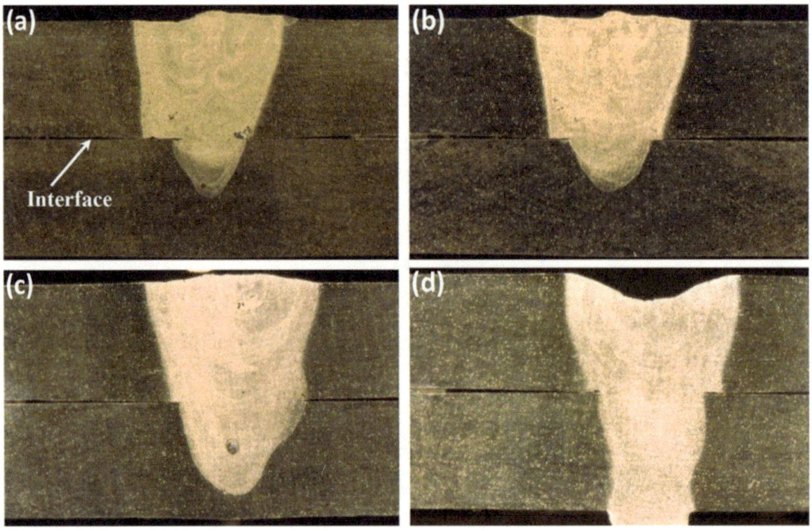

Figure 6. Welds cross-sectional views of PA-treated samples with different laser powers: (a) 800 W, (b) 900 W, (c) 1000 W, and (d) 1100 W.

Magnesium oxide is categorized under porous metal oxides by the P-B ratio and absorbs moisture from the air [6, 29, 32–35] as we discussed earlier in this section. Due to P-B ratio [1], magnesium is categorized under the alloys that form a nonprotective oxide layer on its surface. When magnesium is in atmosphere with moisture, magnesium hydroxide ($Mg(OH)_2$) is formed [6, 36, 37], as shown in Eq. (1). The decomposition temperature of magnesium hydroxide is about 200°C [36]. The water molecules will be released from magnesium hydroxide at this temperature [38]. The released water molecules have no escape path through the overlapped sheets. Instead, the water vapor escapes through the molten pool, resulting in pore formation inside the weld bead in the AR samples. In the PA-treated samples, preheating is performed prior to laser welding in order to decompose the magnesium hydroxide. Water molecules are then vaporized prior to laser welding at the faying surface of the two overlapped sheets. The stable keyhole in the laser welding acts as a chimney to vent the vaporized water to the atmosphere. Therefore, fewer pores are generated in the weld bead.

$$Mg(OH)_2 \rightleftarrows MgO + H_2O \tag{1}$$

To compare the surface chemical composition of AR and PA-treated samples, an XPS test was performed on the surface of both samples. The XPS survey spectra provided qualitative and quantitative surface composition information on samples of the AR- and PA-treated samples.

The primary detected elemental components in both cases are Mg, O, and C. The XPS spectrums of the two cases are shown in **Figure 7**.

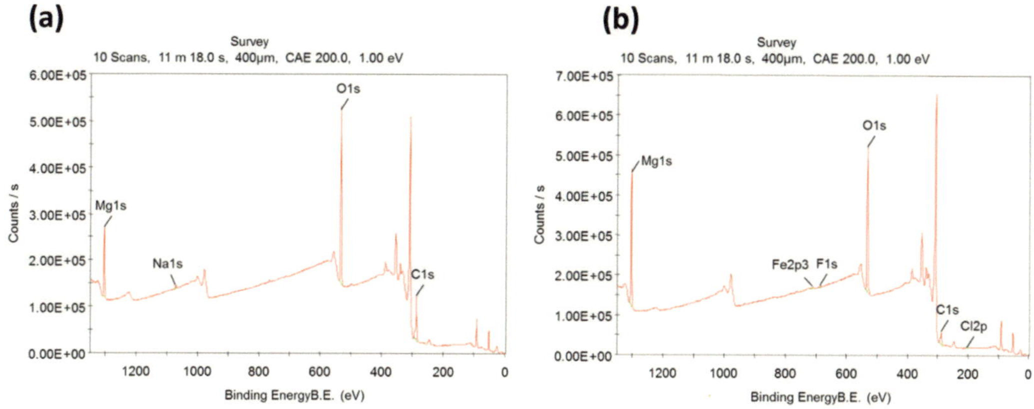

Figure 7. The XPS spectrum of the different cases: (a) AR sample, and (b) PA-treated sample.

The corresponding XPS surface composition results of AR and heat-treated samples are shown in **Table 3**. To quantitatively verify the oxide layer, the O/Mg surface atomic ratios are calculated from the XPS results. The stoichiometric surface composition of the AR sample is characteristic of $Mg(OH)_2$ (AR sample is 1.98, and $Mg(OH)_2$ [theoretical] is 2.00). The stoichiometric surface composition of the heat-treated sample is characteristic of MgO on its surface (PA-treated sample is 0.99, and MgO [theoretical] is 1.00). The above results indicate that the heat treatment converted the original $Mg(OH)_2$ surface layer to MgO on the as-received sample.

Sample	C	O	F	Na	Mg	Fe
AR	29.7	46.3		0.6	23.4	–
Heat-treated	11.5	43.6	0.3	–	44.0	0.3

Table 3. XPS survey spectra results (at.%).

The detection of hydrogen on the surface of AR and heat-treated magnesium alloy is not possible using the XPS or EDS methods due to atomic mass of this element. Reflected electron energy loss spectroscopy (REELS) testing is performed in order to quantitatively detect hydrogen on the surface of the AR and heat-treated samples. The results are shown in **Figure 8**. In **Figure 8**, the main peaks in the REELS spectra for the two samples are normalized to the same maximum intensity. Therefore, the observed differences in the intensity of the shoulders due to hydrogen on the main peaks illustrate true quantitative differences in the amount of surface hydrogen on the two samples. REELS confirmed the presence of hydrogen, presumably as $Mg(OH)_2$, on the surface of both samples. REELS indicated that the as-received sample had more hydrogen present compared to the heated sample, which is consistent with the XPS results, indicating the presence of primarily $Mg(OH)_2$ on the AR sample and $Mg(OH)_2$ plus MgO on the heated sample. This is discussed in more detail in authors' previous work [16].

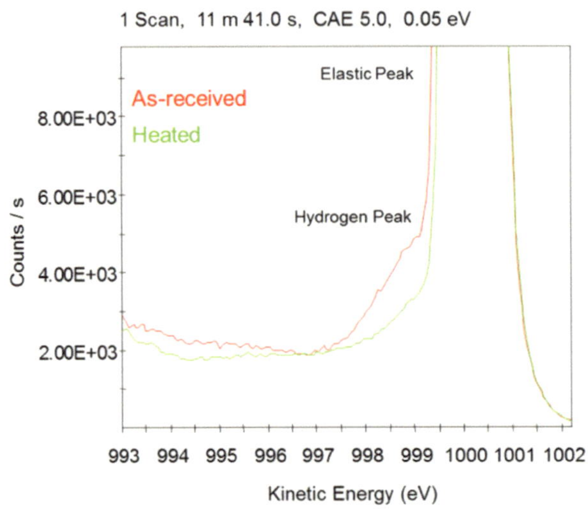

Figure 8. The comparison of REELS spectra for AR sample and heat-treated sample.

To verify if the preheating process provide sufficient temperature and width, a 3D model was developed. A heat source model for the plasma arc is the surface heat flux with a Gaussian distribution [39] and the material properties considered temperature dependent [18]. **Figure 9** shows the 3D view of the temperature field obtained by the numerical model in the plasma arc preheating procedure. It is good to check the isotherm of 200°C (minimum temperature for decomposition of magnesium hydroxide) for the two overlapped sheets. It is shown that the isotherm of 200°C reaches the interface of the two overlapped sheets.

Figure 10 shows the temperature history at the faying surface of the two overlapped sheets. The isotherm of 200°C shows that a width of 3.4 mm (1.7 mm each side) was preheated at the interface. Based on the experimental results achieved in Chapter 2, the maximum width of the weld pool at the faying surface of the two overlapped sheets in laser welding is 1.75 mm. This confirms that the preheated area reaching the temperature of 200°C could cover almost two times the width that is required.

The corresponding tensile test results are presented in **Figure 11** showing that the average tensile-shear loads of PA-treated samples are greater than those of the AR- and MR-treated samples. This is resulted from a lower level of pores in PA-treated samples. The weld bead tensile-shear load improves by 19% with PA-treated samples.

http://dx.doi.org/10.5772/66608

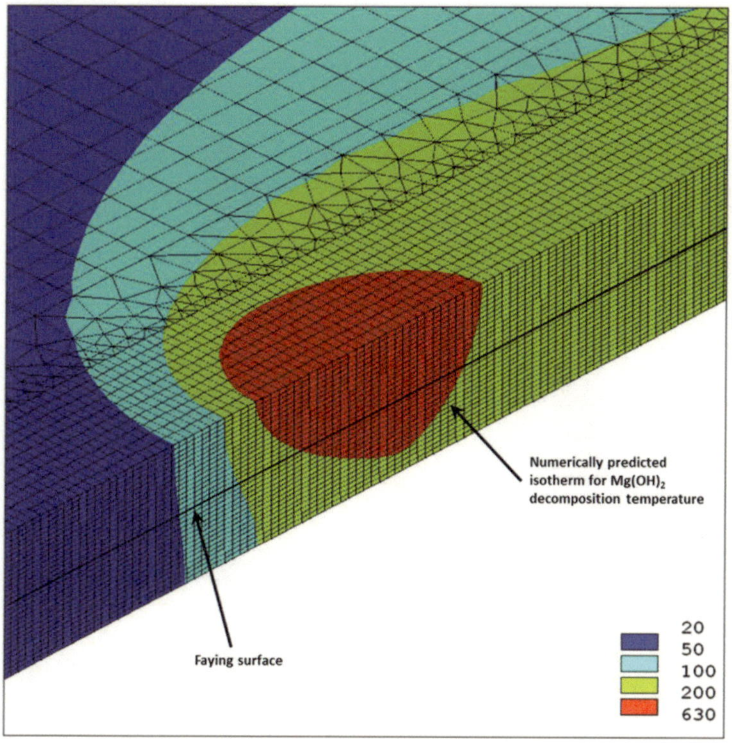

Figure 9. 3D view of temperature history in two overlapped sheets with a plasma arc preheating process with scanning speed of 30 mm/s.

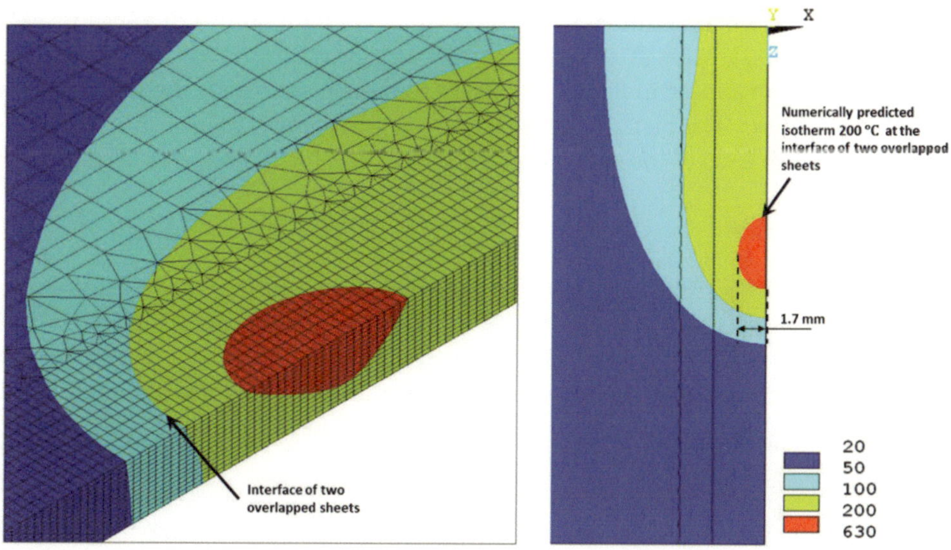

Figure 10. The temperature history of the preheating process at the interface of two overlapped sheets.

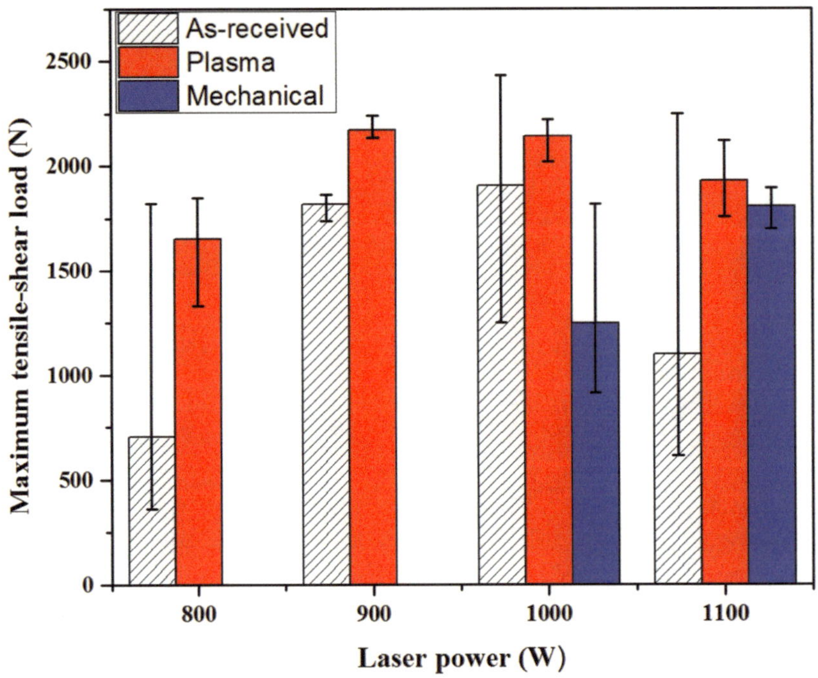

Figure 11. Tensile test results of laser welding AZ31B magnesium alloy.

4.2. Dual-beam laser welding of magnesium alloy

4.2.1. Experimental setup

In order to join AZ31B magnesium sheets in a lap joint configuration, a dual-beam laser was used to weld overlap samples. The single laser beam was changed to a dual-beam laser by using an optical prism as seen in **Figure 12**. An optical prism can divide the single laser beam into two separate spots. The distance between the two split beams was 0.6 mm, and the ratio of the laser power for each of the spots could be adjusted.

4.2.2. Experimental results

Figure 13 presents the profiles of the weld bead obtained during the laser welding process with two beam configurations. It is shown that the surface quality of the weld bead with dual-beam laser is smoother than the top surface of the weld bead obtained by a single laser beam. The improvement of surface quality could be described by a nondimensional hydrodynamic Froude number as a ratio of dynamic force to gravitational force expressed by Eq. (2) [41]:

$$F_r = \frac{V^2}{lg} \tag{2}$$

where V is welding speed, g is gravitational acceleration, and l is the characteristic length of the molten pool. Lower values of the Froude number corresponds to the improved quality of the surface [41]. By having two in-line laser spots in dual-beam laser welding, the length of the molten pool increases, and as a result, the Froude number decreases; that is, the top surface of the weld bead improves. An increase in the welding speed increases the Froude number resulting in a reduction of quality of the top surface of the weld bead, as confirmed by the authors in their earlier work [12].

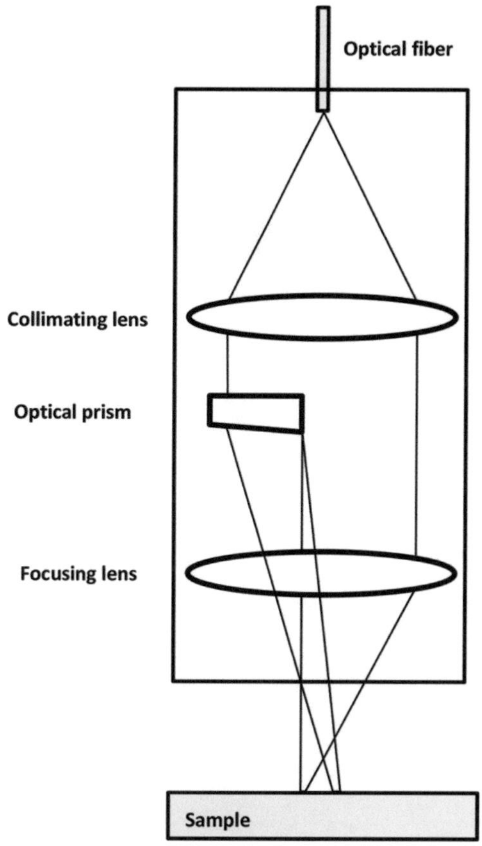

Figure 12. Schematic view of a dual-beam fiber laser head [40].

Four different beam energy ratios were chosen in combination with two laser power and welding speed settings and the results compared to those from a single beam in order to evaluate the effect of the split beam configuration. The baseline cross-sections from a single-beam exhibited pores emanating from the faying interface for both combinations of laser power and welding speed, refer to **Figure 14(i)–(j)**. Introducing the split beam at relatively high ratios of lead to lag beams for process conditions of 900 W at 30 mm/s and 1000 W at 50 mm/s also exhibited pore formation emanating from the faying interface in the weld cross-section; refer to **Figure 14(a)(c)** and **(e)(g)**, respectively. Not until the ratio was reduced to 20:80 was the pore formation at the faying interface of the weld cross-section effectively mitigated; refer to **Figure 14(d)** and **(h)**, for process conditions of 900 W at 30 mm/s and 1000 W at 50 mm/s, respectively. Thus, a split beam ratio of 20:80 for the leading beam power to lagging beam power appears to be an optimum for mitigating pore formation at the faying interface.

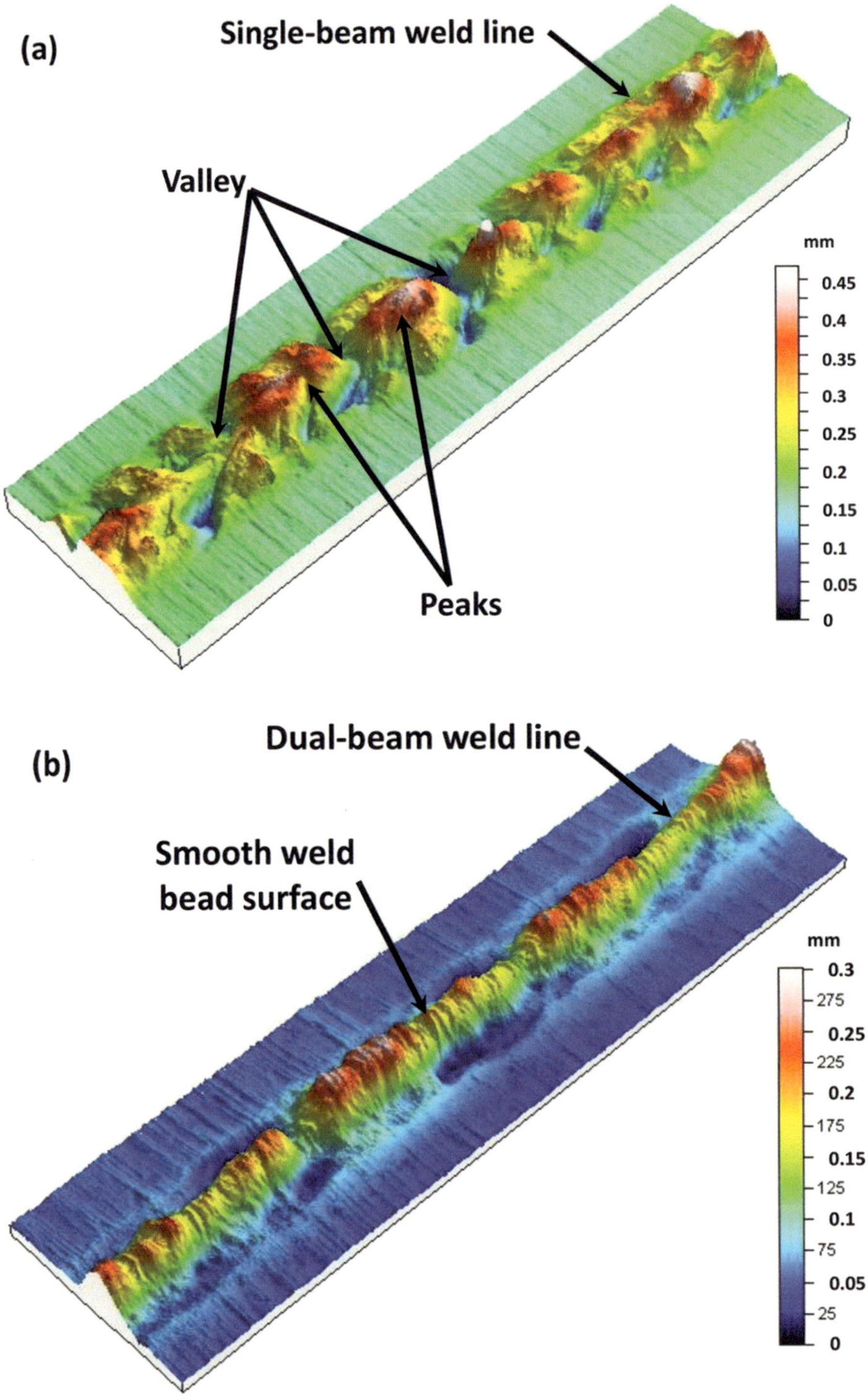

Figure 13. The profiles of the weld bead achieved under a welding speed of 30 mm/s and a laser power of 1000 W with (a) single-beam laser, and (b) dual-beam laser with a 20:80 split beam energy ratio [14].

http://dx.doi.org/10.5772/66608

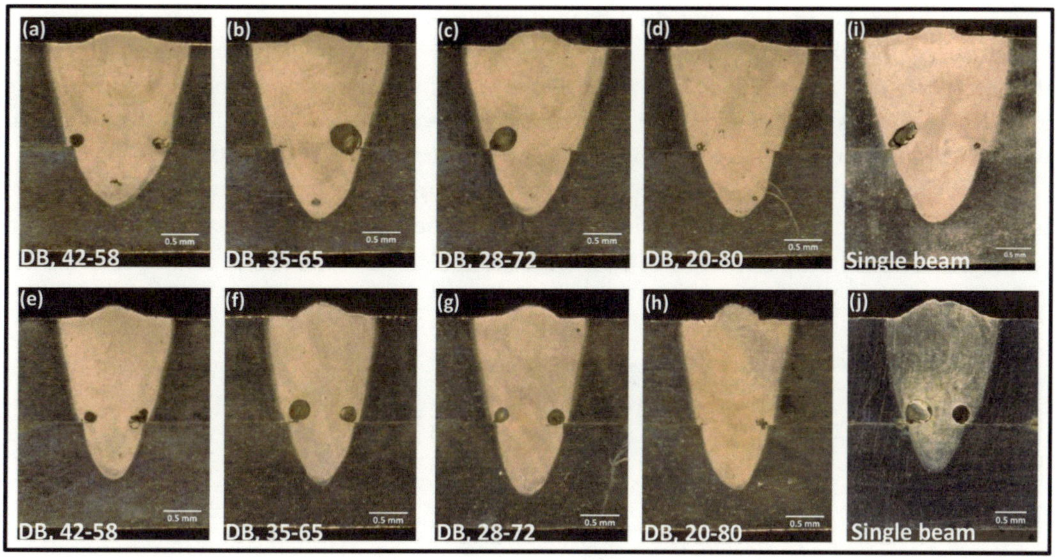

Figure 14. Cross-sectional views of the welds with a different beam ratios and laser process parameters (a)–(d) and (i) P= 900 W, V= 30 mm/s, (e)–(h) and (j) P = 1000 W, V = 50 mm/s.

As shown in **Figure 14**, pores are mitigated by a dual-beam laser for only an optimized beam ratio which indicates that there may be another mechanism involved in the mitigation of pore formation besides increasing of the molten pool size. As discussed previously, when the lead beam contains 20% of the laser power for preheating, the remaining 80% of laser power is used for welding and is sufficient to form a stable keyhole. The stable keyhole then serves as a chimney to vent the hydrogen gas from the decomposed magnesium hydroxide as shown in **Figure 15**. For the other beam ratios, the lead beam provides more than enough energy to preheat the sample; however, the lag beam is not able to form a stable keyhole thus increasing the probability of the hydrogen gas to be entrapped in the solidified weld pool. Thus, two mechanisms are necessary in order to achieve a high-quality weld. The first is the preheating procedure to decompose the magnesium hydroxide accomplished by the lead beam prior to laser welding. The second is the formation of a stable keyhole to vent the hydrogen gas during laser welding by the lagging beam. With a beam ratio of 20:80, the preheating procedure and formation of a stable keyhole provide an optimal distribution of heat resulting in a high-quality weld. **Figure 15** shows the schematic view illustrating the mechanism that mitigates pore formation in the weld bead during dual-beam laser welding.

In order to study the dynamic behavior of the weld pool, a high-speed CCD camera assisted with a green laser as an illumination source is used. **Figure 16** is composed of weld pool images captured by a CCD camera during the laser welding of AZ31B magnesium alloy with a laser power of 1000 W and welding speed of 30 mm/s with both single-beam and dual-beam laser configurations. No keyhole formation is observed during the single-beam laser welding process resulting in an unstable welding process refer to **Figure 16(a)** and **(b)**. In contrast, the images of the dual-beam laser weld pools exhibit a keyhole refer to the shiny dot at the center of the molten pool in **Figure 16(c)** and **(d)**. Although there was some instability in the formation of the keyhole during dual-beam laser welding, the instability was significantly less than for the case of single-beam laser welding. When the single-beam is used, the magnesium hydroxide decomposes and releases the water vapor into the molten pool resulting in collapse of the keyhole, turbulent flow in the molten pool, and porosity in the solidified weld. However, in the case of applying the dual-beam laser welding, the lead beam preheats the interface of the two overlapped sheets resulting in decomposition of the magnesium hydroxide. The lag beam with sufficient laser power then generates a stable keyhole, shown in **Figure 16(c)** and **(d)**, leading to mitigation of porosity.

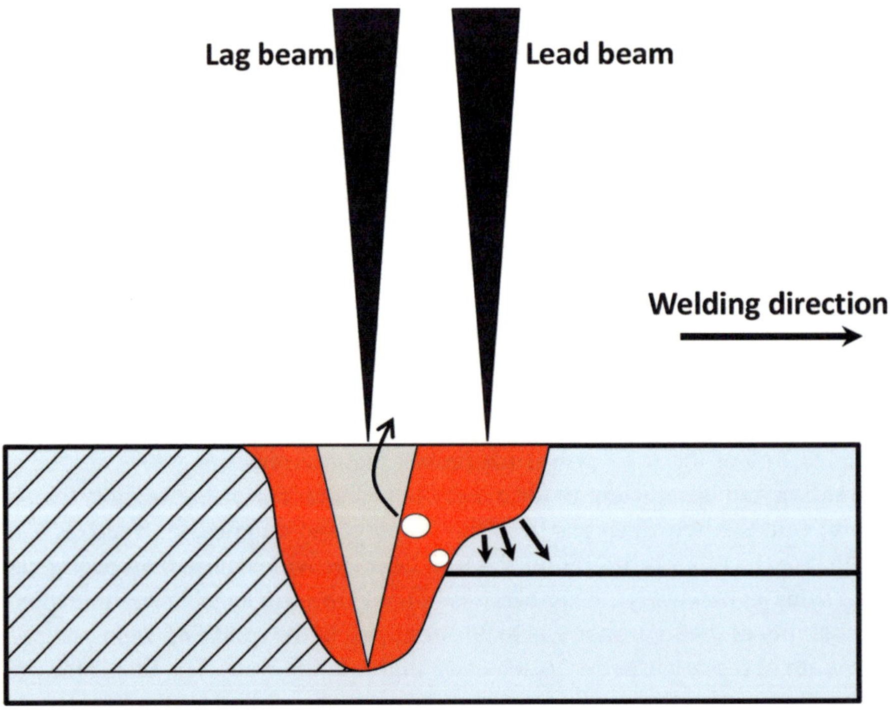

Figure 15. The pore formation mitigation mechanism during the dual-beam laser welding process.

Figure 17 presents the images of the laser-induced plasma plume captured during the laser welding of an AZ31B magnesium alloy with single-beam and dual-beam laser configurations. It is found that the plasma plume is more stable during the dual-beam laser welding process. **Figure 17(a)** presents the images of the plasma plume in the single-beam laser welding process. It is shown that the plasma plume is not stable and fluctuates during the laser welding process caused by water vapor at the faying surface of the two overlapped sheets. The plasma plume was extinguished during the welding process caused by the collapse of the unstable keyhole. The images of the plasma plume captured during the dual-beam laser welding (**Figure 17(b)**) show that the plasma plume exhibited greater stability than that obtained by single-beam laser welding. This is also reported by Xie [42].

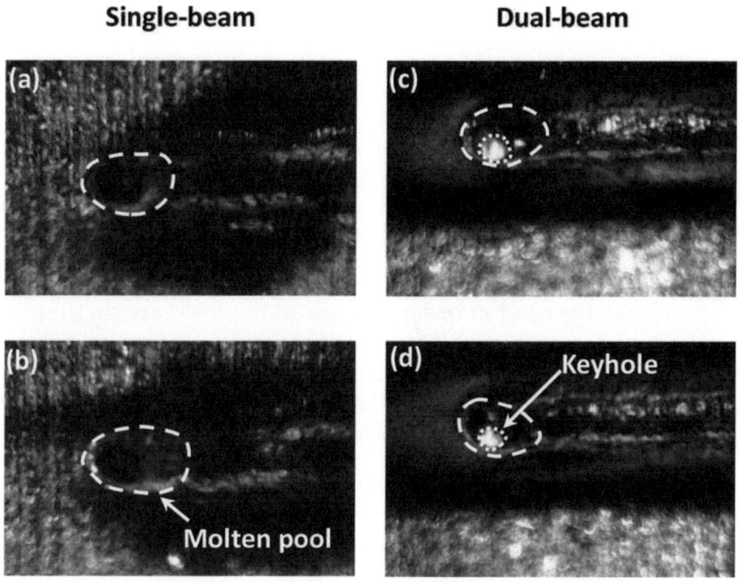

Figure 16. Images of the weld pool obtained with CCD camera with the laser power of 1000 W and welding speed of 30 mm/s (a), (b) single-beam, and (c), (d) dual-beam (lead/lag beam power ratio 20:80).

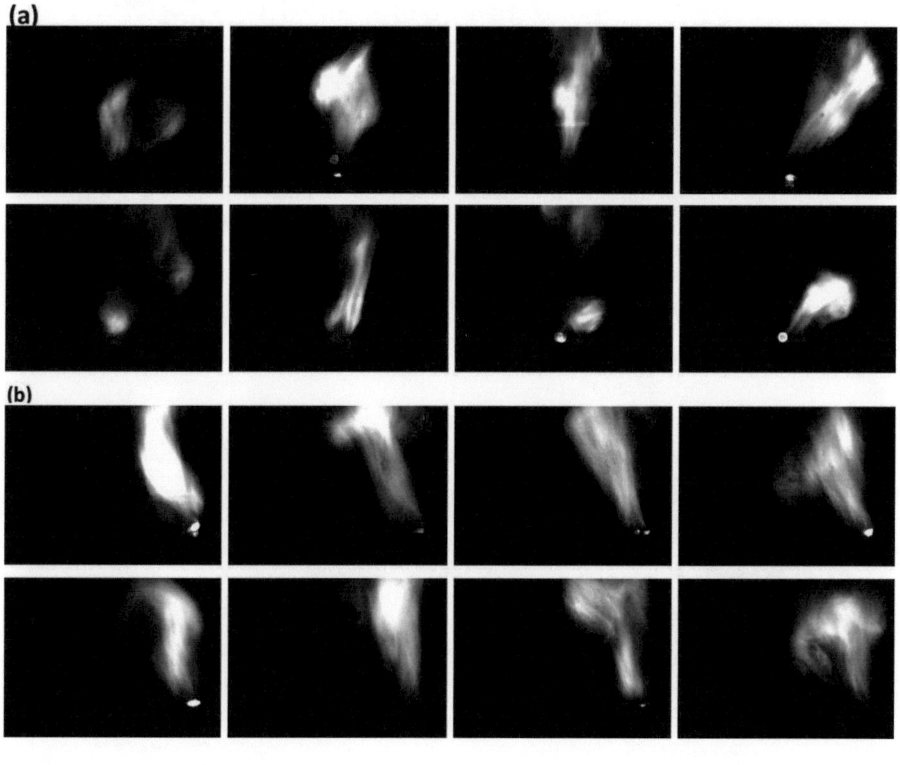

Figure 17. Images of laser-induced plasma plume captured during laser welding with laser power of 1000 W and welding speed of 30 mm/s with (a) single-beam laser, and (b) dual-beam laser with a 20:80 split beam energy ratio.

4.3. Nondestructive detection of defects using a spectrometer

4.3.1. Experimental setup

To observe the presence of pores, the weld bead had to be destructively sectioned in last two sections. In this part, spectroscopy is proposed as a nondestructive means to detect thepresence of pores in a lap joint configuration of a laser-welded AZ31B magnesium alloy as shown in **Figure 18**. The calculated electron temperature by the Boltzmann plot method was correlated to the presence of pores in the weld bead. Utilizing BeamView^{TM} software, a CCD camera with a rate of 15 fps was also used in order to capture the intensity distribution of the laser-induced plasma. This device captures the intensity distribution in the laser and light sources.

In order to ensure that all conditions were the same for each run, the oxide layer was partially removed along the welding path, as schematically presented in **Figure 19**.

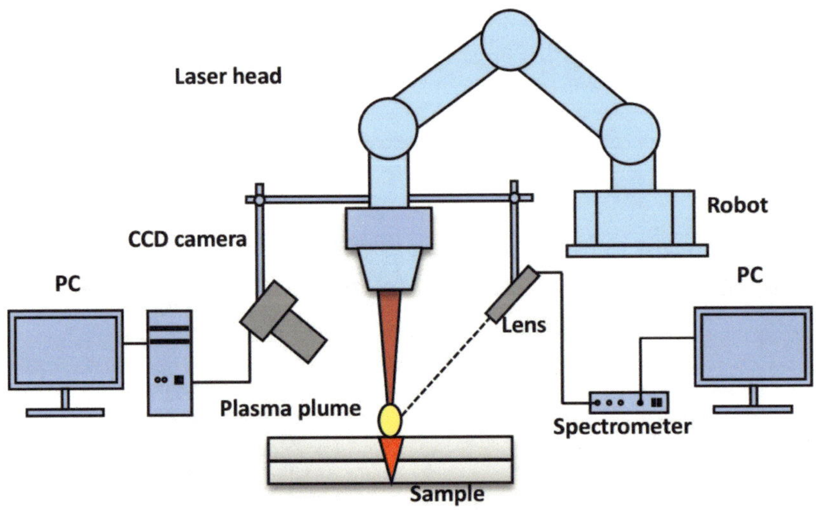

Figure 18. Schematic view of the laser welding process [15].

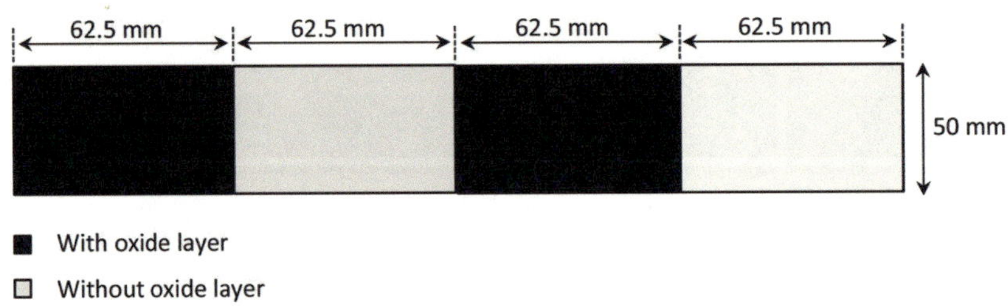

Figure 19. Schematic top view of partially removing the oxide layer on AZ31B magnesium alloy sheets.

In the laser welding process, a collision of free electrons in the plume region produces plasma that contains useful information, as discussed earlier. The plasma electron temperature can be calculated by the Boltzmann plot method. This method includes some important assumptions that should be considered prior to obtaining the final results. It is assumed that the laser-induced plasma with a homogeneous distribution should be optically thin and in local thermal equilibrium (LTE) for laser welding process [43]. By this assumption, the distribution of energy in the plasma region follows Maxwell's equation. Also the collision of particles is dominant to the radiation process [43–45]. The LTE assumption can be satisfied when the population of electrons in the unit volume is sufficiently high [43],

$$N_e \geq 1.6 \times 10^{12} T_e^{\frac{1}{2}} (\Delta E)^3 \ (\text{cm}^{-3}) \tag{3}$$

where N_e is the electron density, T_e is electron temperature, and ΔE is the largest energy gap in the atomic energy level system. The integrated spectrum of emission-line intensities is given as [43],

$$\frac{N_n}{N_m} = \frac{g_n}{g_m} exp(-\frac{E_n - E_m}{kT}) \tag{4}$$

where g is the degeneracy, k is the Boltzmann constant, and E is the emitted energy. The electron temperature can be calculated for a chosen pair of emission lines by the following equation [43]:

$$T = \frac{E_{m_2} - E_{m_1}}{kLn(\frac{\omega_{m_2 n_2} A_{n_2 m_2} g_{m_2} R}{\omega_{m_1 n_1} A_{n_1 m_1} g_{m_1}})} \tag{5}$$

where R is the ratio of two populations.

By knowing the emission-line parameters, Eq. (5) can be applied to calculate the electron temperature of the laser-induced plasma. This method is advantageous because the calculations are relatively straightforward and can be easily implemented for the laser welding process. The upper energy levels for the selected emission lines, E_{m1} and, E_{m2} has to be satisfied by the following expression, $E_{m1} - E_{m2} > KT$ [47] (see **Table 4**).

Selected element	Wavelength (nm)	Energy of the upper level (cm⁻¹)	Statistical weight	Transition probability (S⁻¹)
Mg I	383.83	47957.027	5	4.03e7
Mg I	517.26	41197.403	3	3.37e7

Table 4. Spectroscopic information of selected elements [46].

4.3.2. Experimental results

The oxide layer is partially removed by a mechanical procedure prior to laser welding, as seen in **Figure 19**. As shown in this figure, one part is without an oxide layer followed by the part with the as-received oxide layer intact. **Figure 20** shows the top, bottom, and cross-sectional views of the weld bead achieved with laser power of 1200 W under two conditions: with and without an oxide layer.

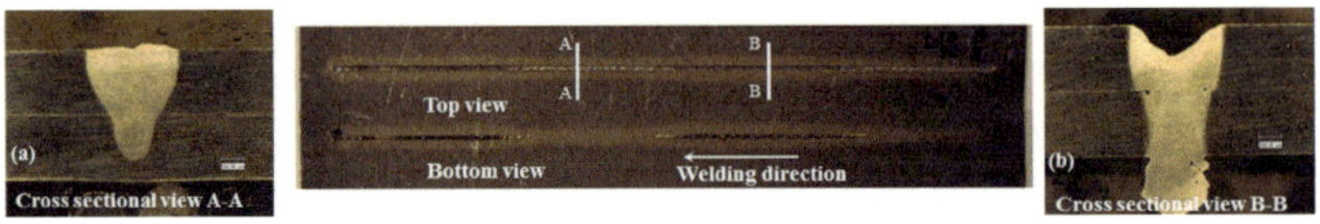

Figure 20. Top, bottom, and cross-sectional views of the weld with laser power of 1200 W for two different conditions at the faying surface of overlapped sheets (a) without oxide layer and (b) with oxide layer.

As clearly shown in **Figure 20**, there are discernable changes on weld bead, top and bottom view between two cases (with and without oxide layer). It is also clear that the presence of the oxide layer at the faying surface results in higher laser energy absorption that is discussed in detail in the work published by author [15] as well as Section 2.4.2. of this chapter.

The spectrum intensity of laser-induced plasma during the laser welding process is shown in **Figure 21**. Comparing the two cases regarding the presence of the oxide layer at the faying surface, it can be observed that the spectrum intensity is slightly less in the case with the oxide layer. As discussed earlier, the presence of the oxide layer at the interface had a dominant effect on the penetration depth. By considering the fact that the laser beam energy input to the metal sheets is constant, having greater penetration or a larger bead width means that more energy was absorbed by the metal and less was available to form a laser plume on top of the weld bead. The spectrum intensity for other laser parameters with more details could be found in the work that was done by authors [15].

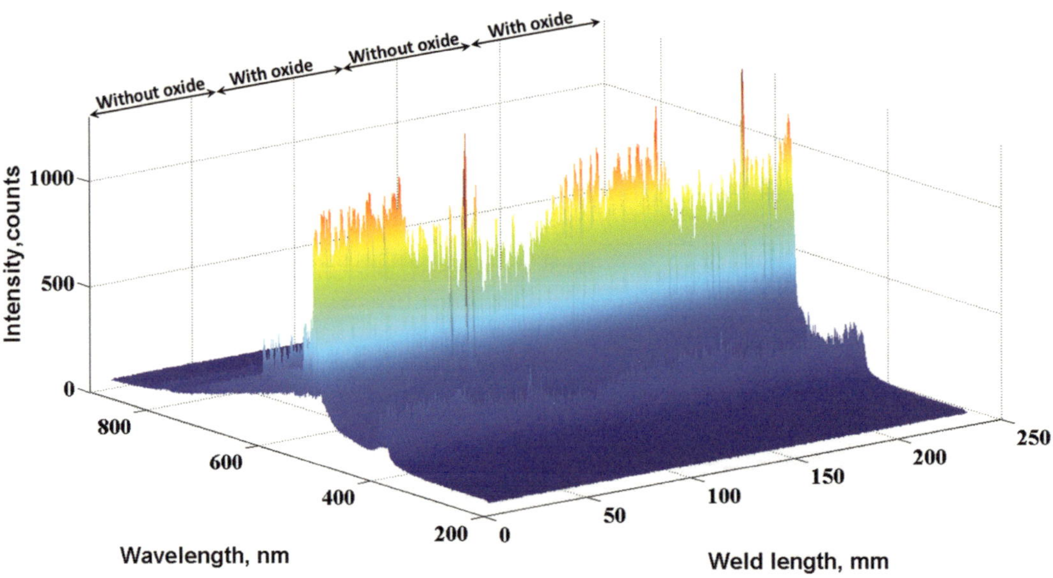

Figure 21. The spectrum intensity of laser induced plasma with laser power of 1200 W in the laser welding process.

In order to verify the intensity of the laser-induced plasma plume for the two surface conditions, with and without an oxide layer, a laser beam CCD device was used to view and capture images of the plasma plume in real time. This device is capable of capturing the intensity distribution images from almost any light or laser source. **Figure 22** shows the images captured during the laser welding of AZ31B magnesium alloy. **Figure 22(a)** and **(b)** presents the images of the plasma plume captured in the section without an oxide layer at the interface. **Figure 22(c)** and **(d)** shows the captured images of the plasma plume in the section with an oxide layer at the faying surface. As shown in this figure, the intensity is higher in the section without an oxide layer. This result confirms the recorded spectrum intensity results presented in **Figure 21**. The captured plasma plume images also revealed that the plasma plume is narrow and tall for the without-oxide section, although it is thicker and shorter for the section with an oxide layer. This result confirms that the laser-induced plasma plume was more stable during the welding of the section without an oxide layer.

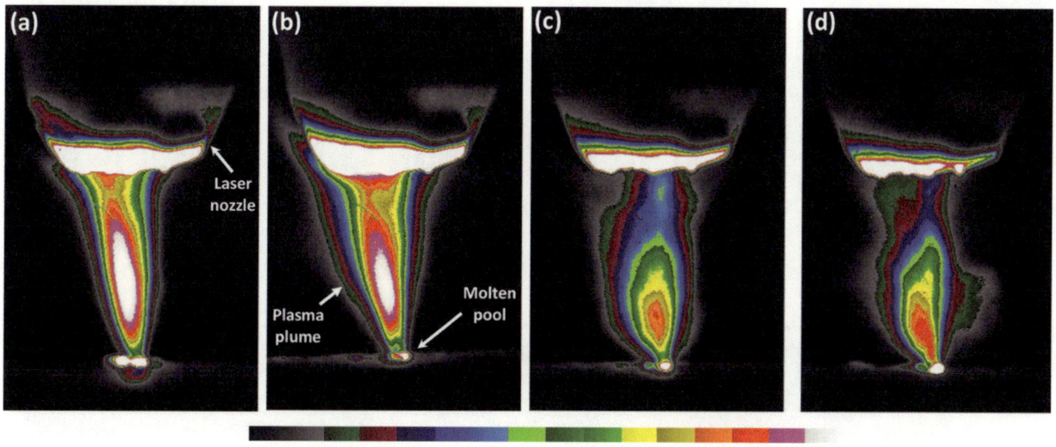

Figure 22. Captured images of laser induced plasma plume in real-time during laser welding with two surface conditions at the faying surface (a), and (b) without oxide layer, and (c), and (d) with oxide layer.

The calculated electron temperature from the spectrum intensity presented in **Figure 21** is shown in **Figure 23**. The electron temperature results exhibits varying levels and amounts of fluctuation along the weld bead. When the oxide layer is removed from the faying surface, the electron temperature shows a lower and uniform value with low fluctuation. The electron temperature shows a higher value with high fluctuation with an oxide layer at the faying surface. With an oxide layer at the faying surface, the keyhole is not stable due to the release of water molecules caused by magnesium hydroxide decomposition. This phenomenon was observed by a high-speed CCD camera, shown earlier in **Figure 3** of the work published previously by the authors [15]. When the keyhole is not stable, the formation of a plume and plasma at the top of the molten pool is not stable. In addition, the presence of spatters in the case with an oxide layer at the faying surface disturbs the plasma plume resulting in a decrease in the spectrum intensity (see **Figure 21**). These spatters increase the magnesium concentration in the plasma plume region causing a higher value for the electron temperature. This shows that basically there is a good correlation between pore formation in the weld bead and the calculated electron temperature.

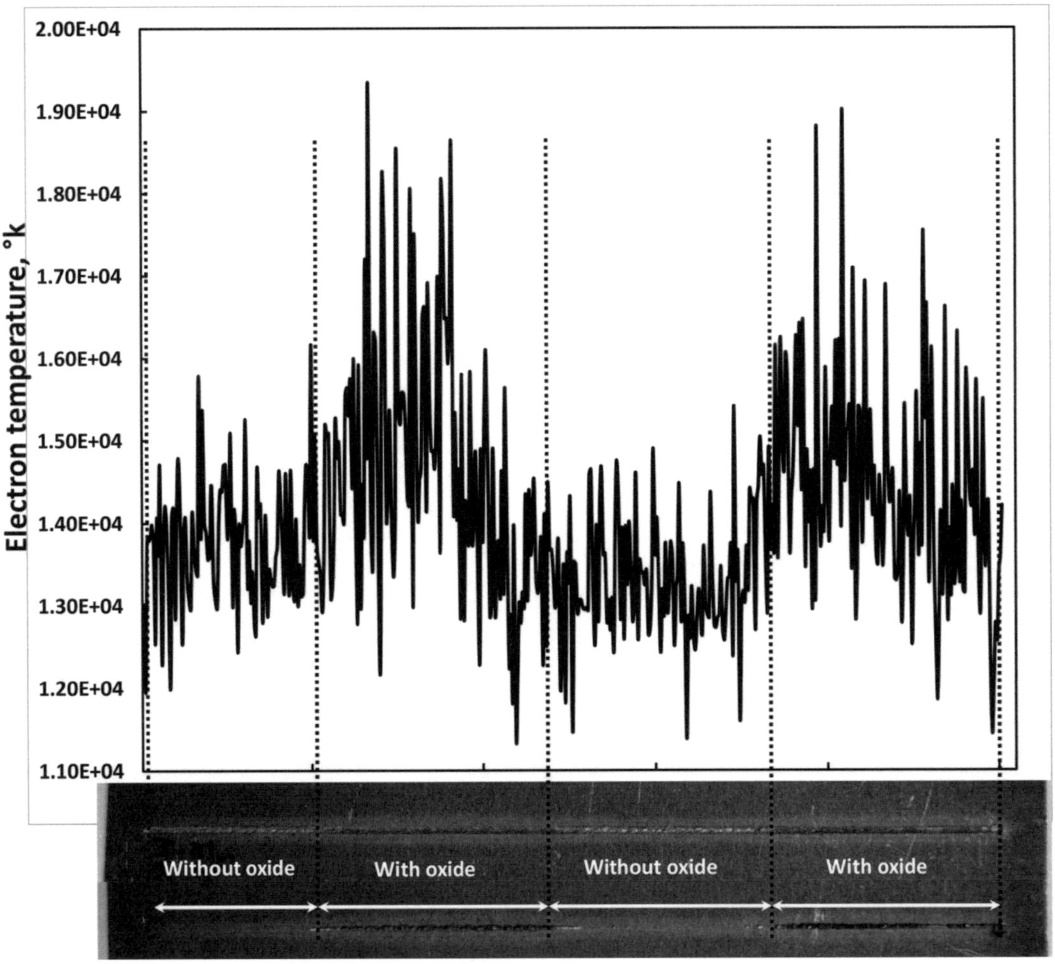

Figure 23. The calculated electron temperature for laser induced plasma with laser power of 1200 W.

4.4. Two-pass laser welding of magnesium alloy

From the earlier discussions, it is understood that welding as-received AZ31B magnesium alloys with a single-beam configuration results in pore formation in the weld bead (see **Figure 2**). In this section, a new method is introduced to use laser for preheating and welding. A defocused laser preheats the surface and in second pass, the focused laser beam welds two magnesium alloy sheets.

4.4.1. Experimental setup

The schematic view of two-pass experimental setup is shown in **Figure 24**. It can be seen that a larger defocused laser beam is applied to perform preheating before laser welding process. After that, a smaller focused beam is used to weld magnesium sheets (**Figure 24(b)**). The time between the first and second pass was about 10 s. **Table 5** shows the process parameters used in preheating and welding processes. The preheating parameter effect was evaluated on the weld quality. The laser welding power and speed were set at 1000 W and 30 mm/s which were used from authors' previous work [16].

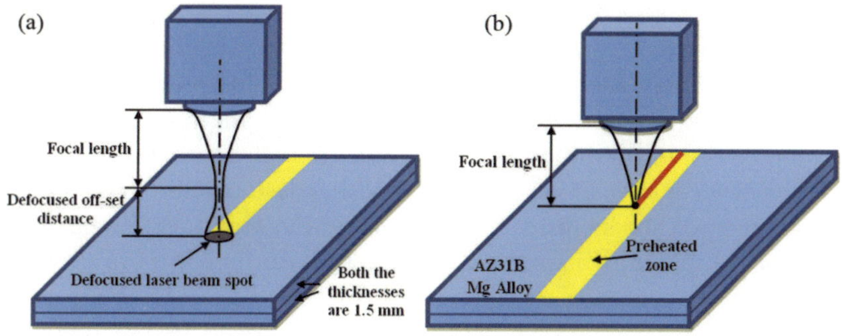

Figure 24. Two-pass laser welding process schematic view (a) preheating by a large defocused beam and (b) welding by a small focused beam.

	Preheating power (W)	Preheating speed (mm/s)	Focal distance (mm)	Laser welding power (W)	Laser welding speed (mm/s)
1	–	–	0	1000	30
2	2000	100	+40	1000	30
3	3000	100	+40	1000	30
4	2000	100	+35	1000	30
5	3000	100	+35	1000	30

Table 5. The preheating process parameters.

A spectrometer was employed also to study the laser induced plasma plume during the laser welding process as it was discussed in detail in Section. 2.3. The detected spectra was then used to calculate electron temperatures for the two different processes, two-pass laser welding (TPLW) and one-pass laser welding (OPLW). The electron temperature was used to compare the stability of the laser welding process for two experimental conditions. The spectrometer wavelength resolution was 0.4 nm, its integration time was 3 ms, and the slit width was 50 μm. The selected elements used as well as all equations for electron temperature calculation are as same as the one in Section 4.3.

4.4.2. Experimental results

Figure 25 presents the cross-sectional views of the welds. The laser power and welding speed (for either single pass or the second pass of a two-pass process) were the same for all cases, refer to **Table 5**. From observation of **Figure 25(a–d)**, it is clear that using an optimum preheating parameter set, a good quality of weld without pores can be achieved. These welds are achieved using a two-pass laser welding process with different power levels and focal distances. We had to make sure that the faying surface of the two overlapped sheets preheated effectively since this was the source of pore formation. As it can be seen in **Figure 25(c)**, pores are formed even though the preheating process was performed. This shows that to get a good weld quality, selecting optimum preheating parameters is critical. The viscosity of the molten magnesium is significantly low [6]. Therefore, when the weld bead was fully penetrated, the molten material dripped out at the bottom of two overlapped sheets caused pore formation. It can be observed from **Figures 25(a–d)** that when a higher preheating power is used, a shallower weld penetration was achieved. This can be discussed due to the difference between laser energy absorption on the surface magnesium alloy. The area of preheating between two overlapped sheets (upper sheet) is shown in **Figure 26**. The preheated region is lighter in color compared to its surrounding. Based on discussion in previous sections, the lighter area and non-preheated area (darker area) were assumed to be magnesium oxide (MgO)and magnesium hydroxide ($Mg(OH)_2$), respectively. Based on previous researches [48] magnesium hydroxide has a higher absorption coefficient in comparison to magnesium oxide [49]. Therefore, using a higher preheating laser power resulted in a greater rate of magnesium hydroxide decomposition at the faying surface. Then, this caused a lower level of power absorption leading to a less weld bead penetration. As a result, **Figure 25(a–d)** preheating with higher laser power resulted in less absorption in the second pass and a shallower penetration.

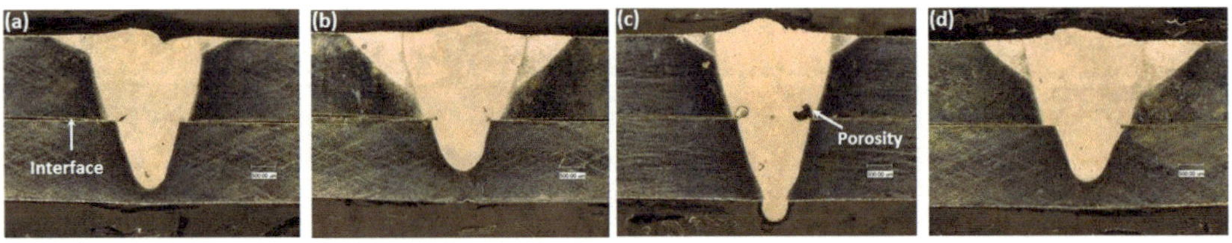

Figure 25. Polished cross-sectional views of magnesium sheet lap-shear laser welds. (a)–(d) Two-pass process with 1000 W and 30 mm/s welding parameters and preheat parameters of 100 mm/s and (a) 2000 W and 40 mm focal distance, (b) 3000 W and 40 mm focal distance, (c) 2000 W and 35 mm focal distance, and (d) 3000 W and 35 mm focal distance.

Figure 26. The preheated area with a lighter color on the upper sheet.

The top view of the two-pass samples acquired with different focal distances is displayed in **Figure 27**. The blowholes were seen in the sample that used less focal distance. The preheating process should not melt the substrate; however, when +20 mm focal distance was used, the energy density was high enough to melt substrate. As mentioned earlier, the viscosity and surface tension of molten magnesium is significantly low, which makes magnesium susceptible to weld defects like humping, or blowholes [6]. At a focal distance of +20 mm, due to higher level of preheating power, the mentioned properties of magnesium lead to have a low surface quality as shown in **Figure 27(b)**.

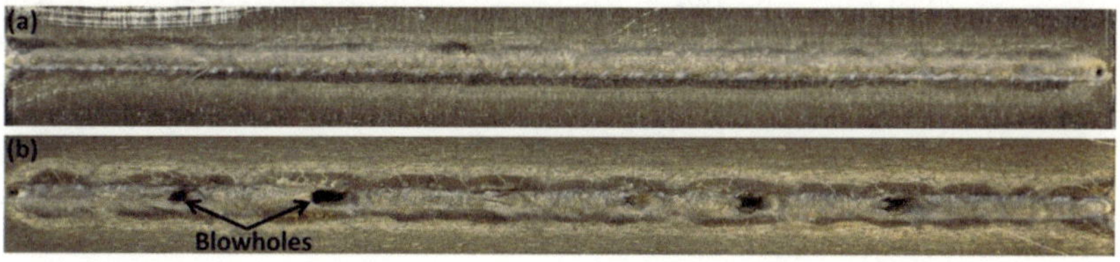

Figure 27. Comparison of focal distance: (a) +35 mm, and (b) +20 mm, and the effect it has upon the upper weld surface quality under the same preheating condition (3000 W, and 100 mm/s), laser power is 1000 W and welding speed is 30 mm/s.

The spectrum intensity from plasma plume single pass and two-pass welding processes is shown in **Figure 28**. Higher plasma plume intensity is achieved during the two-pass laser welding process (see **Figure 28(b)**). This can be due to a more stable plasma plume which can be resulted from a steadier welding condition which results in a higher weld quality. This stability was higher in two-pass compared to single pass. As discussed earlier, the stability is the result of optimum preheating which resulted in decomposition of magnesium hydroxide. Magnesium oxide (MgO) has a lower beam absorption that leads to higher plasma plume intensity since a greater amount of energy was available to form the plasma plume.

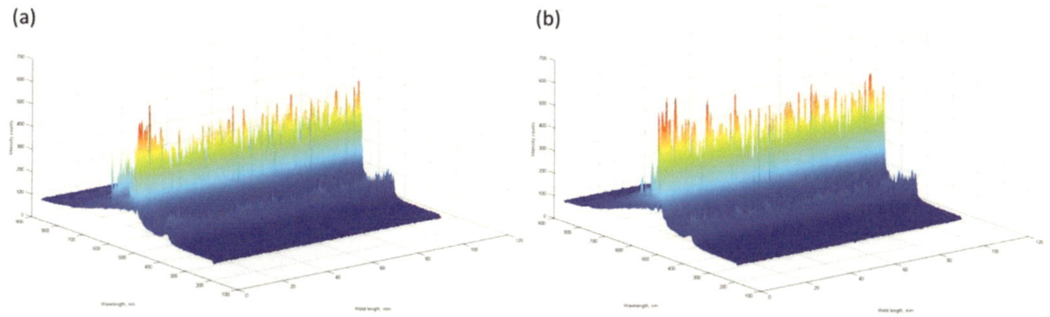

Figure 28. Laser-induced plasma intensity spectrum captured during laser welding under two conditions: (a) single pass and (b) two-pass laser welding.

The electron temperature was calculated for both the single pass and two-pass laser weld spectrum intensities as shown in **Figure 29**. The single pass laser welding process was less stable as expected and discussed in **Figure 25** which resulted in pore formation in the weld. The calculated electron temperature showed higher values when the single pass laser welding was performed. During the single pass laser welding process, the plasma plume intensity was affected by the release of hydrogen into the molten pool caused an unstable welding condition. This unstable condition advanced to the spatter formation as well as molten metal ejection from the molten pool. As also reported by other researchers [50], this can lead to an elevated level of concentration of molten metal in the plasma region leading to an increase in electron temperature. On the other hand, the electron temperature calculated from the two-pass laser welding process was lower. This reveals that the preheating was effective in releasing hydrogen gas prior to laser welding. This was resulted in a better stability in the weld. Less spatter formation and a more stable condition of the weld pool were both contributed to a lower electron temperature values in the two-pass laser welding process.

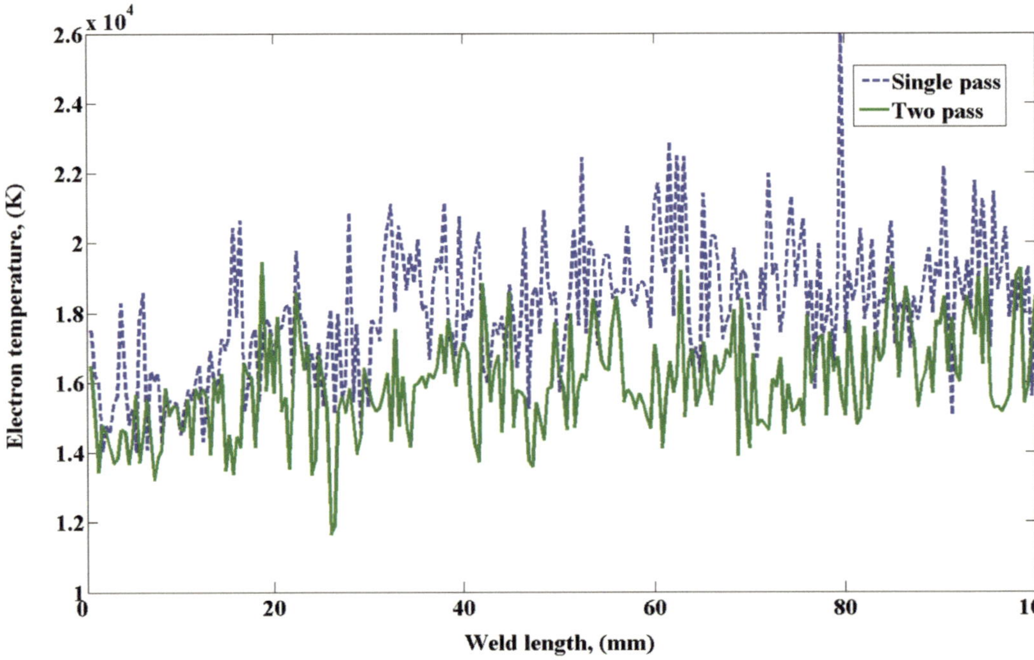

Figure 29. Calculated electron temperature during the laser welding process with different conditions: single pass and two pass.

To find out about the mechanical properties of the welds, tensile strength testing was performed on both single pass and two-pass laser welded sample groups. Five samples were cut for each group. The experimental results of lap-shear strength test are shown in **Figure 30**. The tensile load of was significantly greater for two-pass samples compared to the single pass laser welded samples. This shows that during the two-pass laser welding process, less pores are formed. Also, the greater variation in the as well as lower strength values both reveals that the single pass weld had less stable condition which was caused by release of hydrogen in molten pool as discussed earlier. Therefore, by performing an optimum preheating process during the two-pass welding process, pore formation could be effectively mitigated. More details about this study were published previously in earlier authors' work [17].

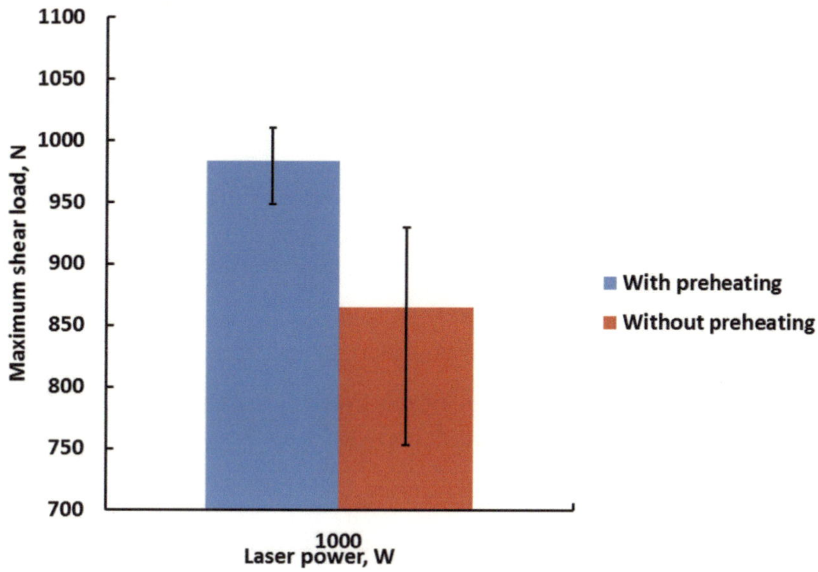

Figure 30. Average tensile shear load of laser welded AZ31 sheet in an overlap configuration with and without preheating.

5. Conclusions

(1) The as-received oxide layer on the surface of magnesium alloy causes pore formation in the weld bead, resulting from decomposition of the magnesium hydroxide at the faying surface of the two overlapped sheets. Pore formation could be mitigated effectively by introducing a plasma arc preheating source to decompose magnesium hydroxide prior to laser welding.

(2) Dual-beam laser welding with a 20:80 split beam energy ratio provides an improved weld bead quality with respect to porosity.

(3) The top surface quality of the weld bead obtained with dual-beam laser welding is smoother and more uniform than the quality obtained from single-beam laser welding as a result of less humping.

(4) The results of on-line monitoring the molten pool and laser-induced plasma plume revealed that the dual-beam laser welding is relatively more stable than single-beam laser welding.

(5) There was a good correlation between the calculated electron temperature detected from the laser-induced plasma and the presence of laser weld bead defects during the laser welding process.

(6) Two-pass laser welding mitigated pore formation in the weld bead. The calculated electron temperature is lower for two-pass laser welding compared to single pass laser welding indicating that the two-pass process is more stable.

Acknowledgements

This work was financially supported by the NSF's Grant No. IIP-1034652.

Author details

Masoud Harooni[1, 2, 3] and Radovan Kovacevic[1*]

*Address all correspondence to: kovacevi@smu.edu

1 Mechanical Engineering Department, Research Center for Advanced Manufacturing, Lyle School of Engineering, Southern Methodist University, Dallas, TX, USA

2 Keihin North America (Honda OEM), Greenfield, IN, USA

3 Center for Additive Manufacturing Research at IUPUI (CAMRI), Indiana University – Purdue University Indianapolis, Indianapolis, IN, USA

References

[1] Liu, L., Welding and joining of magnesium alloys, Woodhead Publishing, Cambridge, UK, 2010.

[2] Chowdhury, S.M., Chen, D.L., Bhole, S.D., Powidajko, E., Weckman, D.C., Zho, Y., (2011). microstructure and mechanical properties of fiber-laser-welded and diode-laser-welded AZ31 magnesium alloy. Metallurgical and Materials Transactions A 42: 1974–1989.

[3] Shrestha, S., (2010). Magnesium and surface engineering. Surface Engineering 26: 313–316.

[4] Blawert, C., Hort, N. and Kainer, K.U., (2004). The science, technology, and application of magnesium. Transactions of Indian Institute of Metals, Orlando, FL, USA, 57: 397–408.

[5] Yu, Z.H., Yan, H.G., Gong, X.S., Quan, Y.J., Chen, J.H., Chen, Q., (2009). Microstructure and mechanical properties of laser welded wrought ZK21 magnesium alloy. Materials Science and Engineering A 523: 220–225.

[6] Cao, X., Jahazi, M., Immarigeon, J.P., Wallace, W., (2006). A review of laser welding techniques for magnesium alloys. Journal of Materials Processing Technology 171: 188–204.

[7] Gao, M., Tang, H.G., Chen, X.F., and Zeng, X.Y., (2012). High power fiber laser arc hybrid welding of AZ31B magnesium alloy. Materials and Design 42: 46–54.

[8] Cao, X., and Jahazi, M., (2009). Effect of welding speed on the quality of friction stir welded butt joints of a magnesium alloy. Materials and Design 30(6): 2033–2042.

[9] Shi, H., Qiu, R., Zhu, J., Zhang, K., Yu, H., and Ding, G., (2010). Effects of welding parameters on the characteristics of magnesium alloy joint welded by resistance spot welding with cover plates. Materials and Design 31: 4853–4857.

[10] Berlin, A., Nguyen, T.C., Worswick, M.J., Zhou, Y., (2011). Metallurgical analysis of magnetic pulse welds of AZ31 magnesium alloy. Science and Technology of Welding and Joining 16: 728–734.

[11] Harooni, M., Kong, F., Carlson, B., Kovacevic, R., (2012). Mitigation of pore generation in laser welding of magnesium alloy AZ31B in lap joint configuration. ASME International Mechanical Engineering Congress & Exposition, Houston, TX.

[12] Harooni, M., Kong, F., Carlson, B., and Kovacevic, R., (2012). Studying the effect of laser welding parameters on the quality of ZEK100 magnesium alloy sheets in lap joint configuration. International Congress on Applications of Lasers & Electro-Optics (ICALEO), Anaheim, CA.

[13] Harooni, M., Carlson, B., and Kovacevic, R., (2013). Effect of process parameters on the weld quality in laser welding of AZ31B magnesium alloy in lap joint configuration, International Congress on Applications of Lasers & Electro-Optics (ICALEO), Miami, FL.

[14] Harooni, M., Blair, C., Kovacevic, R., (2014). Dual-beam laser welding of AZ31B magnesium alloy in zero-gap lap joint configuration. Optics and Laser Technology 56: 247–255.

[15] Harooni, M., Carlson, B., and Kovacevic, R., (2014). Detection of defects in laser welding of AZ31B magnesium alloy in zero-gap lap joint configuration by a real-time spectroscopic analysis. Optics and Lasers in Engineering 56, 54–66.

[16] Harooni, M., Carlson, B., and Kovacevic, R., (2014). Pore formation mechanism and its mitigation in laser welding of AZ31B magnesium alloy in lap joint configuration. Materials and Design 58, 265–276.

[17] Harooni, M., Ma, J., Carlson, B., and Kovacevic, R., (2015). Two-pass laser welding of AZ31B magnesium alloy. Journal of Materials Processing Technology 216, 114–122.

[18] Harooni, M., Experimental and numerical studies on the issues in laser welding of light-weight alloys in a zero-gap lap joint configuration. PhD thesis, Southern Methodist University, Dallas, TX, USA, 2014.

[19] Mikucki, B.A., and Shearouse, J.D., (1993). Interdependence of hydrogen and microporosity in magnesium alloy AZ91. Proceedings of Magnesium Properties and Applications for Automobile, Detroit, MI.

[20] Katayama, S., Seto, N., Kim, J.D., Matsunawa, A., (1998). Formation mechanism and suppression procedure of porosity in high power laser welding of aluminum alloys. Laser Institute of America, ICALEO, Orlando, FL.

[21] Liu, L., Song, G., Liang, G., and Wang, J., (2005). Pore formation during hybrid laser-tungsten inert gas arc welding of magnesium alloy AZ31B—mechanism and remedy. Materials Science and Engineering: A 390, 76–80.

[22] Pastor, M., Zhao, H., Martukanitz, R.P., Debroy, T., (1999). Porosity, underfill and magnesium loss during continuous wave Nd:YAG laser welding of thin plates of aluminum alloys 5182 and 5754. Welding Journal 78: 207s–216s.

[23] Zhao, H. and Debroy, T., (2001). Pore formation during laser beam welding of die cast magnesium alloy AM60B—mechanism and remedy. Welding Journal 80: 204s–210s.

[24] Matsunawa, A., Kim, J.D., Mizutani, M., Katayama, S., (1998). Dynamics of keyhole and molten pool in laser welding. Journal of Laser Applications 10: 247–254.

[25] AHC Surface Coating [Online]. Available: https://www.ahc-surface.com/en/surface-treatment/processes/magoxid-coat. [Accessed 27 June 2016].

[26] Keronite [Online]. Available: http://www.keronite.com/. [Accessed 27 June 2016].

[27] Tagnite [Online]. Available: http://www.tagnite.com/tagnite_coating/. [Accessed 27 June 2016].

[28] Zhu, J., Li, L., and Liu, Zh., (2005). CO_2 and diode laser welding of AZ31 magnesium alloy. Applied Surface Science 247: 300–306.

[29] Marya, M. and Edwards, G.R., (2000). The laser welding of magnesium alloy AZ91. Welding in the World 44: 31–37.

[30] Powidajko, E.J.M., Weldability of AZ31B magnesium sheets by laser welding processes. PhD thesis, University of Waterloo, Waterloo, Canada, 2009.

[31] Pilling, N.B., and Bedworth, R.E., (1923). The oxidation of metals at high temperatures. Journal of Institute of Metals 29: 529–591.

[32] Wegrzyna, J., Mazurb, M., Szymańskib A., and Balcerowskab B., (1987). Development of a filler for welding magnesium alloy GA8. Welding International 1: 146–150.

[33] Liu, L.M., Ren, D.X., (2010). Effect of adhesive on molten pool structure and penetration in laser weld bonding of magnesium alloy. Optics and Lasers in Engineering 48: 882–887.

[34] Otsuki, S., Fukuzawa, S., (2004). Organic electroluminescent device and method for fabricating same. U.S. patent 6737176.

[35] Wakelin, S., (2007). System and method for getting gas-phase contaminants within a sealed enclosure. U.S. patent 7160368.

[36] Fournier, V., Marcus, P., and Olefjord, I., (2002). Oxidation of magnesium. Surface and Interface Analysis 34: 494–497.

[37] Kuroda, Y., Yasugi, E., Aoi, H., Miura, K., Morimoto, T., (1988). Interaction of water with the surface of magnesium oxide. Journal of the Chemical Society, Faraday Transactions: Physical Chemistry in Condensed Phases 84: 2421–2430.

[38] Halikia, I., Neou-Syngouna P., and Kolitsa D., (1998). Isothermal kinetic analysis of the thermal decomposition of magnesium hydroxide using thermogravimetric data. Thermochimica Acta 320: 75–88.

[39] Pavelic V., Tanbakuchi, R., Uyehara, O.A., Myers, P.S., (1969). Experimental and computed temperature histories in gas tungsten arc welding of thin plates. Welding Journal Research Supplement 48: 295s–305s.

[40] Haboudou, A., Peyre, P., Vannes, A., and Peix, G., (2003). Reduction of porosity content generated during Nd:YAG laser welding of A356 and AA5083 aluminium alloys. Materials Science and Engineering A 363, 40–52.

[41] Ready, J.F., (2001). LIA handbook of laser materials processing. Laser Institute of America Orlando, FL, USA.

[42] Xie, J., (2002). Dual beam laser welding. Welding Journal 81: 223s–230s.

[43] Griem, H.R., (1997). Principles of plasma spectroscopy. Cambridge University Press, Cambridge, UK.

[44] Liu, L. and Hao, X., (2008). Study of the effect of low-power pulse laser on arc plasma and magnesium alloy target in hybrid welding by spectral diagnosis technique. Journal of Physics D: Applied Physics, 41, 1–10.

[45] Rizzi, D., Sibillano, T., Calabresa, P.P., Ancona, A. and Lugara, P.M., (2011). Spectroscopic, energetic and metallographic investigations of the laser lap welding of AISI 304 using the response surface methodology. Optics and Lasers in Engineering 49: 892–898.

[46] NIST Atomic Spectra Database [Online]. Available: http://www.nist.gov/pml/data/asd.cfm. [Accessed 01 February 2013].

[47] Lacroix, D., Jeandel, G., and Boudot, C., (1997). Spectroscopic characterization of laser induced plasma created during welding with a pulsed Nd: YAG laser. Journal of Physics D: Applied Physics, 81: 6599–6606.

[48] Ren, D., Liu, L., and Li, Y., (2012). Investigation on overlap joining of AZ61 magnesium alloy: laser welding, adhesive bonding, and laser weld bonding. International Journal of Advanced Manufacturing Technology 61(1–4): pp. 195–204.

[49] Magnesium oxide (MgO), Crystran [Online]. Available: http://www.crystran.co.uk/optical-materials/magnesium-oxide-mgo. [Accessed 5 March 2014].

[50] Kong, F., Ma, J., Carlson, B., Kovacevic, R., (2012). Real-time monitoring of laser welding of galvanized high strength steel in lap joint configuration. Optics and Laser Technology 44: 2186–2196.

Chapter 2

Internal Friction in Magnesium Alloys and Magnesium Alloys-Based Composites

Zuzanka Trojanová, Peter Palček, Pavel Lukáč and Mária Chalupová

Additional information is available at the end of the chapter

http://dx.doi.org/10.5772/67028

Abstract

In practice, some problems connected with undesirable mechanical vibrations or interruption of acoustic bridges may be solved using high damping materials. Especially, transport industry needs high damping light materials with proper mechanical properties. Magnesium alloys and magnesium alloys-based metal matrix composites may be considered as materials exhibiting such behaviour. Damping of mechanical vibrations and their conversion to the heat (internal friction) is conditioned by the movement and redistribution of various defects in the crystal lattice. Generally, internal friction depends on the material microstructure and conversely changes in the material microstructure may be studied using the internal friction measurements. The strain amplitude-dependent internal friction was investigated at room temperature in commercially available Mg alloys and Mg alloys-based composites with the aim to identify changes in the microstructure invoked by thermal and mechanical loading. The temperature-dependent internal friction indicated the following effects: (a) mechanisms connected with dislocations and grain boundaries in the microcrystalline pure Mg, (b) precipitation and phase transformations in alloys and (c) generation as well as relaxations of thermal stresses in composites. The internal friction was measured in the bending mode in two frequency regions: I.: units and tens of Hz and II.: units of kHz.

Keywords: magnesium alloys, magnesium alloys-based composites, amplitude-dependent internal friction, temperature-dependent internal friction, dislocations, solute atoms, thermal stresses

1. Introduction

Internal friction is an anelastic relaxation connected with dissipation of the mechanical energy carried by the sonic or ultrasonic wave in a material and their conversion mostly into heat. In a solid material, it is manifested by deviations from Hooke's law and stress-strain hysteresis in the case of cyclic loading. Defects in the microstructure as point defects, grain boundaries, phase transformations, electrons, phonons or redistribution of the heat are responsible for internal friction. The general quantities describing the internal friction is the *specific damping capacity* $\Psi = \Delta W_{diss}/W_{max}$, or the *loss factor* $\eta = \Delta W_{diss}/2\pi W_{max}$, where ΔW_{diss} is the mechanical energy dissipated in a material during one cycle of the applied stress, and W_{max} is the maximum mechanical energy stored in it. In real experiments depending on the apparatus used, various quantities are measured: the mechanical loss factor $\tan\varphi$, the logarithmic decrement δ or the reciprocal value of the quality factor Q. These quantities are simply proportional:

$$\Psi = \eta/2\pi = \tan\varphi = \delta/\pi = Q^{-1}. \tag{1}$$

It should be mentioned that such proportionality is valid for small values of damping $\Psi \ll 1$, which is valid for most metallic materials. Very effective sources of internal friction are dislocations. If a material containing dislocations is submitted to a harmonic applied stress $\sigma = \sigma_0 \sin\omega t$ with an angular frequency $\omega = 2\pi f$, the lost energy ΔW_{diss} depends only on the anelastic strain. In the case of an anelastic dislocation strain

$$\Delta W_{diss} = \oint \varepsilon_d \, d\sigma. \tag{2}$$

The maximum stored energy can be well approximated by the maximum elastic stored energy

$$W_{max} = \int_0^{\sigma_0} \sigma \, d\varepsilon_{el} = \frac{1}{2} J_{el} \sigma_0^2, \tag{3}$$

where J_{el} is the elastic compliance related to the shear modulus $G^{-1} = J_{el}$. Mechanical loss factor due to the presence of dislocations in the material may be written as

$$\tan\varphi = \frac{1}{\pi J_{el} \sigma_0^2} \oint \varepsilon_d \, d\sigma. \tag{4}$$

The logarithmic decrement, δ, as another damping quantity, is given as

$$\delta = \frac{1}{n} \ln \frac{A_i}{A_{i+n}}, \tag{5}$$

where A_i and A_{i+n} are the amplitudes of the *i*th cycle and $(i + n)$th cycle, respectively, separated by *n* periods of the free vibrations of the specimen.

In experiments, three types of dependences can be estimated:

(i) Amplitude (strain/stress) dependence of damping capacity;

(ii) Temperature dependence of the damping capacity;

(iii) Frequency dependence of the damping capacity.

Because the damping capacity is in many cases connected with the atomic jumps, Arrhenius equation may be used

$$f = f_0 \exp\left(-\Delta G/kT\right), \tag{6}$$

where *f* is frequency, f_0 a constant, ΔG the Gibbs free energy, *k* the Boltzmann constant and *T* the absolute temperature. In this way, the frequency dependence of the damping capacity may be converted to the temperature dependence.

The dynamic strain dependence of the damping capacity can be divided into a strain-independent and a strain-dependent component. If the logarithmic decrement is measured, the experimental finding can be expressed as

$$\delta = \delta_0 + \delta_H(\varepsilon), \tag{7}$$

δ_0 is the amplitude-independent component, found at low amplitudes. The component δ_H depends on the strain amplitude and it is usually caused by the presence of dislocations in the material. The strain amplitude dependence of the logarithmic decrement suggests dislocation-unpinning processes. Dislocations-solute atoms interactions may be explained using the Granato-Lücke (G-L) theory of dislocation internal friction. In the G-L theory [1, 2], the dislocation substructure exhibits longer segments with the lengths of L_N along which weak-pinning points (solute atoms) are randomly distributed. The mean distance between two weak-pinning points is ℓ assuming that $\ell \ll L_N$. The mean total density of dislocations is ρ. Shorter dislocation segments vibrate under the periodic stress $\sigma = \sigma_0 \sin\omega t$ carried with the sonic or ultrasonic wave. At $T = 0$ K and at sufficiently high stress, the dislocation is able to break away from the weak-pinning points which may be point defects (solute atoms) or their small clusters. The longer segments L_N are assumed to be determined by unbreakable pinning points, for example, the nodes of the Frank-Read sources. Dislocations break away from weak-pinning points under stress that is determined by the largest double loop in a segment. The length of this double loop is strongly dependent on the distribution of the solute atoms. The stress necessary for the break-away decreases with increasing temperature because the break-away process is thermally activated [3]. At higher temperatures, the break-away can occur at lower stresses than necessary in pure mechanical process, but higher activation energies are required because the break-away is simultaneous from several neighbouring pinning points. In the high temperature and low frequencies approximation, the stress dependence of the decrement component δ_H can be expressed as [3]

$$\delta_H = \frac{\rho L_N^2}{6} \frac{\nu_d}{\omega} \left(\frac{3\pi kT}{2U_0}\right)^{1/2} \left(\frac{\ell^3 \sigma_0^2}{U_0 G}\right)^{1/2} \exp\left[-\frac{4}{3} \frac{U_0}{kT} \left(\frac{U_0 G}{\ell^3}\right)^{1/2} \frac{1}{\sigma_0}\right], \tag{8}$$

here G is the shear modulus, σ_0 is the amplitude of the applied stress and ω its angular frequency, ν is the dislocation frequency, U_0 is the activation energy, kT has its usual meaning. This relationship has a similar form as the original formula given by Granato and Lücke [1, 2]. The δ_H component depends exponentially on the stress amplitude. The amplitude-independent component δ_0 may be in the frame of G-L theory expressed as

$$\delta_0 = \frac{\Omega B \rho \ell^4 \omega}{E_L b^2}, \tag{9}$$

where Ω is an orientation factor, B is the coefficient of dislocation friction, E_L is the tension in a dislocation line and b is the Burgers vector of dislocations. The critical strain ε_{cr} at which the internal friction becomes amplitude dependent may be used to calculate the effective critical stress amplitude corresponding to the micro-yield stress according to the following equation:

$$\sigma_{cr} = E\, \varepsilon_{cr}, \tag{10}$$

where E is Young's modulus. The stress necessary for a thermal break-away of dislocation loops σ_T at a finite temperature is given by [3]

$$\sigma_T = \sigma_{cr} = \sigma_M \left[1 - \left(\frac{kT}{U_1} \ln A\right)^{2/3}\right] \tag{11}$$

with

$$A = \frac{2}{3} \frac{\nu}{\omega} \frac{\sigma_M}{\sigma_0} \left(\frac{kT}{U_1}\right)^{2/3}. \tag{12}$$

σ_M is assumed to be break-away stress in pure mechanical process. For a double loop with the loop length ℓ_1 and ℓ_2, it occurs at the stress

$$\sigma_M = \frac{2F_m}{b(\ell_1 + \ell_2)}. \tag{13}$$

Here, F_m is the maximum force between the dislocation and the pinning point. $U_1 = 4/3(F_m^{\,3}/\Phi)^{1/2}$, where Φ is a constant.

While amplitude-dependent component of internal friction is only due to the presence of dislocation in the material, amplitude-independent part is influenced by several contributions as dislocations, grain boundaries, precipitates, interfaces, or heat transfer. The damping capacity of materials is a function of the microstructure, stress, temperature, or frequency. The anelastic strain can result from the motion of structural defects, such as point defects, dislocation,

http://dx.doi.org/10.5772/67028

grain boundaries, and phase transformations, and conversely the internal friction can be used to study such motions. Two basic dependencies were measured: amplitude-dependent internal friction (ADIF) and temperature relaxation spectrum of internal friction (TDIF).

It is often stated in the literature that magnesium and magnesium alloys have good damping properties. Damping depends on many factors: purity, grain size, alloy composition, thermomechanical treatment of magnesium materials; their microstructure and substructure, temperature, frequency and the stress amplitude. Therefore the universal statement that magnesium and magnesium alloys have high damping capacity is not true. Experiments are necessary to establish to what extent damping occurs for each particular case. The first internal friction effect observed in magnesium was established by Kê [4]. One relaxation peak found at 490 K was attributed to the grain boundary sliding. Similar results of polycrystalline magnesium and magnesium alloys were found by several authors [5–11]. Internal friction peaks measured at low temperatures in magnesium by Fantozzi et al. [12] and Seyed Reihani et al. [13] were attributed to the so-called Bordoni relaxation, that is, the intrinsic mobility of dislocations through the Peierls barriers. Relaxation peaks found in the vicinity of 0.4 T_m observed by Nó et al. and Trojanová et al. were attributed to dislocation glide in the non-compact slip planes [14–16]. Ageing effects in magnesium alloys were studied using ADIF [17, 18] and also TDIF [19]. The influence of microplastic and cyclic deformation on the damping capacity was studied in Mg and Mg alloys using ADIF [8, 20, 21]. Formation of new interfaces in magnesium alloys via precipitation [22], mechanical twinning [23] or an adding of ceramic particles or fibres [24, 25] may substantially increase the damping capacity. Materials with the high damping capacity are known as HIDAMETS (HIgh DAmping MEtalS) or HIDAMATS (HIgh DAmping MAterialS) among them magnesium materials have an important place. Thermal cycling of magnesium alloys reinforced with ceramic fibres or particles may substantially deteriorate microstructure and mechanical properties [26].

2. Amplitude-dependent internal friction

2.1. Cast alloys

Gravity cast magnesium alloys AM20, AX41 and AZ31 were annealed step by step for 20 min at increasing temperature of the thermal cycle. The temperature step was 20°C. Immediately after thermal cycling, amplitude dependence of internal friction was measured. Amplitude dependences of decrement estimated for AM20 alloy are introduced in **Figure 1a–c** for various upper temperatures of the thermal cycle. The amplitude dependences of decrement are practically identical for thermal cycling with the upper temperatures up to 200 °C. Thermal cycling with higher temperatures between 200 and 260 °C causes a decrease in the amplitude dependent component while the amplitude independent component remains unchanged. Thermal cycling with the upper temperatures between 260 and 400 °C causes an increase in the amplitude dependent component. The amplitude dependence of decrement obtained at 400°C achieved similar values as in the case of the as-cast sample.

According to Eq. (8), the δ_H component of decrement depends exponentially on the stress amplitude. The experimental curves were fitted according to formula (8) in the form

$$\delta = \delta_0 + C_1 \varepsilon \exp(-C_2/\varepsilon), \tag{14}$$

Here, the C_1 parameter is proportional to $\ell^{2/3}$ and the C_2 is proportional to $\ell^{-2/3}$. Discussion of the C_2 parameter is simpler because it depends only on the length of the small dislocation segments ℓ and the activation energy U_0. Assuming that the activation energy is in the first approximation the same (all measurements were performed at room temperature), we may attribute observed temperature dependence of the C_2 parameter to changes in the microstructure. The typical feature of the microstructure of AM20 magnesium alloy is the formation of $Mg_{17}Al_{12}$ intermetallic phase particles as it is obvious from **Figure 2a**. The estimated parameters C_2 are shown in **Figure 1d**.

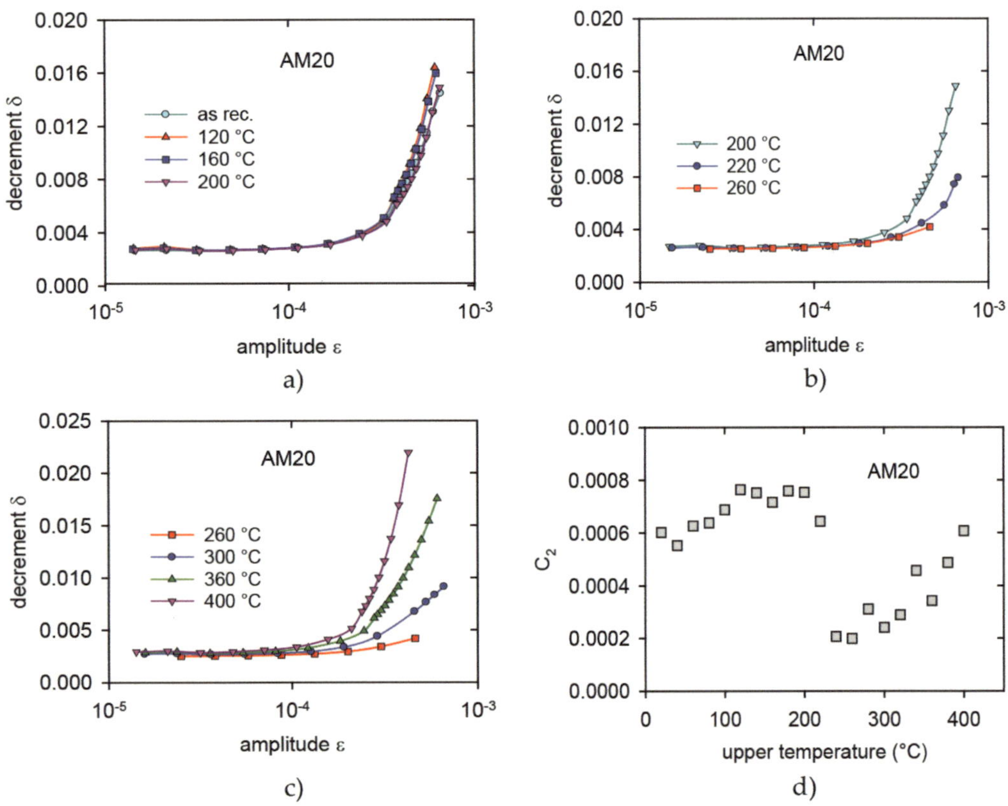

Figure 1. Amplitude dependence of decrement measured in the AM20 alloy after thermal cycling: lower (a), medium (b), higher (c) upper temperatures of the thermal cycle, temperature dependence of the C_2 parameter (d).

The temperature dependence of the C_2 parameter slightly increases up to 200°C, then rapidly decreases with increasing upper temperature of the thermal cycle, and starting from a temperature of 260-280°C, it again increases. The length of dislocation segments ℓ is done by the interaction of the solute atoms with the dislocation line. Although the concentration of solute atoms in the alloy is lower than the solubility limit, solute atoms in the as-cast alloy are not

http://dx.doi.org/10.5772/67028

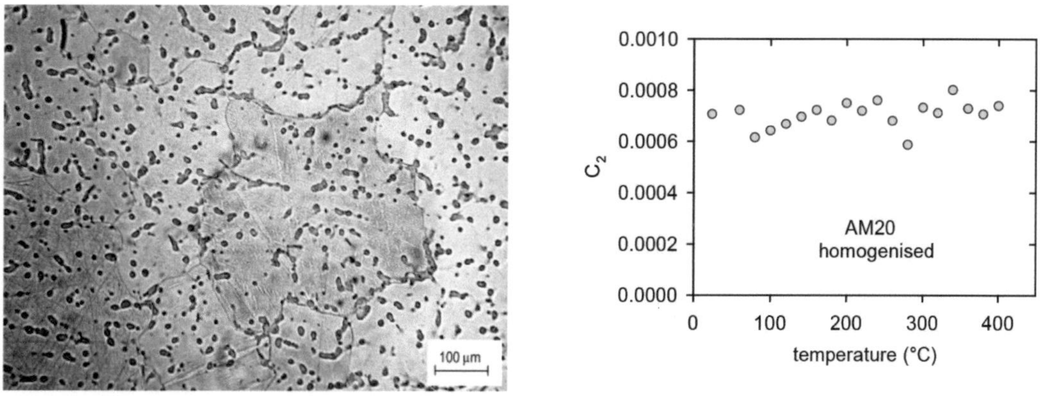

a) b)

Figure 2. Microstructure of the as-cast alloy (a); temperature dependence of the C_2 parameter in the homogenised alloy alloy (homogenised is also in the picture above) alloy (b).

homogeneously dispersed. At temperatures between 200 and 320°C, solute atoms are movable and the solute solution is purified due to precipitation process in the alloy. Concentration of weak pinning points (solute atoms) decreases, and if the dislocation density is constant, the effective length ℓ of dislocation segments increases. The C_2 parameter is proportional to $C_2 \propto \ell^{-2/3}$, thus with increasing annealing temperature, the values of C_2 parameter decreases. Above 320°C, the C_2 slightly increases; this can be caused by several reasons: a decrease in dislocation density, ρ, thermal stresses and the interaction of dislocations with precipitates. The observed increase of C_2 parameter at temperatures up to 160°C could be caused by segregation of solute atoms at dislocations. Such processes were observed during plastic deformation of various Mg alloys [27, 28]. The strain-ageing phenomena are due to dislocation-solute atom interaction. After a homogenisation treatment (413°C for 18 h), free solute atoms occupy dislocation lines. In the G-L model, dislocations are pinned at some place and the free elastic segments move under the oscillating stress. Damping of these oscillations is controlled by interactions with phonons and electrons. In concentrated alloys, the dislocation line is strongly pinned by high number of solute atoms and their small clusters, which occupy the dislocation line. Because of very short distance between the weak pins (solute atoms), the force necessary for break-away of dislocation is very high. At $T = 0$, the dislocation starts to glide when the Peach-Koehler force achieves the value of the binding force between the dislocation line and solute atoms; the dislocation segment L_c (L_c is the pinning correlation length) is released. Then, the critical stress σ_c is done [29] as

$$\sigma_c = \left(\frac{n_p^2 f_p^4}{\gamma}\right)^{1/3} \quad if \quad f_p << \gamma\, n_p b^3, \tag{15}$$

where n_p is the density of dislocation pinning points (solute atoms or their small clusters), f_p is the elementary pinning force and γ is the dislocation elastic energy $\gamma = \kappa^{-}\, Gb^2$ where b is the Burgers vector magnitude, κ^{-} is a constant of the order of 0.5. The collective pinning energy may be estimated as

$$U_c = (\gamma\, n_p f_p^{\,2} b^6)^{1/3}. \tag{16}$$

When the stress (strain) amplitude reaches its critical value, the dislocation motion has features typical for avalanche motion. Long free dislocation segments may operate in the slip plane and damping rapidly increases. Such collective pinning is obvious from **Figure 3a** for AX41 alloy. The critical strain (stress) for break-away of dislocations depends only slightly on the thermal treatment, that is, the microstructure is very stable due to the formation of Mg_2Ca precipitates which are more thermally stable in comparison with $Mg_{17}Al_{12}$. The constant length of shorter dislocation segments is manifested by a nearly constant C_2 parameter as it was observed in the homogenised AM20 and AX41 alloys (**Figures 2b** and **3b**).

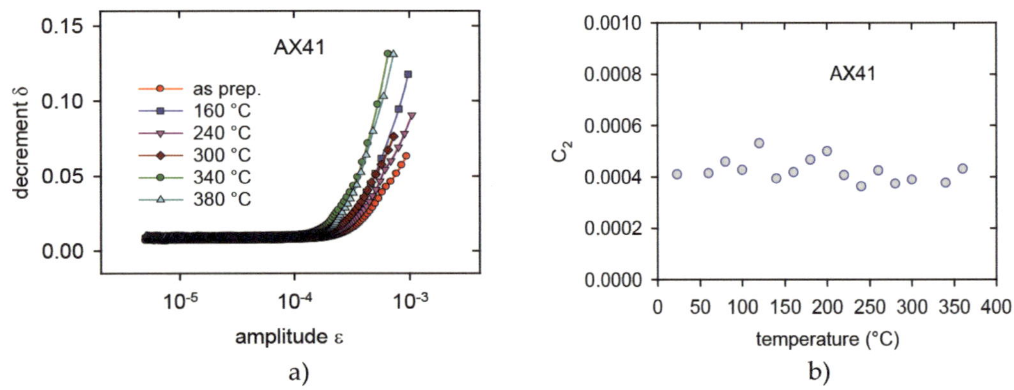

Figure 3. Amplitude dependence of decrement measured in AX41 alloy after thermal cycling (a) and temperature dependence of the C_2 parameter (b).

Different results were found in the case of an AZ31 alloy in the as-cast state (see **Figure 4a, b**). The alloy was thermally treated in the similar way as the AM20 alloy. ADIF measured in the as-cast sample and after the thermal cycling up to 240°C showed no changes in the microstructure (**Figure 4a**). Thermal treatment at higher temperatures influenced both the amplitude-independent as well as amplitude-dependent components of the decrement (**Figure 4b**). Both increased with increasing upper temperature of the thermal cycle. A rapid increase of the δ_0 component indicates some microstructural changes in the sample. The microstructure of the sample after the measurement series is shown in **Figure 5**. Tiny precipitates situated in the grain boundaries are the characteristic feature in the microstructure (**Figure 5**). The formation of precipitates is connected with the purification of the grains interior. A decrease of the solute atom concentration inside of grains increases the effective length of dislocation segments ℓ. The δ_0 component is very sensitive to ℓ, as it follows from Eq. (8); it means the internal friction rapidly increases in the amplitude-independent component.

http://dx.doi.org/10.5772/67028

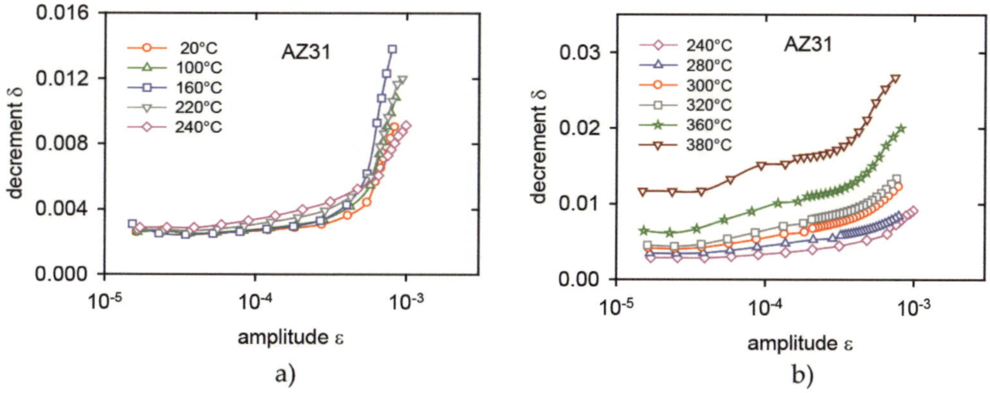

Figure 4. Amplitude dependence of decrement measured in AZ31 alloy for various upper temperatures of the thermal cycle: lower temperatures (a) and higher temperatures (b).

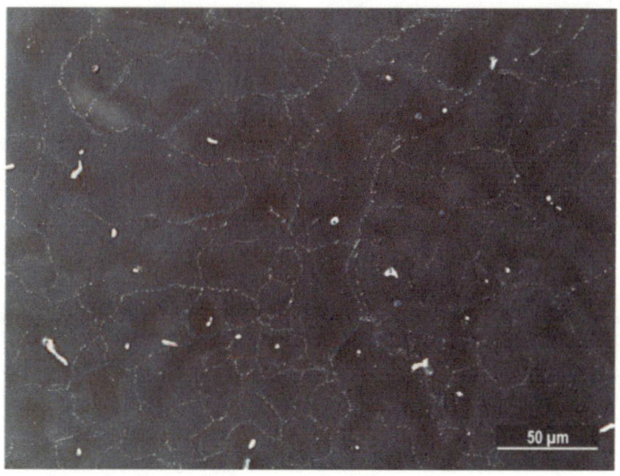

Figure 5. Microstructure of the AZ31 alloy after the thermal cycling.

2.1.1. Summary

From the ADIF investigations in AM, AX and AZ magnesium alloys, it can be concluded that changes in the microstructure of alloys may be detected using the internal friction measurement. Thermodynamic processes occurring in the alloys change the effective length of the dislocation segments. These processes take place mainly at temperatures higher than 200°C. Free solute atoms occupy the dislocation line and the stress necessary for the break-away of dislocations is very high. Avalanche release of dislocation lines (and also their length) leads to a rapid increase in the internal friction.

2.2. Magnesium alloys-based composites reinforced with short fibres

Metal matrix composites (MMCs) are materials combining two or more components with different physical and chemical properties. The commercial Mg alloys AZ91, ZC63 were reinforced by δ-alumina short fibres (Saffil®, 3 μm in diameter with the mean length of about 87 μm) applying the squeeze-cast technology. The preform consisted of short Saffil fibres and a binder system (containing Al_2O_3 and starch). Fibres exhibited a planar isotropic fibre distribution. The preform was preheated to a temperature higher than the melt temperature of the alloy and then inserted into a preheated die. The pressure was applied in two stages. Composites based on AZ91, ZC63 alloys were exposed to the T4 heat treatment (AZ91: 413°C/18 h; ZC63: 440°C/8 h). Microstructure of the AZ91+14.6% Saffil fibres composite is introduced in **Figure 6a**. The sample was cut so that the fibres plane was parallel to the section plane. It can be seen that the distribution of fibres is not ideal and many fibres were broken during the squeeze-casting process.

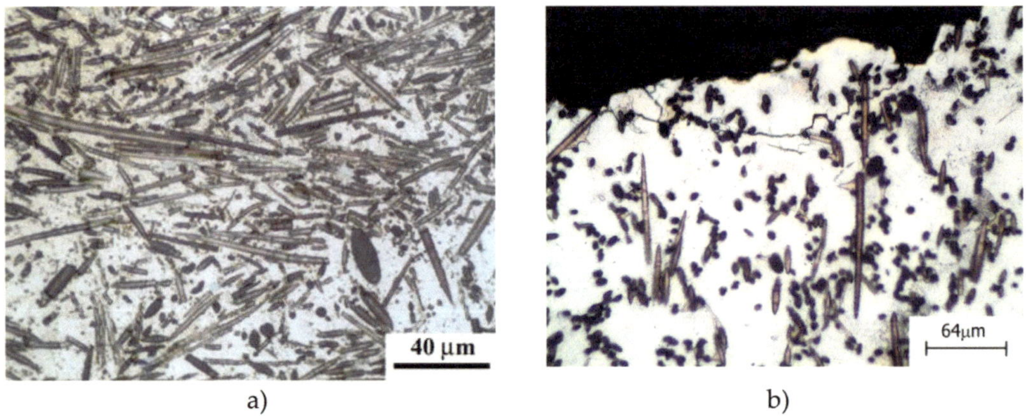

Figure 6. AZ91/Saffil composite before mechanical cycling (a) and cracks in the vicinity of the fatigue fracture plane after mechanical cycling (b).

Figure 7a shows the plots of the logarithmic decrement against the strain amplitude for the ZC63 magnesium alloy reinforced with 24.9 vol.% Saffil fibres before and after thermal cycling between room temperature and an increasing upper temperature of the thermal cycle. **Figure 7b** shows results obtained for higher temperatures. From **Figure 7a,b**, it can be seen that the strain dependencies of the logarithmic decrement exhibit two regions in good accord with Eq. (7). The values of δ_H in the strain amplitude-dependent region increase very strongly with increasing upper temperature of the thermal cycle up to 260°C, and then above 260°C, the values of δ_H again decrease with the upper temperature. Similar results were found for other composites based on magnesium alloys; only the critical temperature T_C at which the original increase of the amplitude-dependent decrement component turns to the decrease is different for various magnesium alloys. Similar results were found for the AZ91+14.9 vol.% Saffil short fibres (**Figure 8a, b**). The critical temperature T_C was in this case found to be 280°C. This critical temperature was found for ZC63/Al_2O_3 260°C (temperature 280°C was not measured), ZE41/ Al_2O_3 300°C. The critical temperature for QE22 was about 360°C and for AX41/ Al_2O_3 310°C, depending on the melting point of the matrix alloy.

In the lower-strain amplitude region, the decrement is only weakly dependent on the strain amplitude. In the second region, for higher strains (or stresses, according to Hooke's law they are proportional), the decrement increases strongly with increasing strain amplitude.

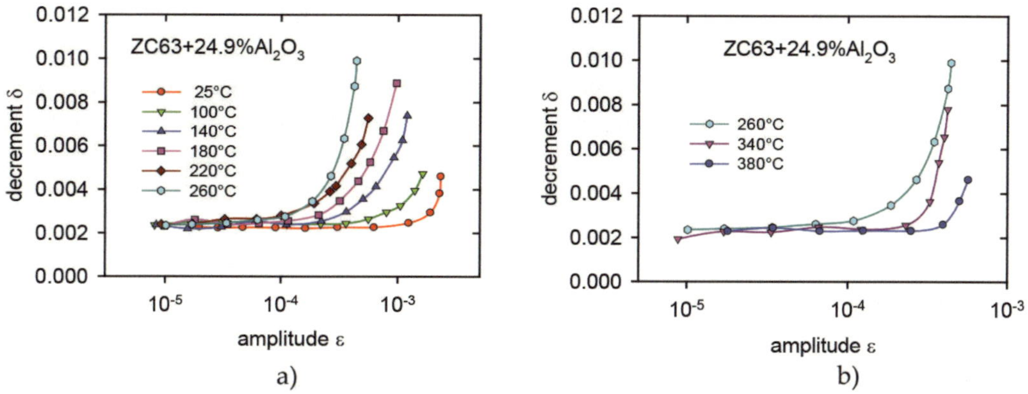

Figure 7. Amplitude dependence of decrement measured for ZC63 alloy reinforced with 24.9 vol.% of Saffil fibres after thermal cycling: lower upper temperatures of the thermal cycle (a) and higher temperatures of the thermal cycle (b).

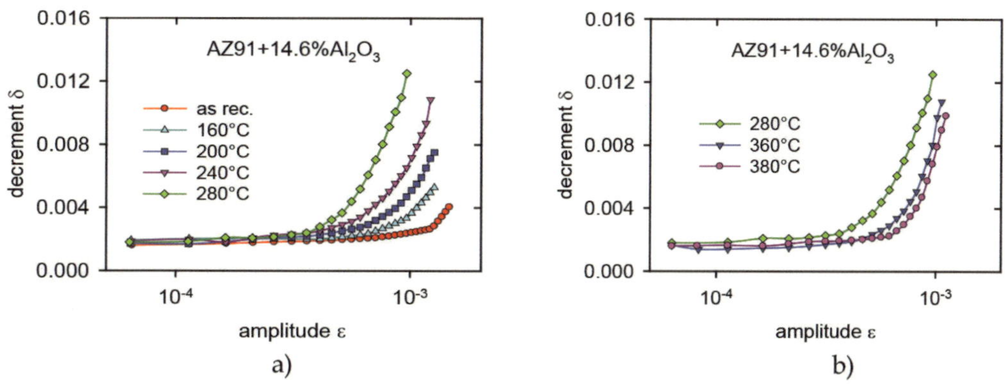

Figure 8. Amplitude dependence of decrement measured for AZ91 alloy reinforced with 14.6 vol.% of Saffil fibres after thermal cycling: lower upper temperatures of the thermal cycle (a) and higher temperatures of the thermal cycle (b).

A number of possible damping mechanisms can be identified in MMCs. Interfaces present in composites are very effective source of damping (interfacial frictional sliding, local dissipative interfacial processes and interfacial diffusion). Grain boundary sliding can occur in the same way as in the unreinforced materials. A finer grain size, which is typical for most MMCs, means that there this process may occur with the higher intensity. On the other hand, ceramic reinforcements exhibit low dislocation density, then dislocation motion and grain boundary sliding are limited; the damping effect from these mechanisms will be reduced in proportion to the remaining volume fraction of the matrix. All these effects (dislocation damping, interface and grain boundary damping) may influence the amplitude-independent δ_0 component, while the amplitude-dependent component δ_H is caused by dislocation motion in the material.

The experimental data were analysed using relationship (14). Values of the C_2 parameter are introduced in **Figure 9** for ZC63 (**a**) and AZ91 (**b**) alloys reinforced with Saffil fibres. Note that good correlation with the theory presented was found only for temperatures higher than

50–100°C. Decreasing values of the C_2 parameter with the upper temperature of the thermal cycle up to temperatures 240–260°C indicate that the length of shorter dislocation segments ℓ increases with increasing upper temperature. Further cycling at temperatures higher than 240–260°C again increased C_2 parameter, that is, the length of dislocation segments ℓ decreased.

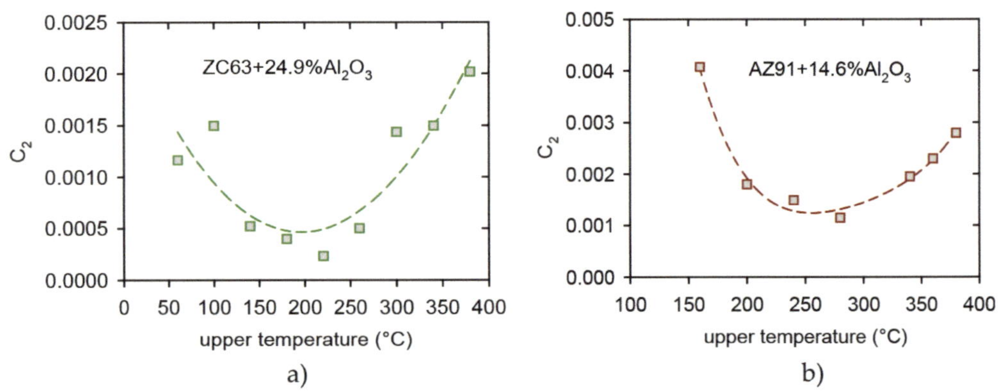

Figure 9. Temperature dependence of the C_2 parameter in the AZ63 alloy with 24.9% of Saffil fibres (a) and AZ91 alloy with 14.6% of Saffil fibres (b).

Metallic matrices and ceramic reinforcement are in the equilibrium at the manufacture temperature. Cooling to the room temperature induces thermal stresses [30]. The coefficient of thermal expansion (CTE) of ceramic reinforcement is lower than that of most metallic matrices. This means that when the composite is subjected to a temperature change, thermal stresses will be generated. Generally, these thermal stresses may be formulated in the following form:

$$\sigma_t = \mathrm{f}(C, r_i)\,\Delta\alpha\Delta T, \tag{17}$$

where $\mathrm{f}(C,r_i)$ is a function of the elastic constants C, and geometrical parameters r_i, $\Delta\alpha$ is the absolute value of the difference in the expansion coefficients of the components, and ΔT is the temperature change. Typically there is a large difference between the thermal expansion coefficient (CTE) of the matrix and the reinforcing phase. Even small temperature changes may generate thermal stresses. These internal stresses generated by thermal cycling of composites can be reduced by various relaxation mechanisms: dislocations production, their glide, by decohesion or sliding of the matrix-reinforcement interface, by diffusion of solute atoms in the matrix. Density of newly created dislocations created in the vicinity of the reinforcement is significantly higher than elsewhere inside of the matrix. The dislocations generated during cooling from an elevated temperature below a certain temperature T_{cr} can glide inducing plastic deformation into the matrix.

The observed results may be explained if we consider that during cooling and also during thermal cycling new dislocations are created due to the difference in the CTE. This process may be influenced by the thermodynamic processes occurring in the matrix during the thermal cycling. Chmelík and co-workers [31] measuring the acoustic emission from the thermally treated composite (without any applied stress) have shown that new dislocations are formed in the cooling part of the thermal cycle. Higher dislocation density decreased the effective length of dislocation segments considering that the number of pinning points is constant. Number of free foreign atoms or their small clusters can be modified by thermodynamic processes in the matrix. A redistribution of solute atoms may be studied by electrical resistivity measurements. Residual resistivity ratio RRR ($RRR^{-1} = \rho_e(77\text{ K})/\rho_e(293\text{ K})$) measured for step-by-step annealed sample at increasing temperature is introduced in **Figure 10a** for ZC63/Al_2O_3 MMC. A sharp drop in RRR^{-1} detected in the temperature range from 160 to 240°C may be explained by the solute redistribution. This decrease of the resistivity reflects very complex processes; precipitates of various types can be formed in the matrix. The absence of the expected ageing effect might be due to the massive CuMgZn consuming a high amount of alloying elements. Electrical resistivity measurements have shown that the precipitation processes occur in some Mg-based MMCs approximately between 200 and 300°C [32–34]. Similarly, the matrix alloy is purified from the free solute atoms also in the AZ91 alloy as it will be shown in Section 3.3.

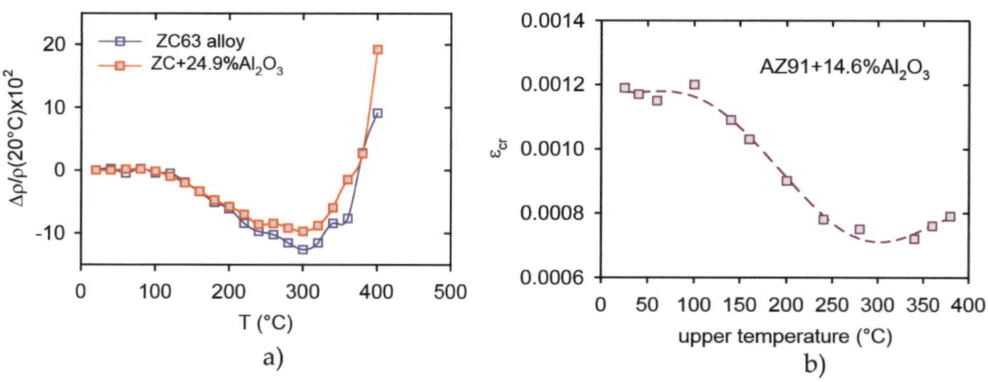

Figure 10. Temperature dependence of the residual resistivity ratio in ZC63 alloy and composite (a) and critical strain in the AZ91+14.6% of Saffil fibres (b).

From **Figure 9a, b**, it is obvious that the C_2 constant decreases with increasing upper temperature of the thermal cycle. This tendency finished at temperatures between 200 and 300°C (depending on the matrix alloy) and then it again increases with temperature. Such behaviour may be explained by the formation of plastic zones which may accommodate thermal stresses produced at the matrix/ceramic fibre interfaces.

The radius of this plastic zone is given by the following approximate relationship [35]:

$$r_{\text{plz}} = r_f \left[\frac{4\Delta\alpha\, E_M}{(5-4\nu)\ \sigma_y} \cdot \Delta T \right]^{1/2}, \tag{18}$$

where E_f and E_M are Young's moduli of fibres and matrix, respectively, v is Poisson constant and σ_y the yield stress in the matrix, r_f is the radius of fibres. Similarly, it is possible to express the volume fraction of the plastically deformed matrix [35]

$$f_{plz} = f\left[\frac{4\Delta\alpha E_M}{(5-4v)\sigma_y} \cdot \Delta T - 1\right]. \tag{19}$$

If the volume fraction of the plastified matrix increases above a certain value, the plastic zones may overlap. In the overlapped zones, the dislocation loops formed near the interfaces have the opposite sign on both sides of the fibre. At higher temperatures, the yield stress in the matrix is lower than internal stress and at temperatures higher than 260–340°C, the tensile stresses change to compression ones. At this moment, dislocations move in the plastic zones and annihilation of dislocations can occur and hence the dislocation density decreases.

Chmelík and co-workers [31] have measured acoustic emission of samples thermally cycled between room temperature and an increasing upper temperature of the thermal cycle. At elevated temperatures from 200°C (in the case of AZ91 MMC) to 280°C, permanent elongation of the sample was detected, followed by rapid shortening of the sample after cycling at temperatures higher. These characteristic temperatures were different for various magnesium alloys used as the matrix alloys, but in each case, the critical temperature at which the sample became to be shorter accords with the critical temperature obtained in the internal friction experiments (AZ91/Al_2O_3 280°C, ZC63/Al_2O_3 260°C (temperature 280°C was not measured), ZE41/Al_2O_3 300°C, QE22/Al_2O_3 360°C). A decrease of the amplitude-dependent component of the decrement is observed in the region where the sample is plastically deformed by thermal stresses. Acoustic emission signal was detected only in the cooling parts of the thermal cycle while high-temperature strain of the sample was completely noiseless.

The temperature dependence of the critical strain ε_{cr} is shown in **Figure 10b**. All measurements were performed at ambient temperature. Thus, we can assume that parameter A and the activation energy U_1 in Eq. (11) are, in the first approximation, independent of thermal treatment. Then, the critical strain ε_{cr} depending on the length of dislocation segments ℓ follows the same tendency as the C_2 parameter.

2.2.1. *Summary*

Thermal stresses are generated after the thermal cycling of magnesium alloys reinforced by short Saffil fibres. These stresses may be relaxed by anelastic as well as plastic strain. Internal friction measurements can detect newly created dislocations in the vicinity of fibres ends.

Internal friction measurements showed that changes in the microstructure start at temperatures above about 200°C. Precipitation processes and solute atoms migration are responsible for these changes. Thermal internal stresses generated at temperatures higher than

http://dx.doi.org/10.5772/67028

220–280°C are high enough to invoke motion of newly created dislocations. Thermal cycling at temperatures higher than ~300°C causes movement and annihilation of new dislocations (which are only slightly pinned and well movable) in the matrix, under appearing compressive thermal stresses, which leads to a decrease of internal friction.

2.3. Influence of mechanical cycling on ADIF

The same composite as in Section 2.2, that is, AZ91 alloy with 14.6% of Saffil fibres, was mechanically cycled in the same apparatus in which the ADIF was measured. The sample vibrated certain number of cycles at the resonant frequency, and then the amplitude dependence of decrement was measured.

The end of the fatigue life of the sample was manifested by the rapid increase of the resonant frequency. The sample was cycled up to the sample failure. It was reached at $N = 2.19 \times 10^8$ cycles, the maximum strain amplitude was 1×10^{-3}.

The amplitude-dependent internal friction curves determined in the as-received state and after definite numbers of cycles are introduced in **Figure 11a**. A rapid increase of the decrement was observed between the first measured curve for $N = 0$ and the second measured curve after 1 min of cycling, corresponding to $N = 4.02 \times 10^3$ cycles. The logarithmic decrement estimated in the amplitude-independent region for the strain amplitude $\varepsilon = 2 \times 10^{-4}$ is shown in **Figure 11b**. A saturated region, where the decrement is more independent of the number of cycles, was found from $N = 4 \times 10^3$ to approximately 1×10^6 cycles. Further cycling led to a rapid decrease of the decrement and then again an increase followed with the fracture of the sample. A rapid increase of the amplitude-independent component of decrement δ_0 was observed for number of cycles $N = 4 \times 10^7 - 9 \times 10^7$ as it is demonstrated in **Figure 12a**. After further cycling, local maxima appeared in the amplitude-independent part of the curves as it can be seen in **Figure 12b**.

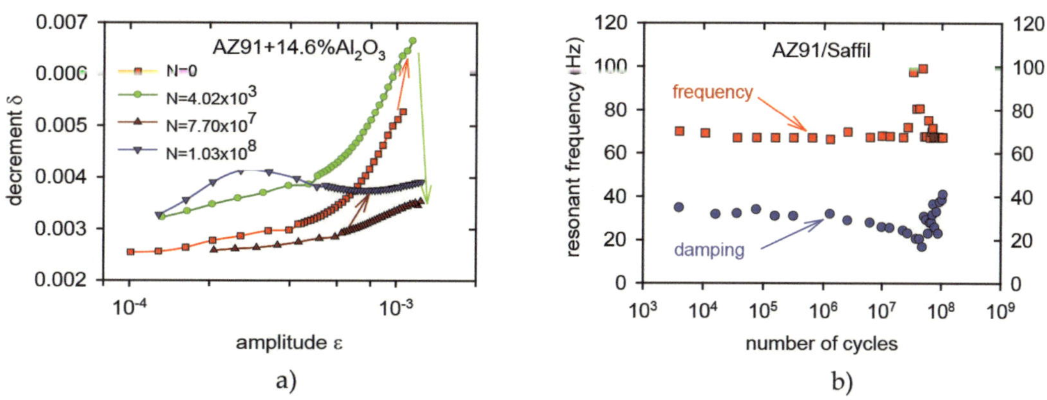

Figure 11. Strain amplitude dependences of the logarithmic decrement for as-received state and after certain number of cycles (a) and resonant frequency and damping measures after N cycles (b).

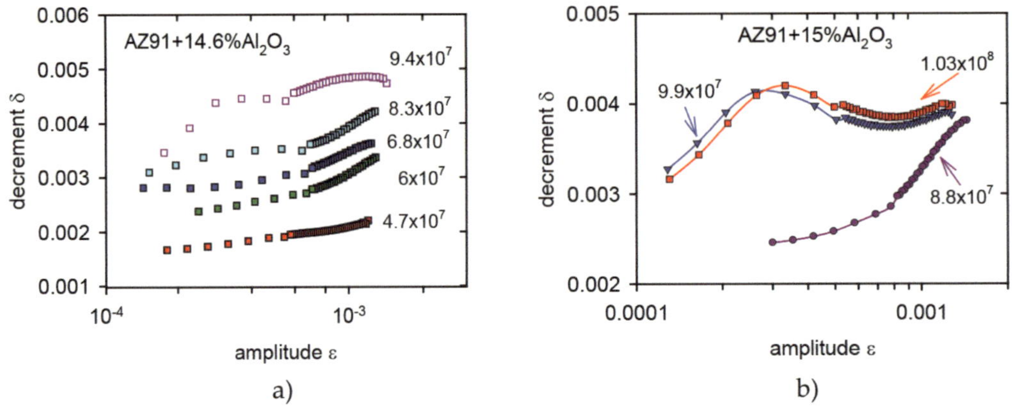

Figure 12. Rapid increase of decrement observed after cycling (a); local maximum in the amplitude independent part of decrement (b).

Changes in the composite response upon cyclic loading are primarily caused by plastic deformation in the matrix [36]. The cycling response in the material containing particles impenetrable for dislocations leads to a rapid increase in the dislocation density during cyclic loading until a saturated state is reached. Composites typically exhibit a higher dislocation density compared with their monolithic counterparts. The dislocations have to circumvent the impenetrable ceramic fibres and in the case of the AZ91 alloy also precipitates $Mg_{17}Al_{12}$. New dislocations are generated at the matrix/obstacles interfaces. According to Eq. (9), the amplitude-independent component of the logarithmic decrement depends on the dislocation density and the length of the shorter dislocation segments ℓ. The observed increase of the decrement at the onset of the cycling process is a result of the increase in the dislocation density. Because the number of the pinning points is constant and the effective length of the shorter dislocation segments ℓ is higher, the decrement increases. During further cycling, the dislocations are trapped by the interface, and induce disordered three dimensional dislocation structures with a few dislocations extending in the matrix. The density of movable dislocations is reduced. Shuttling of these dislocations accommodates the imposed strain but the damping decreases as it is obvious from **Figure 11a, b**. A rapid increase of the logarithmic decrement for the number of cycles higher than 4.7×10^7 (**Figures 11b** and **12a**) indicates nucleation and growth of fatigue cracks. We observed also influences of the number of cycles on the sample resonant frequency. A rapid increase of the damage parameter of the specimen after N cycles, $D(N)$, can be defined [37] as

$$D(N) = 1 - \frac{E_N}{E_o} = 1 - \frac{f_N^2}{f_o^2}, \qquad (20)$$

where E_N is an effective Young's modulus of the specimen after N cycles and E_0 is Young's modulus for $N = 0$ and f_N and f_0 are the resonant frequencies of the specimen after N cycles and for $N = 0$, respectively. Microstructural investigations showed that the initiation sites of cracks were fibres, clusters of fibres and large-size precipitates (see **Figure 6b**).

http://dx.doi.org/10.5772/67028

A smooth local maximum in the amplitude dependence of the logarithmic decrement was observed after very high numbers of cycles (see **Figures 11a** and **12b**). This maximum is caused by the formation of cracks in the fatigued matrix. A simple rheological model taking into account the crack origin of damping was developed by Göken and Riehemann [38]. According to the model, one elementary crack is assumed to be represented by a frictional grip that is attached in series to a spring E_1 representing the loss of modulus due to opening of the crack. This frictional grip is absolutely firm for stresses lower than its critical stress $\sigma<\sigma_{ci}$. In this stress region, the spring E_1 and frictional grip σ_{c1} attached parallel to another spring E_r are elongated by an amount of ε_{c1} with $\sigma_{c1}=(E_r+E_1)\varepsilon_{c1}$. The grip is separated at $\sigma=\sigma_{c1}$ where the spring E_1 is released and the strain increases according to $\sigma_{c1}=E_r\varepsilon$ because only E_r representing the relaxed modulus is elongated. In this process, the elementary crack opens at the critical strain and the mechanical energy ΔW is converted to heat by the displacement of the two crack surfaces, by emission of dislocations at the crack tips, or by crack growth. The mechanical loss can be written as

$$\eta = \frac{1}{2\pi}\frac{\Delta W}{W} \cong \frac{E_i}{E_r}\left(\frac{\sigma_c}{\sigma}\right)^2 \propto \left(\frac{\varepsilon_c}{\varepsilon}\right)^2 \tag{21}$$

for $\varepsilon\geq\varepsilon_c$ and $\eta = 0$ for $\varepsilon<\varepsilon_c$. Introducing n combinations of grips σ_{ci} and springs E_i with $I = 1\ldots n$, the damping behaviour of n cracks can be modelled. The local maximum in the amplitude dependence of the logarithmic decrement was caused by the distribution of cracks length and their orientations leading to a nearly continuous distribution of the critical strains σ_c. All occurring local maxima could be fitted with good accuracy when a lognormal distribution of critical stresses σ_c was assumed. In **Figure 12b**, the local maximum after 9.9×10^7 cycles is shifted to slightly higher damping and higher strain amplitude after additional cycles leading to a total number of 1.02×10^8 cycles. This may indicate that new smaller cracks are created or dislocation damping increases again during this period. The latter could already be observed in the high amplitude region from 8.8×10^6 to 9.9×10^7cycles. Fatigue cracks were found after mechanical cycling in the composite microstructure as it is demonstrated in **Figure 6b**.

2.3.1. Summary

Fatigue can successfully be studied by the measurement of strain amplitude-dependent damping after increasing numbers of mechanical cycles. For mechanical cycling of AZ91/Saffil composite mainly dislocation effects could also be found for cycle numbers up to 4.7×10^7. For further cycling, a strong increase of damping can be attributed to nucleation and growth of cracks leading to fracture of the sample. Smooth relative maxima appearing on the damping versus strain curve can be explained and evaluated by a rheological model of the effect of cracks on damping.

2.4. Magnesium zirconia nanocomposite

Magnesium with 3 vol% of zirconia nanoparticles was prepared by powder metallurgical route. Powder from Mg was mixed with ZrO_2 nanoparticles (the mean size of 14 nm) and milled together for 1 h in a planetary ball mill. The mixture was subsequently pre-compressed followed by hot extrusion at a temperature of 350°C under a pressure of 150 MPa. After extrusion, the originally more or less equiaxial grains changed into elliptical grains with the long axis parallel to the extrusion direction. The grain size in the cross section was about 3 μm and in the extrusion direction 10 μm. ADIF was measured after mechanical cycling at room temperature. Experiments were performed in similar way as in the case of AZ91/Saffil composite. Decrement estimated for two strain amplitudes depending on the number of cycles is shown in **Figure 13a**.

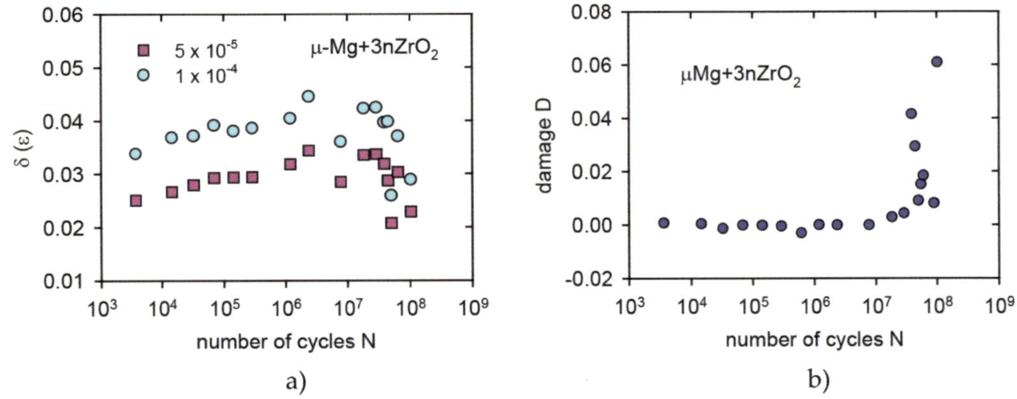

Figure 13. The logarithmic decrement measured for two amplitudes depending on number of cycles (a) and damage parameter versus number of cycles (b).

Mechanical cycling in the region from 3 × 10^3 to 2.4 × 10^6 leads to the decrement increase. Observed increase of decrement is very probably a consequence of the dislocation density increase and also an increase of the effective length between weak pinning points. Further cycling up to 2.9 × 10^7 cycles decreased the decrement. This drop is caused by a decrease of the dislocation density, which is a consequence of interactions between dislocations. The higher dislocation density limited the slip length of vibrating dislocation segments. Rapid increase of the decrement is caused by the cracks creation. Rapid decrease of the damage parameter shown in **Figure 13b** observed for the number of cycles of $N>10^7$ and the decrease of the resonant frequency at the end of the fatigue process indicates loss of the stiffness as a consequence of the creation and propagation of cracks.

2.4.1. *Summary*

Magnesium zirconia nanocomposite was cycled at room temperature and after certain number of cycles the logarithmic decrement was measured. Plastic deformation during cycling increased the dislocation density and also the measured decrement. The loss of stiffness at the end of the fatigue process is a consequence of the cracks formation and propagation.

2.5. Ultra-fine-grained magnesium

Ultra-fine-grained magnesium (UFG-Mg) samples were prepared by ball milling of powder in an inert atmosphere and subsequent consolidation and hot extrusion. The linear grain size of specimens estimated by X-ray line profile analysis was about 100–150 nm. Samples were cyclically loaded until failure with bending vibrations in the apparatus also used for damping measurements. The resonant frequency of the system ranged from 59 to 62 Hz. The end of the fatigue life of a sample was manifested by a rapid decrease of the resonant frequency. The sample was cycled up to failure, which was reached at $N = 2.74 \times 10^8$ cycles.

The amplitude-independent component, δ_0, and the damping value at the maximum amplitude, δ_{max}, plotted against the cycle number are shown in **Figure 14a**. It can be seen that the decrement is more or less constant at the onset of the cycling process. Cycling in the region between 10^7 and 10^8 cycles leads to an increase of the decrement. The observed increase of the decrement indicates a rapid increase of the dislocation density and also an increase in the distance between weak pinning points. This increase of the decrement, on the other hand, is accompanied by a decrease of the resonant frequency. **Figure 14b** shows the damage D (defined by Eq. (20)) plotted against the cycle number N. A rapid increase of D occurred for $N > 10^8$. The measured decrease of the resonant frequency at the end of fatigue process indicates a stiffness loss due to the formation of cracks and may also be caused by an increase of the number of bowable or moveable dislocations as a function of cycling. A rapid increase of the amplitude-dependent component of the decrement at the end of the sample life is due to rapid nucleation or growth of cracks and a resulting strong local increase in the free dislocation density. Newly created dislocations at cracks tips accommodate these stresses.

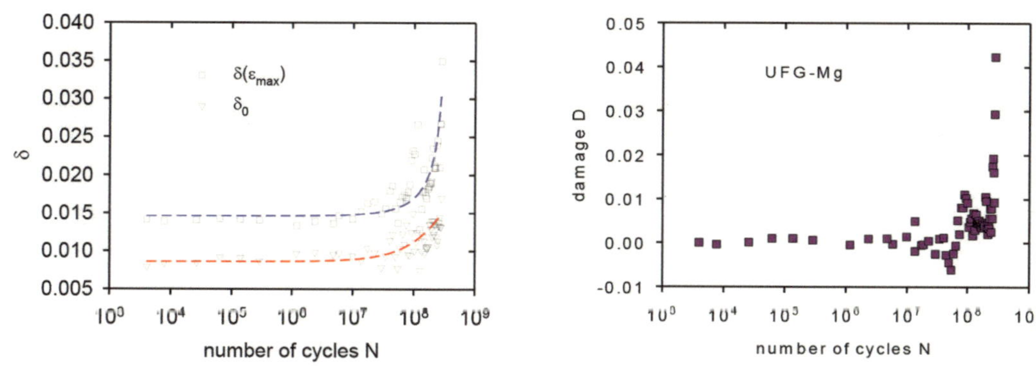

a) b)

Figure 14. Amplitude dependence of decrement measured in UFG-Mg for two amplitudes (a); damage parameter depending on the number of cycles (b).

2.5.1. Summary

The accumulation of plastic deformation mainly due to an increase of the dislocation density is the significant feature of the fatigue process. Cycling higher than 10^8 cycles led to nucleation and growth of cracks. Newly created and propagated cracks increased the logarithmic decrement and damage parameter.

3. Temperature relaxation spectrum of internal friction

3.1. Microcrystalline Mg

Microcrystalline magnesium was prepared using the powder metallurgical route. The temperature dependence of the loss angle tan φ estimated over a wide range from room temperature up to 480°C is given in **Figure 15a** for three measuring frequencies of 0.5, 5 and 50 Hz. Two peaks showing different intensities have been found: the weak low-temperature peak and the more intense high-temperature peak. It can be seen that both peaks shift to a higher temperature with increasing frequency, indicating that they may be relaxations peaks. Positions of peaks in the temperature scale are for three measuring frequencies introduced in **Table 1**. The curves shown in **Figure 15a** were measured during one heating course in the multi-frequent mode. Repeating of the thermal measuring cycles did not change the height and position of peaks. The internal friction peaks are assumed to be imposed by a background IF_b expressed by

$$IF_b = A + B\exp\left(\frac{-C_b}{kT}\right), \tag{22}$$

where k is the Boltzmann constant and T is the absolute temperature, A, B and C_b are constants. After subtracting the background by using a fitting programme PeakFit, the maximum temperature T_P was estimated for both the relaxation peaks. Exact temperatures of both peaks estimated for various frequencies are given in **Table 1**. The peak widths are broader than that for a Debye peak, characterised by single relaxation time. The internal friction peak appears at the condition $\omega\tau = 1$ [39], with

$$\omega\tau = \omega\tau_0\exp(\Delta H/kT), \tag{23}$$

where ω is the angular frequency ($=2\pi f$, f is the measuring frequency), τ is the mean relaxation time, τ_0 the pre-exponential factor, and ΔH is the mean activation enthalpy. **Figure 15b** shows the semilogarithmic plot of the frequency versus reciprocal value of the peak temperature T_P—the so-called Arrhenius plot.

From the slope and intercept of the straight line, the mean activation enthalpy ΔH for the low-temperature peak has been obtained and its value is 1.02 ± 0.05 eV. Internal friction in magnesium has been studied by several authors. Their results are summarised in **Table 2**. The peaks found above room temperature up to 100°C were described to dislocation effects. As far as the peak observed may have a dislocation nature, prestraining of the sample at a low temperature should influence the height of the peak. Deformation by rolling at room temperature ($\varepsilon \approx 2\%$) increased the strength of the low-temperature peak. This result is different from an estimation presented in the paper [14], where probably the identical peak P_D vanished after annealing at 570 K. Similar peak was found by Trojanová and co-workers [40, 41] for ultra-fine-grained magnesium prepared by ball milling and hot extrusion.

In the case of a relaxation processes linked to dislocation motion, the activation area $A = V/b$ (b is the Burgers vector and V activation volume) is the area that a dislocation has to move in order to activate the process, this means to overcome the barrier. Therefore, the product $V\sigma^*$

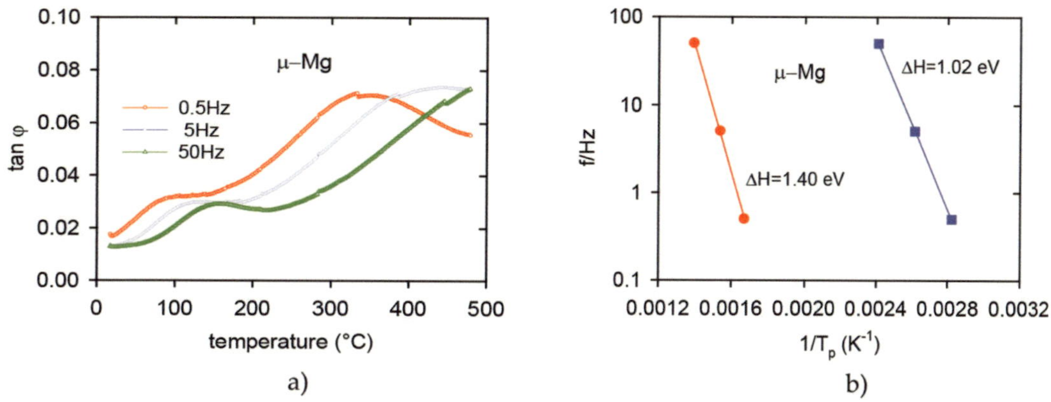

Figure 15. Temperature dependence of the loss angle measured at three frequencies (a) and Arrhenius plots constructed for low- and high-temperature peaks (b).

f/Hz	*P*1 (°C)	*P*2 (°C)
0.5	84.7	325
5	133.9	377.4
50	142.2	445.2

Table 1. Peak temperatures observed for various frequencies.

Material	*T* (°C)	*f* (Hz)	Δ*H* (eV)	References
Mg99.999%	76.8	1–2		Seyed Reihani [42]
Mg99.97%	219.8	0.46		Kê [4]
Mg99.9999%	67	1		Nó et al. [14]
Mg99.9999%	157	1		Nó et al. [14]
High purity Mg	146.8			Fantozzi et al. [43]
Mg99.9999%	148.8	1	1.00	Nó [44]
μ-Mg	84.8	0.5	1.02	This work
μ-Mg	325	0.5	1.40	This work
UFG-Mg	102	1	0.95	Trojanová et al. [40]
UFG-Mg	358	1	1.77	Trojanová et al. [40]
UFG-Mg+3nGr	78	1	1.08	Trojanová et al. [45]

Table 2. Occurrence of the internal friction peaks in magnesium prepared by various routs .

represents the work done by the applied stress to promote the dislocation motion over the local barrier. Neglecting the entropy term, the activation volume can be expressed as

$$V = -\frac{\partial \Delta H}{\partial \sigma} = -k \ln\left(\omega\tau_0\right) \frac{\partial T_p}{\partial \sigma}, \tag{24}$$

using Eq. (23) and the condition $\omega\tau = 1$ for peak appearance. **Figure 16a** shows the internal friction spectrum measured at different strain amplitudes ε (from 8.3×10^{-5} to 2.3×10^{-4}). As the strain amplitude increases the peak shifts towards higher temperatures and its height increases while the background increases moderately. Taking for $\ln(\omega\tau_0) = -31.33$ ($\tau_0 = 7.9 \times 10^{-15}$ s, $\omega = \pi$ Hz), and for $\partial T_p/\partial\sigma = (1/E)1.84 \times 10^4$ (linear regression from **Figure 16b**, $\sigma = E\varepsilon$, E is the Young modulus), we obtain for the activation volume $V(\text{exp}) = 1.5 \times 10^{-28}$ m^3 and dislocation activation volume $V_d = \psi V_{exp} = 7.2 \times 10^{-28}$ m^3 = 22 b^3 (E = 42 GPa, and the Taylor factor ψ = 4.8). Considering geometrical interpretation of the activation volume

$$V = bd\,\ell_D = bA, \tag{25}$$

where the energy barrier has a width d. ℓ_D is the length of the dislocation segment between barriers. Parameters of the thermally activated process (activation enthalpy and activation volume) are typical for certain barriers and their experimental estimation gives a possibility to identify the main barriers relevant for dislocation motion. Usually, the thermally activated process is studied in plastic deformation. On the other hand, the inelastic behaviour of metallic materials can be considered as the thermally activated dislocation motion.

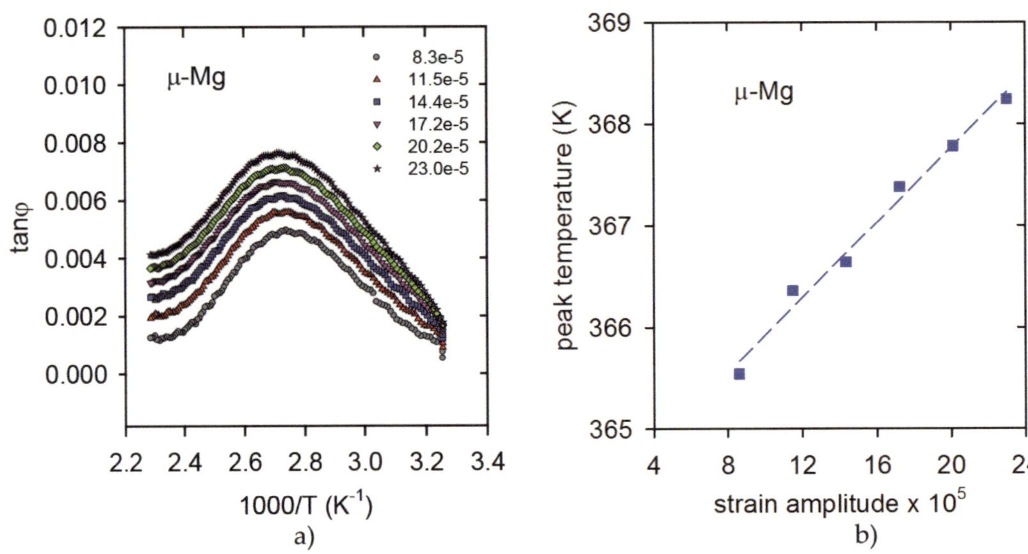

Figure 16. Temperature dependence of the low-temperature peak measured for various strain amplitudes (a) and amplitude dependence of peak temperatures (b).

The main deformation mode in magnesium is basal slip of <a> dislocations. The secondary conservative slip may be realised by <a> dislocations on prismatic and pyramidal planes of the first kind. Couret and Caillard studied by TEM prismatic glide in magnesium in a wide temperature range [46, 47]. They showed that the screw dislocations with the Burgers vector of $\frac{1}{3}$ $[11\bar{2}0]$ are able to glide on prismatic planes and their mobility is much lower than the mobility of edge dislocations. The deformation is controlled by thermally activated glide of these screw dislocation segments. A single controlling mechanism has been identified as the Friedel-Escaig mechanism. This mechanism assumes a dissociated dislocation on a compact plane (0001) that joins together along a critical length L_r producing double kinks on non-compact plane. The activation energy for this process is

$$U = 2U_K + 4U_C + 2U_R \tag{26}$$

where $2U_K$, $4U_C$ and $2U_R$ are the formation energies of the kink pair, of the four constrictions and of the two recombined segments, respectively. The activation area is proportional to the critical length between two kinks and it should be small.

$$A \propto L_r b. \tag{27}$$

The theoretical prediction leads to a value of $2U_K$ superior to 1.2 eV and to an activation area of about 15 b^2. This mechanism should give an internal friction peak similar to the Bordoni relaxation. The Friedel-Escaig mechanism is also sometimes called as pseudo-Peierls mechanism. Additionally, in this model the maximum of the internal friction peak must increase with increasing oscillating stress (strain amplitude) from a certain critical value. Thus, the comparison between the theoretical predictions and the experimental results leads to the conclusions that the observed relaxation is very probably pseudo Bordoni relaxation based on glide of screw dislocations on non-compact planes controlled by the Friedel-Escaig mechanism.

The high-temperature peak has been observed at temperature ~325°C (0.5Hz). The cold prestraining of the sample did not affect the strength of the high-temperature peak. Position and height of the peak are very stable during heating as well as cooling. The activation energy was obtained from the frequency dependence of the peak temperature (Arrhenius plot) to be 1.40 ± 0.05 eV (see **Figure 15b**). Existence of the high-temperature peak in magnesium has been reported by Kê [4] and other authors [5, 6, 10, 11]. They described this peak as being related to the grain boundary relaxation. Grain boundary sliding is realized by the slip and climb (providing a maintenance of vacancy sources and sinks) of grain boundary dislocations. Since both modes of dislocation motion must occur simultaneously, the slower one will control the grain boundary sliding. The climb mode involves jog formation and grain boundary diffusion and both modes may be affected by the interaction with impurities segregated in grain boundaries. According to ref. [48], grain boundary dislocation segments vibrate under applied cyclic stress, restoring force K is assumed to be from the elasticity of the limiting grains at triple junctions. According to this theory, the relaxation strength is for small grains not depending on the grain size:

$$(\tan\varphi)_{max} = \frac{G\rho_s b^2}{C}, \tag{28}$$

where ρ_s is the dislocation grain boundary surface density (total dislocation length per unit of grain boundary area), G is the shear modulus and C is a constant. The relaxation time in this model is

$$\tau = \frac{kT}{b^2 C_j D K} \exp\left(\frac{\Delta H_j + \Delta H_{GB}}{kT}\right) = \tau_0 \exp\left(\frac{\Delta H}{kT}\right), \tag{29}$$

where C_j is the linear density of jogs along the dislocation line, D is the diffusion coefficient, ΔH_j, ΔH_{GB} are the activation enthalpies for the jog formation and the grain boundary diffusion, respectively. The activation enthalpy for grain boundary diffusion in the coarse-grained magnesium is 0.95 eV [49]. Comparing with the estimated activation enthalpy value of 1.40 eV, the possible jog formation energy should be approximately 0.45 eV, which seems to be a reasonable value. Similar peaks describing the grain boundary sliding are introduced in **Table 2**. Peak observed by Kê [4] was found at 219.8°C and by Trojanová [41] at 358 °C. This is very probably due to the fact that microcrystalline and ultra-fine magnesium were prepared by milling in the protective Ar atmosphere containing 1% of oxygen. MgO particles in the grain boundaries together with the alumina nanoparticles may influence grain boundary diffusion characteristics.

3.1.1. *Summary*

Microcrystalline magnesium was prepared by milling and hot extrusion. Internal friction was measured as a function of temperature with three frequencies 0.5, 5 and 50 Hz. Two relaxation peaks were observed: a peak at 85 (0.5 Hz) with activation energy of 1.02 eV, and a peak at 325°C (0.5 Hz) with activation energy of 1.40 eV. The estimated small values of the activation volume and the peak sensitivity to the stress amplitude lead us to conclude that the low-temperature peak is due to the screw dislocation motion on the prismatic plane controlled by the Friedel-Escaig mechanism. Grain boundary sliding is probably the reason for appearing at the high-temperature peak.

3.2. Mg-Gr nanocomposite

The Mg with 3 vol% of Gr nanoparticles was prepared by the similar procedure as the microcrystalline Mg (see Chapter 3.1). The Mg micropowder with a median particle diameter of about 40 μm was mixed for 8 h with 3 vol.% of graphite powder and milled for 8 h in the planetary ball mill in a sealed argon atmosphere. The composite was encapsulated in an evacuated Mg container, degassed at 350°C, and extruded by the preheated (350°C) 400 t horizontal extrusion press. The material was studied in a transmission electron microscope; the mean grain size of the material was found to be about 200 nm. Temperature relaxation spectra of internal friction were measured in a DMA 2980 (TA Instruments) apparatus while heating and cooling. In a temperature range from room temperature up to 480°C, the measurements were performed in the multi-frequency mode at three frequencies 0.5, 1 and 2 Hz. Throughout the measurements, the strain amplitude was 1.2×10^{-4}. The heating as well as cooling rate was 1 K/min. No grain growth during heating up to 480°C was observed.

http://dx.doi.org/10.5772/67028

Figure 17a shows the temperature dependence of the mechanical loss angle tanφ measured at three frequencies during first heating sub-cycle. Two types of peak have been found: the weak low-temperature peak and the more pronounced two high-temperature peaks. While the low-temperature peak position in the temperature scale depends on the measuring frequency, position of both high-temperature peaks does not depend on the frequency. **Figure 17b** shows the temperature dependence of mechanical loss angle measured during cooling. It is seen that the two high-temperature peaks developed into one peak. Position of this peak is not depending on frequency and it remained constant in the next heating-cooling cycles. The low-temperature peaks extracted from the internal friction spectrum subtracting the background damping are introduced in **Figure 18a, b** which shows the Arrhenius plot for the low-temperature peak. An estimated activation enthalpy of 1.08 ± 0.05 eV is close to activation enthalpies found for μMg and UFG-Mg as it is given in **Table 2.** This fact and together with found peak temperature from a range of 67–102°C indicates that the low-temperature peak is due to the reversible screw dislocation motion on the prismatic plane controlled by the Friedel-Escaig mechanism as it was described in paragraph 3.1.

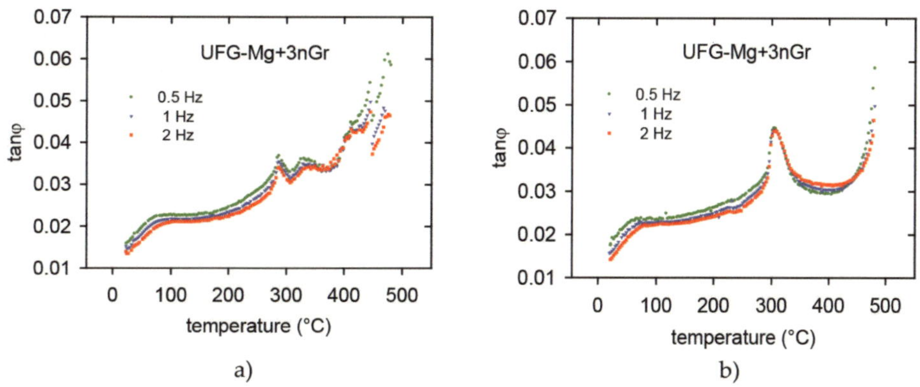

Figure 17. Temperature relaxation spectrum of internal friction measured for three frequencies while heating (a) and cooling (b).

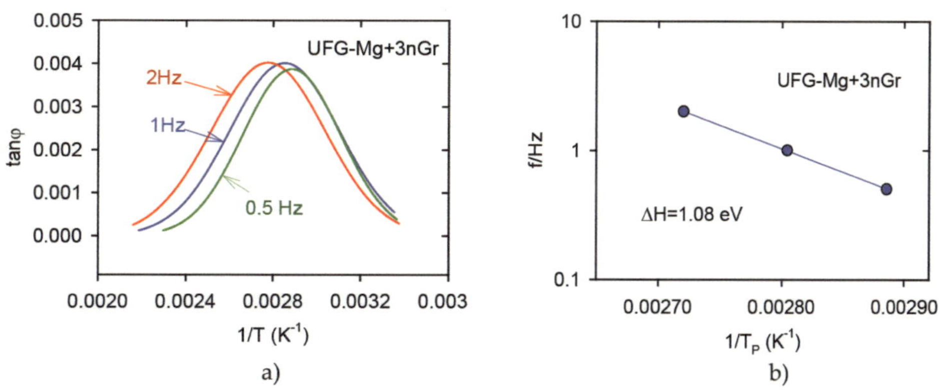

Figure 18. Subtracted low-temperature peaks (a), Arrhenius plot and activation enthalpy obtained for the low-temperature peak (b).

Observed independence of the high-temperature peak on the frequency indicates that this peak has no Debye nature. Its existence is due to newly created dislocations and their motion. An increase in the dislocation density near reinforcement with the different CTE compared with the matrix has been calculated according to Arsenault & Shi [50]

$$\rho = \frac{B\,V_f \Delta\alpha \Delta T}{b\left(1 - V_f\right)} \frac{1}{t_{\min}}, \tag{30}$$

where V_f is the volume fraction of the reinforcing phase, t_{min} its minimum size, b the magnitude of the Burgers vector of dislocations, B a geometrical constant, $\Delta\alpha$ the absolute value of the CTE difference between the matrix and the reinforcement, ΔT the temperature difference. Rawal [51] studied the interface in the cast Mg-Gr composite. He found, using the transmission electron microscopy, enhanced dislocation density near the interface. Dislocations were generally linear and corresponded to available basal slip systems in hcp Mg matrix. Dislocation density measurements made using the weak beam imaging technique indicate an average dislocation density of $1.6 \times 10^{13} - 6 \times 10^{14}$ m^{-2}. High density of dislocations may be attributed to the residual stress state generated because of differential contraction between the Gr and matrix during the fabrication process. In the Mg/Gr composite studied in the present work, the CTE value of high purity Mg matrix is 26×10^{-6} K^{-1} and that of the Gr is approximately 4.0×10^{-6} K^{-1} (the value for carbon fibres is -1.4×10^{-6} K^{-1} [52]). It is expected that such large difference could generate a high density of dislocations during heating as well as cooling of Mg/Gr system. The residual strain, or strain accumulation, produced by the thermal mismatch may be calculated from

$$\varepsilon = \Delta\alpha\Delta T. \tag{31}$$

In the absence of thermal stresses, the vibrating dislocation segments do the dislocation damping IF_d. If the composite material is heated, compressive thermal stresses σ_{th} are generated in the matrix and induce a long-range movement of dislocation, which is superimposed on the oscillatory motion induced by the apparatus. Several models are reported in the literature to describe the mechanism(s) responsible for damping in composites due to dislocation motion in the vicinity of reinforcing phase. Mayencourt and Schaller [24] have reported a non-linear dependence of the internal friction in MMCs with the cooling rate, measuring frequency and strain amplitude. Dislocation motion near the interface is controlled by the solid friction mechanism and break-away of dislocation from solutes. Vincent and co-workers [53] developed a model where a coupling between the thermal-induced strain and the alternating shear stress of the apparatus is considered. Carreño-Morelli and co-workers [25] developed a model based on the growing plastic zones around the reinforcement. Wei et al. [54] studied internal friction in Al with Gr particulates. An internal friction peak was found at about 260°C. The internal friction peak is attributed to the thermal dislocations, the thermal stress coupled to the exciting stress, and the interaction between the dislocations and the cyclic applied stress.

3.2.1. Summary

Mg with 3 vol.% of Gr nanoparticles prepared by milling and hot extrusion has been characterised by mechanical spectroscopy during continuous heating as well as cooling. Internal friction was measured as a function of temperature with three frequencies ranging from 0.5 to 2 Hz. Two internal friction peaks were observed in the temperature relaxation spectrum. The peak observed in the vicinity of ≈ 80°C, having the activation energy of 1.08 eV, has been described to the reversible screw dislocation motion in the Mg grains on the prismatic plane controlled by the Friedel-Escaig mechanism. Thermal stresses generated due to thermal mismatch during heating as well as cooling are very probably the reason for the peak observed at ≈ 300°C. Position of this peak is insensitive to the measuring frequency. Increased dislocation density induced by thermal stresses and movement of dislocations under the thermal stress as well as the applied stress is the reason for the existence of this peak.

3.3. AZ magnesium alloys

TDIF was measured in cast magnesium alloys AZ31, AZ63 and AZ91 after homogenisation treatment (22 h at 390°C). The resonant frequency and damping analyser (RFDA) was used to determine the damping and resonant frequency. The measurements were performed in a wide temperature range from room temperature up to 400°C. The prism-shaped samples 75 × 20 × 10 mm^3 were excited to vibrations in the resonant frequency using a small striker. Damping of the sample-free vibrations was registered with a microphone and processed using special software. The resonant frequency exhibiting ~8 kHz was estimated by means of Fourier transform. The linear heating rate was chosen in the range from 1 to 4 K/min. The AZ alloys are formed by the solid solution of Al and Zn atoms in Mg (δ-phase) and the $Mg_{17}Al_{12}$ intermetallic compound (γ-phase). Precipitates existing in these alloys may be of two kinds: discontinuous precipitate (DP), which is formed by lamellae $\gamma + \delta$, and continuous precipitate (CP) having usually a form of small discs or spheres. Microanalysis of all three alloys showed the presence of the γ-phase in the AZ31 alloy while the discontinuous precipitate was observed only rarely. In the AZ63 alloy, electron compound $Mg_{17}Al_{12}$ was found beside discontinuous precipitate. AZ91 alloy exhibits similar microstructure as AZ63 alloy, only with the different proportional representation of individual phases.

The temperature dependences of the logarithmic decrement measured at the heating rate of 2 K/min for AZ31, AZ63 and AZ91 alloys are introduced in **Figure 19a**. The temperature dependence of the logarithmic decrement was measured in the AZ91 sample after the homogenisation treatment while heating is introduced in **Figure 19b**. The heating rate was 60 K/h, the cooling rate was not controlled. The temperature record showed the linear decrease of temperature up to 120–140°C. When room temperature was reached, a new measurement was started. The measurements were repeated three times. It is obvious from **Figure 19b** that the decrement is more or less constant up to approximately 220°C, and then it increases with increasing temperature. At temperature ~320°C, a local maximum of the logarithmic decrement is observed. The height of this maximum decreases in the second and third run of the measurement. The position of maximum in the temperature axis is the same for all runs.

The temperature dependence of the logarithmic decrement was measured for three heating rates as it is shown in **Figure 20a**. Note that before each measurement, a new homogenisation treatment was performed. Only the first runs of the measurements are given in **Figure 20a**. The higher heating rate, the higher temperature of the local maximum is observed. The height of the maximum simultaneously decreases with the increasing heating rate. To reveal the physical nature of the observed local maximum in the temperature dependence of the logarithmic decrement, a sample with the smaller thickness was tested. Maximum position in the temperature scale was not shifted for the measurement with the resonant frequency of ~7 kHz (see **Figure 20b**). This fact indicates that the observed maximum is not of the Debye type. Some transitory effects must be considered.

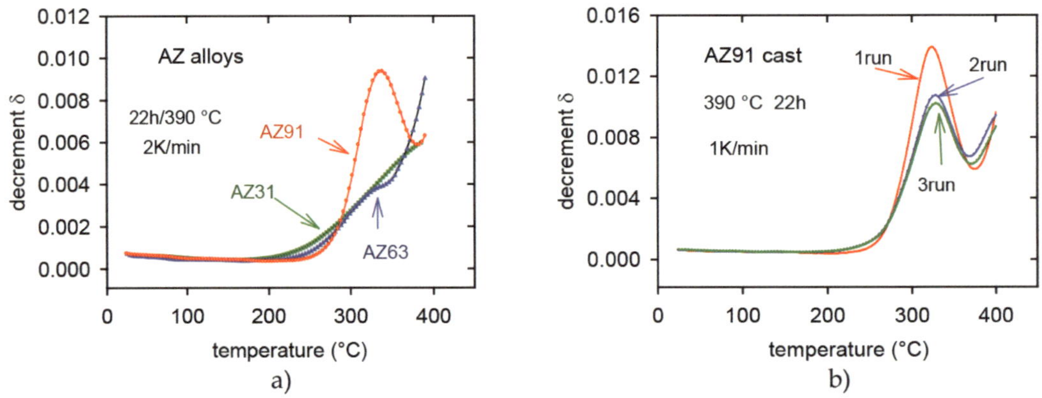

Figure 19. Temperature dependence of the decrement measured for three AZ alloys (a) and AZ91 alloy in three runs (b).

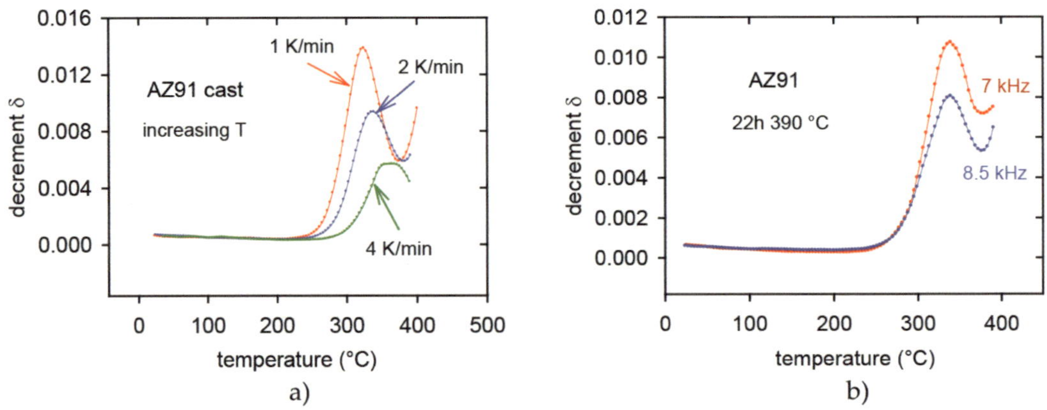

Figure 20. Temperature dependence of the logarithmic decrement while heating performed in three runs for the heating rate 1 K/min (a) and for three heating rates (b).

In the temperature dependence of the logarithmic decrement, four characteristic points were selected: 220, 320, 360 and 390°C. Samples were subjected to the homogenisation treatment and then put into a furnace with the predetermined heating rate of 1 K/min. Reaching the chosen temperature, samples were step by step removed from the furnace and quenched into water of 60°C. Scanning electron microscopy (SEM) observations were performed at room temperature. After heating of the sample with the heating rate of 1 K/min up to 220°C supersaturated solid solution decomposed; DP appeared in the vicinity of grain boundaries as it is obvious from **Figure 21a**, small precipitates visible as light points are CP, their size slowly increased. Tiny CPs are situated primarily in the annealing twins, the matrix in the vicinity of twins is purified, density of CPs is in these places lower (**Figure 21b**).

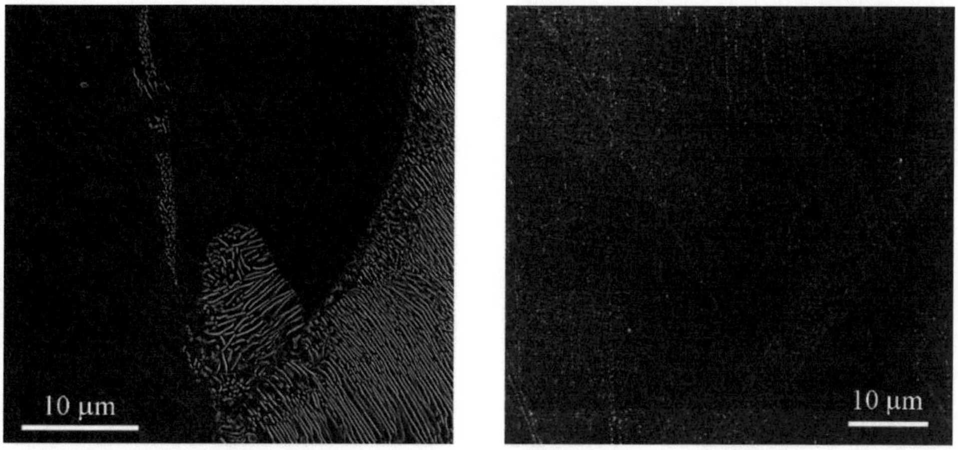

a) b)

Figure 21. Microstructure of the alloy observed homogenisation and linear heating up to 220 °C: discontinuous precipitates (a), small continuous precipitates (b).

When temperature increased up to 320°C, decomposition of DPs started, and then the growing of CPs (**Figure 22a**) followed. Original lamellae were reshaped into small discs as it is visible in **Figure 22b**. CPs depicts the twin; small discs of CPs are ordered in twin/grain boundaries. Note that a temperature of 320°C corresponds to the peak temperature in the temperature dependence of decrement.

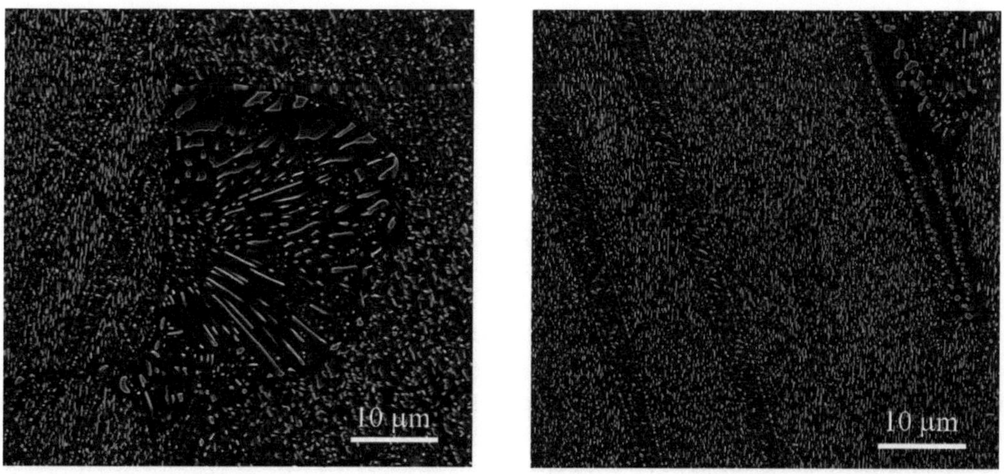

a) b)

Figure 22. Microstructure of the sample subjected to homogenisation treatment and annealed up to 320 °C: decomposition of discontinuous precipitates (a), continuous precipitates situated in twins.

Microstructure of the sample annealed up to 360°C (local minimum in the temperature dependence of decrement) is characterised by the growth of CP (**Figure 23a**). Light points are small Mn particles; small content of Mn (0.24 wt.%) was used for grain refining. In the vicinity of grain boundaries, purified places were established (**Figure 23b**). Meander-like grain boundaries are decorated with larger precipitates (**Figure 23a, b**).

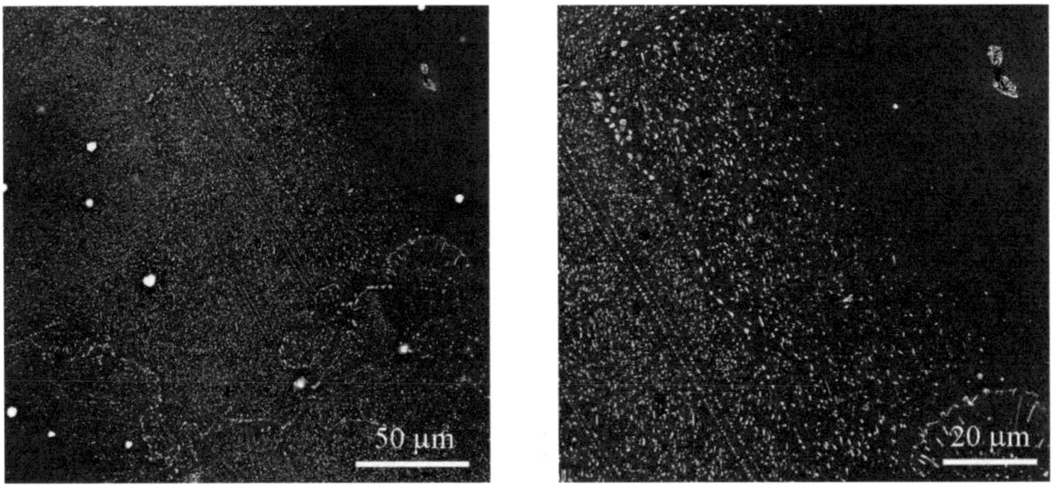

Figure 23. Microstructure of the sample after homogenisation treatment and annealed up to 360 °C: growing continuous precipitates (a), purified place free of precipitates in the upper right corner (b).

At 390°C, the transformation of DP into CP is finished. The CPs decorate the grain boundaries (**Figure 24a**), DP vanished as it is obvious from **Figure 24b** and the size of CP increased. It is possible to conclude from the microstructure observations that, in the temperature range from 220 to 360°C, a transformation of DP originally formed up to 220°C stepwise passed off. Disc-shaped CP changed into small spheres and at the highest temperatures into larger particles with a complex shape. Increasing the heating rate, time for the transformation of DP into CP is shorter. It is very probably the reason why the maximum height decreases with increasing heating rate and at the same time, the maximum is shifted to higher temperatures. DPs are formed when the grain boundary diffusion is dominant, whereas the CPs are formed at higher temperatures from solid solution when volume diffusion becomes faster. As it was shown in previous paragraphs, internal friction is very sensitive to the material microstructure, volume fraction and distribution of structural defects. Any modification in the microstructure induces changes in the internal friction course. Thus, the internal friction measurements may provide information about processes occurring at the atomic scale. Internal friction maxima were found in the AZ magnesium alloys by several authors. Lambri and co-workers [11] found the internal friction peak at about 152°C using the measuring frequency of 1 Hz. Authors ascribed this peak to the grain boundary sliding. The peak height may be controlled by a decrease in solute atoms due to precipitation process. The activation energy estimated from the Arrhenius plot was evaluated as 1.13 eV. Similarly, Liu and co-workers [55] found the internal friction peak situated at 165°C and a frequency of 1 Hz. They determined the corresponding activation energy of 1.26 eV. The peak was ascribed to the grain boundary relaxation. Two internal friction peaks were observed in [19]. P_1 peak estimated at 165°C was characterised by the activation energy of 1.31 eV and the high temperature P_2 peak was found to be sensitive to the heating rate and insensitive to the measuring frequency. Analogous peak to the P_2 internal friction peak was estimated also by Soviarová et al. [28] at 320°C, frequency of 20.4 kHz and a heating rate of 1 K/min. Theoretical models of the internal friction consider that transformations in the solid state may be manifested by internal friction peaks. Damping mechanism can be associated with the kinetics of atoms diffusing to evolving precipitates [56]. Observed independence of the maximum position on frequency indicates that the Arrhenius equation cannot be considered in this case, that is, the interface relaxation and diffusion processes may be excluded as the reason for the maximum occurrence. During the IF measurement, the movable γ-$Mg_{17}Al_{12}$/δ-Mg interface consumes the work. When a new interface forms, defects (solute atoms and vacancies) diffuse and accumulate in the interface which leads to the gradual pinning of the interface. The movable space is a function of heating time, that is, the heating rate. The mobile interface stabilises, so that the internal friction decreases.

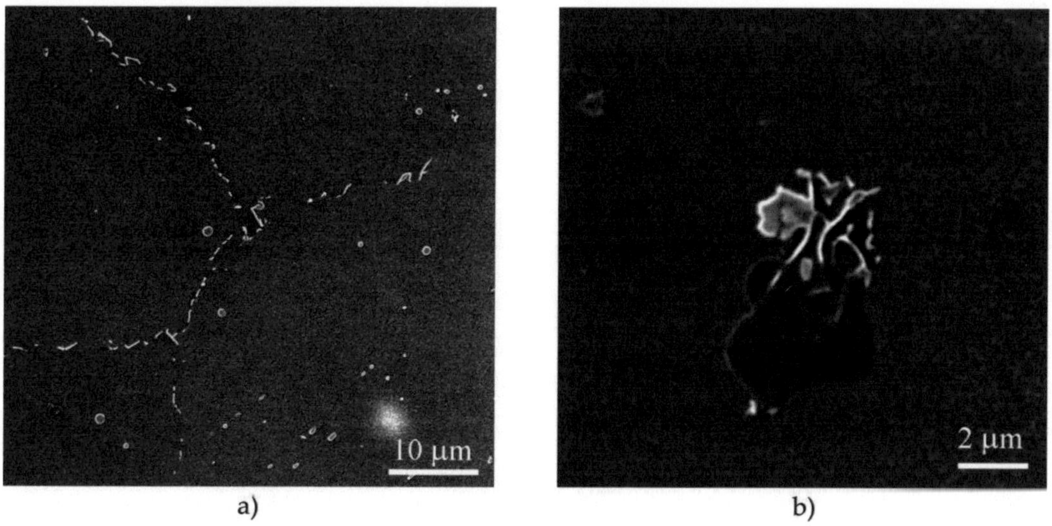

Figure 24. Microstructure of the sample after homogenisation treatment and annealed up to 390 °C: continuous precipitates in grain boundaries (a), particle of γ-$Mg_{17}Al_{12}$ phase (b).

3.3.1. *Summary*

Internal friction measurements of the AZ31, AZ63 and AZ91 alloys showed an exponential increase of the logarithmic decrement with temperature. In AZ63 and AZ91 alloys, this increase is modulated by a local maximum at about of 320°C. The maximum height decreased in the second and third run while the maximum temperature remained the same. The position of the local maximum was sensitive to the heating rate and insensitive to the frequency. Increasing heating rate shifted the peak temperature to higher temperatures. The occurrence of the internal friction maximum is connected with the transformation of discontinuous precipitates to continuous ones. Strain (stress)-supported flux of solute atoms and movement of the γ-$Mg_{17}Al_{12}$/δ-Mg interface are very probably the reason for the absorption of mechanical energy carried by an ultrasonic wave.

Acknowledgements

Z.T. and P.L. gratefully acknowledge the financial support from the Czech Science Foundation under Grant 107-15/11879S. P.P. and M.C. are grateful to the Ministry of Education and Academy of Science of the Slovak Republic for the support by the project VEGA No. 1/0683/15. Parts of this chapter are reproduced from authors' recent publications [34, 41] and conference contributions [15, 36]. We thank our co-authors for permission to reuse parts of the papers. This chapter is dedicated in the memory of Professor Werner Riehemann from the Clausthal University of Technology, who passed away in January 2016.

Nomenclature

Quantities (selected)

σ, ε stress, strain amplitude;

δ, logarithmic decrement;

tanφ loss factor;

Q quality factor;

ρ dislocation density

ΔH activation enthalpy;

V activation volume;

λ, L_N length of dislocation segments

b Burgers vector of dislocations

τ relaxation time

ω angular frequency

α thermal expansion coefficient

Materials

AZ31 Mg-3wt%Al-1wt%Zn

AZ63 Mg-6wt%Al-3wt%Zn

AZ91 Mg-9wt%Al-3wt%Zn

AM20 Mg-2wt%Al-0.4wt%Mn

AX41 Mg-4wt%Al-1wt%Ca

ZC63 Mg-6wt%Zn-3wt%Cu

QE22 Mg-2wt%Ag-2wt%RE

Saffil® 96-97%Al_2O_3, 4 to 3%SiO_2

μMg microcrystalline Mg

UFG-Mg ultrafine-grained Mg

nGr graphite nanoparticles

nZrO_2 zirconia nanoparticles

Author details

Zuzanka Trojanová[1]*, Peter Palček[2], Pavel Lukáč[1] and Mária Chalupová[2]

*Address all correspondence to: ztrojan@met.mff.cuni.cz

1 Department of Physics of Materials, Faculty Mathematics and Physics, Charles University, Prague, Czech Republic

2 Department of Materials Engineering, Faculty of Mechanical Engineering, University of Žilina, Žilina, Slovakia

References

[1] Granato AV, Lücke K: Theory of mechanical damping due to dislocations. J. Appl. Phys. 1956; **27**:583–593. http://dx.doi.org/10.1063/1.1722436.

[2] Granato AV, Lücke K: Application of dislocation theory to internal friction phenomena at high frequencies. J. Appl. Phys. 1956; **27**:789–805. http://dx.doi.org/10.1063/1.1722485

[3] Granato AV, Lücke K: Temperature dependence of amplitude-dependent dislocation damping. J. Appl. Phys. 1981;**52**: 7136–7142. http://dx.doi.org/10.1063/1.328687

[4] Kê TS: Stress relaxation across grain boundaries in metals. Phys. Rev. 1947;**72**: 41–46.

[5] Hu XS, Zhang YK, Zheng MY, Wu K: A study of damping capacities in pure magnesium and Mg-Ni alloy. Scr. Mater. 2005;**52**:1141–1145.

[6] Hu XS, Wang XJ, He XD, Wu K, Zheng MY: Low frequency damping capacities of commercial pure magnesium. Trans. Nonferrous Met. Soc. China 2012;**22**:1907–1911. DOI: 10.1016/S1003-6326(11)61406-4

[7] Fan GD, Zheng MY, Hu XS, Xu C, Wu K, Golovin IS: Effect of heat treatment on internal friction in ECAP processed commercial pure Mg. J. Alloys Compd. 2013;**549**:38–45. doi: 10.1016/j.jallcom.2012.09.040

[8] Fan GD, Zheng MY, Hu XS, Wu K, Gan WM, Brokmeier HG: Internal friction and microplastic deformation behavior of pure magnesium processed by equal channel angular pressing. Mater. Sci. Eng A 2013; **561**:100–108. doi: 10.1016/j.msea.2012.10.083

[9] Esnouf C, Fantozzi G: Medium temperature internal friction in high purity f.c.c and h.c.p metals. J. Phys. C5, 1981;445–450. Doi: 10.1051/jphyscol:1981566

[10] Jiang WB, Kong QP, Magalas LB, Fang QF: The internal friction of single crystals, bicrystals and polycrystals of pure magnesium. Archives of Metallurgy and Materials 2015; **60**:371–375. doi:10.1515/amm-2015-0061

[11] Lambri OA, Riehemann W, Trojanová Z: Mechanical spectroscopy of commercial AZ91 magnesium alloy. Scripta Mater. 2001;**45**:1365–1371. http://dx.doi.org/10.1016/S1359-6462(01)01171-X DOI:10.1016/S1359-6462(01)01171-X#doilink

[12] Fantozzi G, Esnouf C, Benoit W, Ritchie I: Internal friction and microdeformation due to the intrinsic properties of dislocations: The Bordoni relaxations. Prog. Mater. Sci.1982;**27**:311–451. doi:10.1016/0079-6425(82)90003-2

[13] Seyed Reihani SM, Fantozzi G, Esnouf C, Revel G: Frottement intérieur d'un magnésium de haute pureté déformé plastiquement (Internal friction in high purity magnesium plastically deformed). J. Phys. Lett. 1978;**39**:L-429–432. http://dx.doi.org/10.1051/jphyslet:019780039022042900

[14] Nó ML, Oleaga A, Esnouf C, San Juan J: 1990, Internal friction at medium temperatures in high purity magnesium. Phys. Stat. Sol. (a) 1990;**120**:419–427. doi:10.1002/pssa.2211200214

[15] Trojanová Z, Weidenfeller B, Lukáč P, Riehemann W, Staněk M: Anelastic properties of nanocrystalline magnesium, in: M.J. Zehetbauer, R.Z. Valiev editors Nanomaterials by severe plastic deformation, Weiheim: Wiley VCH-Verlag GmBH; 2004. 413–419.

[16] Trojanová Z, Lukáč P, Száraz Z, Drozd Z: Mechanical and acoustic properties of magnesium alloys based (nano) composites prepared by powder metallurgical routs. In: Light Metal Alloys Applications, InTech, Rijeka 2014, 163. ISBN 980-953-307-1127-9, 163–198. http://dx.doi.org/10.5772/57454

[17] González-Martínez R, Göken J, Letzig D, Timmerberg J, Steinhoff K: Influence of heat treatment on damping behaviour of the magnesium wrought alloy AZ61. Acta Metall. Sin. 2007;**20**:235–240. doi: 10.1016/S1006-7191(07)60033-7

[18] González-Martínez R, Göken J, Letzig D, Steinhoff K, Kainer KU: Influence of aging on damping of the magnesium-aluminium-zinc series. J. Alloys Compd. 2007;**437**:127–132. http://dx.doi.org/10.1016/j.jallcom.2006.07.085 DOI:10.1016/j.jallcom.2006.07.085#doilink

[19] Hao GL, Han FS, Wang QZ, Wu J: Internal friction peaks associated with the precipitation in AZ91 magnesium alloy. Phys. B 2007;**391**:186–192. http://dx.doi.org/10.1016/j.physb.2006.09.018 DOI:10.1016/j.physb.2006.09.018#doilink

[20] Fan GD, Zheng MY, Hu XS, Xu C, Wu K, Golovin IS: Improved mechanical property and internal friction of pure magnesium processed by ECAP. Mater. Sci. Eng. A 2012;**566**: 588–594.http://dx.doi.org/10.1016/j.msea.2012.07.031DOI:10.1016/j.msea.2012.07.031#doilink

[21] Zhou H, Wang J-F, Pan F-S, Xu D-D, Tang A-T, Hao LH: Influence of rolling on internal friction peak of Mg3Cu1Mn alloy. Trans. Nonferrous Met. Soc. China 2013;**23**:1610–1616. doi:10.1016/S1003-6326(13)62638-2

[22] Schoeck G: Internal friction due to precipitation. Phys. Stat. Sol. (b) 1969;**32**:651–658. doi: 10.1002/pssb.19690320216

[23] Cui Y, Li Y, Sun S, Bian H, Huang H, Wang Z, Koizumi Y, Chiba A: Enhanced damping capacity of magnesium alloys by tensile twin boundaries. Scr. Mater. 2015;**101**:8–11. http://dx.doi.org/10.1016/j.scriptamat.2015.01.002

[24] Mayencourt C, Schaller R: A theoretical approach to the thermal transient mechanical loss in Mg matrix composites. Acta Mater. 1998;**46**:6103–6114. http://dx.doi.org/10.1016/S1359-6454(98)00274-2

[25] Carreño-Morelli E, Urreta SE, Schaller R: Mechanical spectroscopy of thermal stress relaxation at metal-ceramic interfaces in aluminium-based composites. Acta Mater. 2000;**48**:4725–4733. http://dx.doi.org/10.1016/S1359-6454(00)00264-0

[26] Trojanová Z, Weidenfeller B, Riehemann W: Thermal stresses in a Mg-Ag-Nd alloy reinforced by short Saffil fibres studied by internal friction. Mater. Sci. Eng. A 2006;**442**:480–483. http://dx.doi.org/10.1016/j.msea.2006.04.137

[27] Lukáč P, Trojanová Z: Hardening and softening in selected Mg alloys, Mater. Sci. Eng. A 2007;**462**:23–28. http://dx.doi.org/10.1016/j.msea.2006.01.170

[28] Soviarová A, Palček P, Trojanová Z: 2014, Monitoring of precipitation process in AZ31 and AZ91 magnesium alloys by internal damping measurement, Manuf. Technol. 2014;**14**:447–451.

[29] D'Anna G, Benoit W, Vinokur VM: Internal friction and dislocation collective pinning in disordered quenched solid solutions. J. Appl. Phys. 1997;**82**:5983–5990. http://dx.doi.org/10.1063/1.366463

[30] Farkas G, Trojanová Z, Száraz Z, Minárik P, Máthis K: Effect of the fiber orientation on the deformation mechanisms of magnesium-alloy based fiber reinforced composite. Mater. Sci. Eng. A, 2015;**643**:25–31. 10.1016/j.msea.2015.07.012

[31] Chmelík F, Kiehn J, Lukáč P, Kainer K-U, Mordike BL: Acoustic emission and dilatometry for non-destructive characterisation of microstructural changes in Mg based metal matrix composites submitted to thermal cycling. Scripta Mater. 1998;**38**:81–88. 10.1016/S1359-6462(97)00416-8

[32] Kiehn J, Z. Trojanová Z, Lukáč P, K. U. Kainer KU: Effect of thermal cycling on the damping behaviour of Mg matrix composites. Key Eng. Mat. 1997;**127–131**:993–1000. 195.113.32.106-03/05/07,11:38:39)

[33] Chmelík F, Trojanová Z, Kiehn J, Lukáč P, Kainer KU: Non-destructive characterisation of microstructure evolution in Mg based metal matrix composites submitted to thermal cycling. Mater. Sci. Eng. 1997;A**234–236**:774–777. PII S0921-5093(97)00401-2

[34] Z. Trojanová, P. Lukáč: Microstructure changes in magnesium alloys base composites studied by internal friction. Acta Univ. Carol. Math. Phys. 2005;**46:**11–25.

[35] Carreño-Morelli E, Urreta SE, Schaller R, In: Proceedings of the International Conference on Fatigue of Composites (S. Degallaix, C. Bathias, R. Fougères, eds.), Société Française de Métallurgie et de Matériaux, Paris, 1997,France, 112–119.

[36] Riehemann W, Trojanová Z, Mielczarek A: Fatigue in magnesium alloy AZ91-gamma Alumina fiber composite studied by internal friction measurements FATIGUE 2010. Proc. Eng. 2010;**2**:2151–2160. 10.1016/j.proeng.2010.03.231

[37] Lemaitre J: A Course on Damage Mechanics. Berlin 1996, Springer. ISBN 978-3-642-18255-6

[38] Göken J, Riehemann W: Damping behaviour of AZ91 magnesium alloy with cracks. Mater. Sci. Eng. A 2004;**370**:417–421. doi:10.1016/j.msea.2003.08.110

[39] Nowick AS, Berry BS: Anelastic relaxations in crystalline solids. Academic Press, 1972; New York, NY/London.

[40] Trojanová Z, Lukáč P, Weidenfeller B, Riehemann W: Mechanical damping in magnesium prepared by ball milling in medium temperature region. Kovove Mater. 2008;**46:**249–256.

[41] Trojanová Z, Lukáč P: Elastic and plastic properties of ultra-fine grained magnesium. Int. J. Mater. Product Technol. 2011;**40**:120–139. doi: 10.1504/IJMPT.2011.037209

[42] Seyed Reihani SM, Fantozzi G, Esnouf C, Revel G: 1979, Internal friction of high purity magnesium after plastic deformation. Scripta Metall. 1979;**13**:1011–1015. doi:10.1016/0036-9748(79)90194-7

[43] Fantozzi G, Esnouf C, Seyed Reihani SM, Revel G: 1984, Anelastic behaviour of plastically deformed high purity magnesium between 10 and 500 K. Acta Metall. 1984;**32**:2175–2183. doi:10.1016/0001-6160(84)90160-3

[44] Nó ML. Dislocation damping at medium temperature. In: Schaller R, Fantozzi G, Gremaud G, editors. Mechanical Spectroscopy Q^{-1} 2001. Trans Tech Publications Ltd 2001, Switzerland, 247–267.

[45] Trojanová Z, Bosse M, Ziegmann G, Mielczarek A, Ferkel H: Anelastic properties of Mg+3vol.%Gr prepared by ball milling. Key Eng. Mater. 2006;**319**:189–195. 195.113.32.106-03/05/07,11:37:04

[46] Couret A, Caillard D: An in situ study of prismatic glide in magnesium—I. The rate controlling mechanism. Acta Metall. 1985;**33**: 1447–1454. doi:10.1016/0001-6160(85)90045-8

[47] Courret A, Caillard D: An *in situ* study of prismatic glide in magnesium—II. Microscopic activation parameters, Acta Metall. 1985;**33**:1455–1462. doi:10.1016/0001-6160(85)90046-X

[48] Lakki A, Schaller R, Carry C, Benoit W: High temperature anelastic and viscoplastic deformation of fine-grained MgO-doped Al_2O_3. Acta Mater. 1998;**46**:689–700. http://dx.doi.org/10.1016/S1359-6454(97)00250-4

[49] Frost JH, Ashby MF: Deformation Mechanisms Maps. Pergamon Press; 1982, Oxford.

[50] Arsenault RJ, Shi N: Dislocation generation due to differences between the coefficients of thermal expansion. Mater. Sci. Eng. 1986;**81**:175–187. doi:10.1016/0025-5416(86)90261-2

[51] Rawal SP: Interface structure in graphite-fiber-reinforced metal–matrix composites. Surf. Interface Anal. 2001;**31**:692–700.

[52] Ibrahim IA, Mohamed FA, Lavernia EJ: Particulate reinforced metal matrix composites—a review. J. Mater. Sci. 1991;**26**:1137–1156. doi: 10.1007/BF00544448

[53] Vincent A, Lormand S, Durieux S, Girard C, Maire E, Fougères R: Transient internal damping in metal matrix composites: Exp. Theory. J. Phys. IV 1996;**6**:C8:719–730.doi: 10.1051/jp4:19968157

[54] Wei LN, Li ZB, Han, F.S. 2002, Thermal mismatch dislocations in macroscopic graphite particle-reinforced metal matrix composites studied by internal friction. Phys. Stat. Sol. (a) 2002;**191**:125–136. doi: 10.1002/1521-396X(200205)191:1<125::AID-PSSA125>3.0.CO;2-2

[55] Liu S-W. Jiang H-C, Li X-Y, Rong L-J: Effect of precipitation on internal friction of AZ91 magnesium alloy. Trans. Nonferrous Met. Soc. China 2010;**20**:453–457. doi:10.1016/S1003-6326(10)60517-1

[56] Song Z, Kishimoto S, Zhu J, Wang Y: Study of stabilization of CuAlBe alloy during martensitic transformation by internal friction. Solid State Commun. 2006;**139**:235–239. http://dx.doi.org/10.1016/j.ssc.2006.05.037

Section 2

Surface Treatment

Chapter 3

Mg Alloy Surface Treatment

Jae-In Yu

Additional information is available at the end of the chapter

http://dx.doi.org/10.5772/64884

Abstract

Before the surface treatment of magnesium alloy, it is necessary to consider the degree of electrical resistance, the apparent surface condition, and the degree of corrosion resistance, so that the most appropriate method could be used for the surface to be treated. Magnesium alloy also follows the basic rule of metal, and the best method is to process after mechanical and chemical cleaning.

Keywords: Mg alloy, surface treatment, plasma electrolytic oxidation, AZ31, AZ91, Aluminum, hard coating, porous

1. Introduction

The application of Mg alloy is expected, in particular involving motion or portability of a component, to increase in the future because of its high strength-to-weight ratio and a relatively high stiffness. Magnesium and its alloys are increasingly used in a number of components where weight reduction is of great concern. However, the poor corrosion resistance of Mg and its alloys has limited their applications in corrosive environments. Recently, the creep behavior of Mg-alloy has attracted much attention due to its useful value in practice. However, the conventional magnesium alloy has a relatively low resistance to creep at elevated temperatures in contrast to Al alloy, which has restricted the development of Mg alloy. In addition, magnesium alloys possess many good properties that make them ideal materials for a large number of applications. However, poor corrosion resistance and high chemical reactivity have hindered the wide applications of magnesium alloys. This chapter describes the process of surface treatment of the magnesium alloy.

2. Mechanical cleaning

First, this process is performed after removing oil from the surface. The processes of mechanical cleaning are as follows:

1. Grinding and rough polishing

   This process uses a belt and a grinding surface. It is used for surface treatment of the product by low-pressure casting.

2. Dry-abrasive blast cleaning

   This method falls in the sandblasting category and uses 25 or 35 AFS silica sand method.

3. Wet-abrasive blast cleaning

   This method is used prior to metal coating.

4. Barrel or bowl abrading

   This process is required for the manufacture of pressure casting.

3. Chemical cleaning

Different types of chemical cleaning methods are as follows:

1. Solvent cleaning and vapor degreasing.
2. Emulsion cleaning.
3. Alkaline cleaning.
4. Acid picking.
5. Fluoride anodizing.

4. Mechanical finishing operations

This method is performed as the last step after the chemical treatment or coating. The different types of mechanical finishing operations are as follows:

1. Mass (barrel) finishing.
2. Vibratory finishing.
3. Polishing and buffing.

5. Chemical conversion treatment

This method is used for improving the conversion of the magnesium surface of the neutral state to the state of an alkali. The primary reason is that takes place between the well is combined with other coating materials. It is used in the current processing methods, except in the light range of chromium. If the corrosion resistance is of excellent quality, the method based on the chromate-treated surface comes to mind as the first method.

1. Chrome-pickle treatment

   This method is used to remove the oxide formed on the surface of the total chromium and magnesium, and it can be expressed in gray and gold colors.

2. Dichromate treatment

   This method forms an oxide layer (2 μm) thicker than that formed after chrome-pickle treatment and this thick oxide layer is formed due to the characteristic high electrical resistance.

3. Chrome-manganese treatment

   - The processing method, including chrome-manganese developed by Magnesium Electron Company, is carried out at a temperature of 93℃.
   - It has good coating efficiency, as in chromate treatment.
   - In addition, this method is conducted in low temperature.

4. Ferritic nitrate picking

   - This method is used for cleaning the magnesium alloy surface before the organic coating process. And the method is used for etching.

5. Phosphate treatments

   - This method uses indium treatment of magnesium alloys.

6. Anodic treatments

Anodic treatment uses the battery or other direct current power source, but corresponds to the negative terminal, while the passive load corresponds to a positive electrode terminal. Electronic tube proceeds from the anode along the tube out of the cathode, and anions in electroplating cell are accumulated on the cathode, the places where the anodes accumulate in the electrochemical oxidation.

1. Chemical treatment nr. 9

   - Applying a low DC voltage to the thin film growth.

   - It can make a black or dark brown coat color.
2. Chemical treatment nr. 17
   - Used for all types of magnesium alloy.
   - Low DC voltage is applied → 5 μm degree of the oxide film layer is formed.
   - And dark green colors used in the amorphous structure are implemented.
3. HAE treatment
   - When you use a two-phase coating on the surface of magnesium alloy.
4. Keronite
   - This is a familiar method of forming a coating layer of electrically oxidized magnesium alloy.
   - The method to form the oxide film layer of the 10–60 μm degree.

The amount of the metal in accordance with industrial requirements has been increasing. However, raw material prices have been rising rapidly due to the limited available resources useful to secure stable resources of these metals and it is recognized as a key element of national competitiveness and sustainable economic growth. In particular, recent increase in the social and industrial requirements for the development and commercialization of resource recycling technology for reuse by recovery of metal from the waste that occurs during the life cycle of the metal resources is strengthening the resource weaponization policy of resource-rich countries, including China and Russia [1–7]. Mg is said to be of the lightest weight among structural metal products and possesses characteristics such as nasal degrees, nonrigid, castability, machinability, impact properties, vibration absorption capability, etc. Weight reduction is required in transport equipment, electronic appliances, sports leisure, medical equipment, military products, industrial equipment, and robotics. Mg is added to the second and third element in order to reduce weight and improve performance during practical use. If you want to increase the strength of the alloy heat resistance, and creep resistance (resistance-creep) is improved. In the case of aluminum, zinc (AZ) series alloy addition of the zinc improves the strength, castability, workability, corrosion resistance, weldability, etc., depending on the addition amount of zinc [8–17]. In particular, the addition of rare earth elements such as Ce and Nd has high strength at 200–250°C. The creep (creep-resistance) characteristics are excellent for heat-resistant alloy.

7. The nonferrous metals (Al, Mg) surface treatment characteristics

7.1. Al anodizing

Al in the air combines with oxygen to produce oxidized layer (Al_2O_3) with naturally irregular arrays. This oxide layer is created artificially by anodic oxidation method than the natural oxide layer made with a regular array. This artificially made oxide layer is called anodic aluminum oxide (AAO), generally called aluminum country [18–21]. AAO barrier-type oxide produced according to the conditions (barrier type film; BTF) and the porous oxide type (porous type film; PTF) is divided into two types. BTF is made to proceed if the anodizing electrolyte is at pH 5 or more, and PTF is created when the anodizing proceeds in an electrolyte pH of less than 5. **Figure 1(a)** shows a passivation oxide layer having a uniform thickness of the same structure. At this time the minimum value decreases in the form of look and measure the current density exponential. This is because the thickness of the oxide layer thickened the ion moving rate of the reduction and reduces the electric field strength. This relationship is shown in Eq. (1):

$$i = A\exp\left(\frac{BE_f}{d}\right) \quad (1)$$

where i (A/cm^2) is the current density, E_f is the voltage difference between the aluminum and the electrolyte, d represents the thickness of the oxide layer, and A and B are the coefficients. The current density (i) and electric field (E_f) are the exponential functions that take place from a high electric field in the process of moving ions in the aluminum oxide layer and the boundary surface. PTF is created in the electrolyte, including sulfuric acid, phosphoric acid, and oxalic acid. PAA is formed to a structure (**Figure 1(b)**). For PAA, the thickness of the oxide layers, unlike the BTF, as well as the anodizing voltage, anodizing time, and current density, is known to be affected by the kind of electrolyte. PTF is not to have a minimum value, continuously decreasing as the current density in the form of exponential function when the predetermined time when the BTF is stabilized after the increase (**Figure 2**).

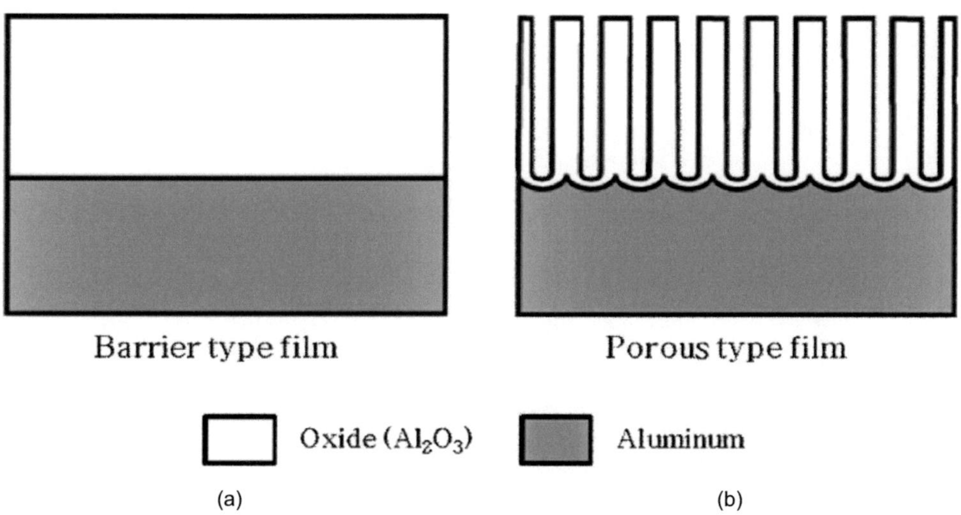

Figure 1. The image of barrier type film and porous type film. (a) Barrier type film; (b) porous type film.

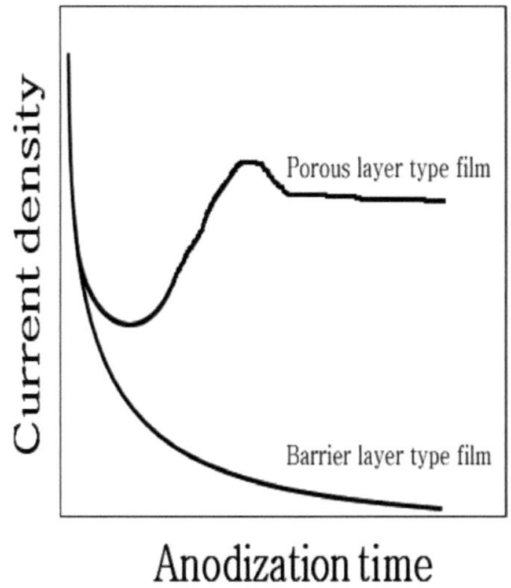

Figure 2. The anodic oxidation current density images. During BTF and PTF are formed over time.

The structure of the PAA was first discovered by F. Keller in 1953, since then many studies have been published. PAA is typically well aligned with a regular hexagonal dense structure (**Figure 3(a)**). Looking at the state of alumina grow in a perpendicular direction on the aluminum surface, there is a cell in a hexagonal shape as shown in **Figure 3(b)** and has a pore in the center of the cell. Pore size of the cells is determined by the temperature and the kind of voltage applied to the electrolyte during the anodic oxidation process, and the size of the pores can be controlled in several tens of nanometers to several hundred nanometers. Observing the cross-section PAA, one end of the structure is blocked in a cylindrical way. A porous oxide layer with a pore layer referred to as a barrier layer of an oxide layer of aluminum is in contact with the hemispherical surface.

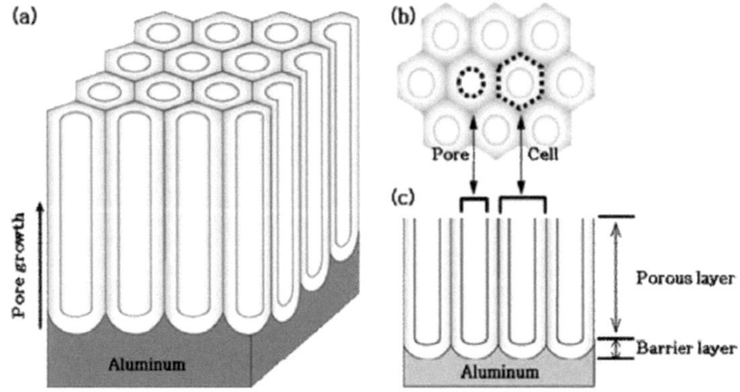

Figure 3. Schematic of Porous Anodic Alumina (PAA). (a) PAA overall appearance, (b) PAA surface appearance, and (c) PAA sided appearance.

Figure 4 shows the surface of the cross-sectional image of the observed PAA.

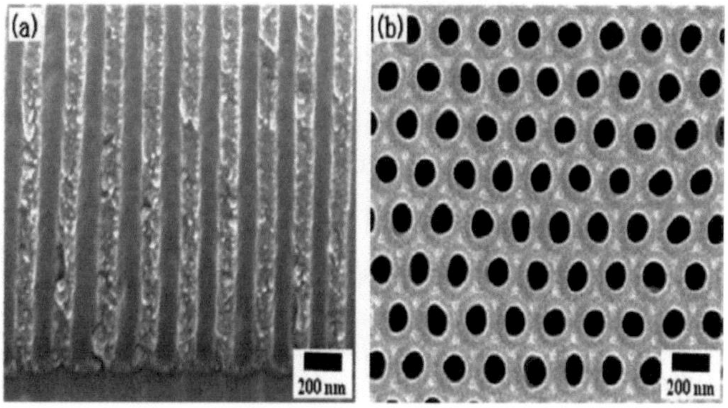

Figure 4. PAA of FESEM image of (a) cross section and (b) surface.

PAA is shown in **Figure 5**. First, as shown in **Figure 5(a)** PAA is made up of pores with irregular arrangement and size. The PAA is thus anodized over time. The size, as shown in **Figure 5(b)**, is constant and the changes in appearance are well aligned. This process is called self-alignment of the information on the magnetic alignment of PAA. During the forming of PAA in the anodizing process, self-alignment, voltage, and the type of electrolyte are known to be affected by the temperature of the electrolyte [22, 23]. In the case of the anodizing voltage expressions (2) and (3) and it is known that as determining the distance between the pore size of the PAA and the pores with the pore, where D-pore, Dint, D-cell the size of each pore, the pore , denotes the size of the cells, U is Empty indicate the anodizing voltage:

$$D_{\text{pore}} = 1.29 \times U \tag{2}$$

$$D_{\text{j}} = D_{\text{cell}} = 2.5 \times \tau \tag{3}$$

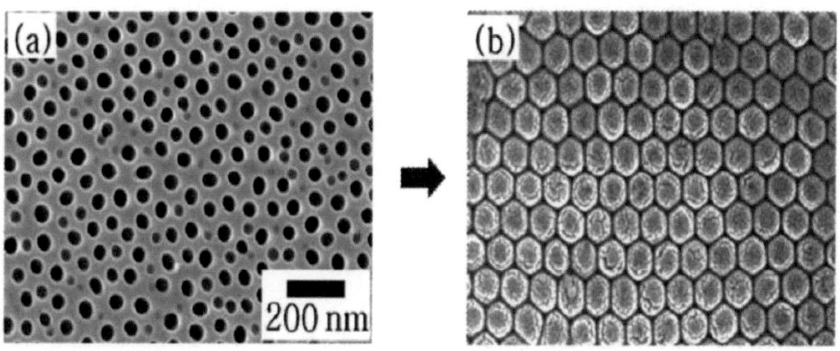

Figure 5. FESEM images of (a) the initial PAA anodized surface and (b) after anodic oxidation under PAA surfaces.

7.2. Mg plasma electrolytic oxidation

The areas that are focused on research and development in each country are different. In Japan, it is used in the development of automation technology, plating equipment, developing speed and partial screen technology, plating, and developing precision and advanced technology in plating bath. In the United States, it is used in purity control, alloy and composite technology development and evaluation, and coating technology, such as high heat-resistant special coating technology, additives development technology, and environmental protection coating technology. There has been rapid progress in the dry surface treatment sector of the developed countries, vacuum plasma technology-based industries, and related high-tech industries. Therefore, the development of surface treatment technology has been applied in the mechanical industry, automobile industry, and electrical electronic industry [24–27]. It is extensively applied to the functional surface treatment for various sectors ranging from the aviation industry to large industrial type. Especially in developed countries, vigorous research has been done in developing highly functional products of high-added value of the industrial upgrading progresses interested in the industry for dry surface treatment technology. Also, dry coating surface treatment process has been applied in LCD, sensor, electronic components, plasma, ion beam, etc. using metal, ceramics, and synthetic resins. High-performance surface treatment technology is also actively applied in the glass and electronics industry. Plasma electrolytic oxidation surface treatment of magnesium alloys are currently only four UK Kero night and henkelsa Germany has a unique blend of solutions. **Figure 6** shows an electrolytic oxidation process diagram of typical magnesium alloys.

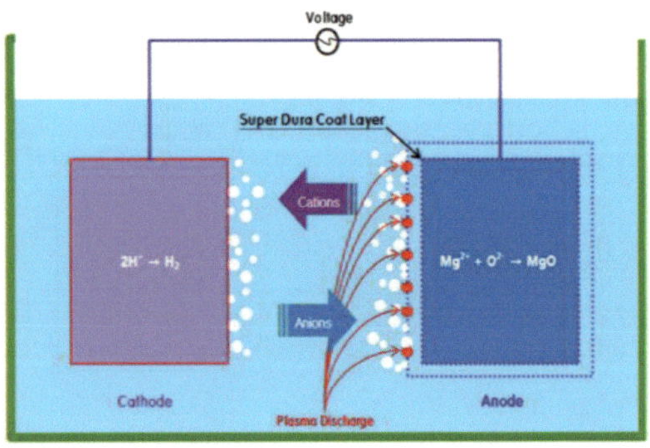

Figure 6. The schematic of PEO device.

8. Surface treatment types

8.1. Chemical treatment (chemical conversion)

Cr(VI)(CrO_3) is generated by the ions in the Cr_2O_7 (DOW1, 7, 20, 21 chromate method). This is a practical solution to many of the chemical conversion treatment method using a magnesium croissant mating. Dow Chemical developed a chromic acid solution for surface coating using

Cr-free chemical conversion solution and its use thereof is restricted. In a typical example using the chemical conversion treatment solution containing $KMnO_4$ amorphous Mg oxide (or hydroxide) and Mn oxide (or hydroxide) it is formed in the surface coating. This reported to also have little in the $KMnO_4$ solution containing HCl $Na_2B_4O_7$ and boron oxide in the coating film produced having a corrosion resistance comparable to the chrome mating method, a change in pH significantly influences the quality of the surface coating. Corrosion resistance for NH_4^+ ions and PO_4^{3-} ions in the Cr^{6+} instead harmless Cr^{3+} treatment using the ion solution purpose utilized for automobile parts of 1 g/L, by 10 g/L, respectively, a (NH_4) been reported $2SO_4$ + H_3PO_4 solution have. Metal phosphate is insoluble and chemically stable, and heat resistance promises a chemical treatment method; however, a drawback is that the corrosion resistance is lower than the croissants mating in the water. As a way to improve this method, forming a composite film by a complex process that combines steam treatment and phosphate treatment of magnesium hydroxide ($Mg(OH)_2$) and phosphate ($NH_4Mg(PO_4)\cdots H_2O$) has been developed [28–31].

The pure Mg, AZ31 and AZ91D alloy is happening initial general corrosion then dipped in molten salt corrosion occurs Phil Reform (filiform corrosion) of the fiber with the generation of hydrogen bubbles caused by the corrosion reaction. As a technique to prevent this corrosion, after treatment in solution 1–10% NaCl + 10% NaOH (pH 12) is performed. If the heat treatment (corrosion-oxidation) in air at 673–773 K generated in the alkali solution $Mg(OH)_2$ in the oxide film is heat-treated, Phil reform corrosion is prevented by changes to MgO.

8.2. Anodizing

It has been switched from a conventional chromium-/fluoride-based electrolyte in terms of environmentally friendly even anodizing with phosphoric acid (H_3PO_4), sodium silicate-based (NaSi), and potassium permanganate-based ($KMnO_4$) electrolyte. In the plasma electrolyte oxidation surface treatment of ammonium phosphate was added, contributing to $Mg_3(PO_4)_2$ improved corrosion resistance that is detected in the positive electrode film and ammonia, alcohols, and amines, and contributes to the densification of the film and flattening by the discharge spark. In addition, because the case of phosphate electrolyte contains magnesium alloy containing Al MgO, $MgAl_2O_4$, is formed $Mg_3(PO_4)$ coating the low melting point composite compounds of two gas discharge easily in the coagulation process. The higher the Al content, a dense coating with low porosity is formed. The phosphate solution containing aluminate ions to form a coating layer similar to the anodized aluminum film shown in the anodic oxidation of aluminum, the corrosion resistance is greatly improved by the addition of small amount of silicates. In addition, the corrosion resistance is greatly improved even when repeated with KOH + $Al(OH)_3$ and then using a mixed electrolytic solution for 5 min, the waveform of a minute unit cell lower voltages. After chemical conversion surface treatment with a phosphoric acid solution as a method of improving the corrosion resistance and coating adhesion at the same time in addition to (K_2SiO_3 + KOH + KF) after anodizing by electrolyte, to form a forsterite (Mg_2SiO_4) as a multistage process for performing 100°C H_2O sealing treatment method by which method, using a mixed solution of phosphoric acid anodizing solution and an aqueous ammonia solution for antistatic and electromagnetic shielding of the electronic circuit of the electronic equipment to obtain a conductive film of 0.4 Ω or less has been developed. Recent use of discontinuous oxidation reaction accompanying the spark discharge at a high voltage of several hundred volts beyond the area of the anodizing electrolytic plasma oxidation method [plasma electrolytic oxidation (PEO)] has been noted. In Applying a high voltage between the electrodes reaches the micro-arc oxidation (MAO) area occurs in the vibration of the breakdown by slight discharge from a material immersed in the electrolyte after the voltage is increased linearly from the anodizing area. The MAO region is formed with a relatively thick porous film and is expected to use as a surface treatment for wear resistance, corrosion resistance, heat resistance, and the like. AZ91D alloy K_2ZrF_6 (10 g/L)-Na_2SiO_3 (10 g/L)-KOH (4 g/L), an electrolytic solution (pH 12.86) at 300 V of voltage for 5 min, was added to form that coating of the surface and cross-section image in the PEO peculiar typical porous as shown in the film. And the oxide layer is grown on only 12.9 μm in only 5 min to form a pore in the vicinity of the surface layer (outer layer) and a dense inner coating layer (inner layer), Mg_2SiO_4, ZrO_2, MgF_2, and SiO_2 composite film consisting of a corrosion resistant component at 60°C or more. PEO dispersing the porous polytetrafluoroethylene (PTEE) nanoparticles in an alkaline electrolyte solution, the addition of a phosphate surfactant in a method of charging a food material within the pores of the film are reported statutes PEO to form a film on the pulsed supply. In addition to the hydrophobic effect and lubricating effect of the PTFE component, improvement in wear resistance is effective in improving the corrosion resistance of magnesium vulnerable to moisture. In a similar way to the method CeO_2 particle-containing coating using the PEO-based electrolyte Na_2SiO_3, self-assembled nanophase particle (SNAP). The PEO + SNAP coating method using a solution has been reported. In This Recent study, on the PEO process various additives on the corrosion resistance of PEO coating in KOH containing alkali silicate electrolyte ($NaAlO_2$, $Na_2B_4O_7$, Na_3PO_4, K_2TiF_6, $CO(NH_2)_2$, etc.), PEO coatings + polymer film composite layer the corrosion resistance, the effect of the current mode and the discharge form on the corrosion resistance of PEO coating, monocrystalline sol gel-forming effect for sealing (pore filling) of the PEO film, the microstructure and corrosion resistance to a wide variety of ideas of preprocessing PEO film by laser the results are presented to verify [32–34].

8.3. Diffusion coating (diffusion coating) and Sol-gel coating method

The Diffusion coating is a process of diffuse coating component and the material component and the excellent adhesive force with a strong metallic bond and thermal resistance between the thermal chemical reaction substrate and the film to a surface treatment method for forming a coating material, while it is in contact with another phase (phase), abrasion resistance, it is possible to control properties such as electrical conductivity, electromagnetic wave shielding ability, and heat conductivity. The Mg diffusion coating of Al is used a lot. Al is used due to its easy generation of the Mg and the corrosion resistance and excellent abrasion resistance between the metal compounds ($Mg_{17}Al_{12}$, Mg_2Al_3, etc.), AZ91 and AZ30, including excellent recycling of large magnesium alloys containing 3–10% of Al. Formation of the intermetallic compound film by diffusion of Al is by processing at temperature 437°C or less. The compound being the diffused surface of Al forms Mg-Al solid solutions, between γ-$Mg_{17}Al_{12}$ metal, and depending on the Al concentration increases the surface of the Mg-Al that has been created.

The β-Mg_2Al_3 is generated when Al is increased above the equilibrium concentration. At 437°C, a two-phase structure of the Mg + $Mg_{17}Al_{12}$ is distributed in a fair response to the $Mg_{17}Al_{12}$ discontinuous phase. Diffusion coating method is an example of magnesium pack cementation (PC) method, a melting of two-stage process has been reported. PC as heated metal powder is provided in the device to perform the simple method of forming a film by spreading the powder onto the substrate surface element and fewer process ratios are easy to apply to the parts of complex shapes. In the diffusion coating of the AZ91D alloy using the Al powder, a 750μm (γ + δ-Mg-$Mg_{17}Al_{12}$) thick film is formed in 1 hour at 450°C, which improves the hardness remarkably. The diameter of the AZ91D alloy capacitive loop (loop capacitive) indicates the size of the polarization resistance as a plot obtained by treatment of 1.5 hours at 420°C; corrosion resistance of the film improves remarkably than the Al diffusion treatment. Using Al + Zn powder mixture as a method for lowering the diffusion temperature is effective to form a film when Al-Zn solid solution is formed with low melting point in the creation of Mg-Al-Zn three alloy in surface Al/Zn powder layer and in the Mg base the concentration gradient between the spread is facilitated. The result is a surface vicinity of excellent corrosion resistance, Al/Zn-rich τ differently, creating different Mg-rich β side of the substrate. The addition of Zn has an effect of preventing sintering of the Al powder and the addition of ethylene glycol in Al powder promotes the adhesion between the base materials.

In halide activators (activator) such as NH_4Cl or $ZnCl_2$ in Al powder to precipitate the activated Al halide activation method for forming a composite film of $Mg_{17}Al_{12}$, $MgZn_2$, Zn_3Mg_7 (Halaide pack cementation; HAPC), by treatment in a vacuum alumina vacuum to lower the diffusion temperature to 400°C, a film containing as a main component a β-$Mg_{17}Al_{12}$ easing laws have been proposed. Melting is a method for forming a corrosion-resistant coating film by immersing the material in the molten salt bath of the halide. When using a low melting point mixed with salt $AlCl_3$ + NaCl, formulas (1) and (2) between the coating compound Mg-Al-based metal is formed by the precipitation and diffusion of the Al produced by the reaction. Eq. (5) presents the spread of Al 523–673 K thermal conductivity differences in the $\Delta G \ll 0$ and it occurs at about 100°C or lower temperature than the pack cementation, where G is the thermal conductivity:

$$AlCl_3 + NaCl \rightarrow NaAlCl_4 \tag{4}$$

$$3Mg + 2NaAlC_{14} \rightarrow 2Al + 3MgC_{12} + 2NaCl \tag{5}$$

Pack cementation activates Al from the Al powder in a low-temperature. It is difficult to produce the film by a high-temperature process since it is a discontinuous process (α-Mg + γ-$Mg_{17}Al_{12}$) and the melt activates even at low temperatures as the Al is generated by interdiffusion occurring and continuing as intermetallic compound of Mg. Two-step method for surface mechanical attrition treatment of the surface by mechanical abrasion is used. After the formation of the nano-crystalline layer on the surface, PC SMAT treatment method has been reported. Plastic deformation is repeated by SMAT processing surface and is formed by deformed layer of about 1500 μm thick. Surface of about 100 μm thickness have been used to produce ultrafine nano-crystalline layer having an average of 30 ± 5 nm hardness and reaches about two times of the bulk. Diffusion coating of AZ91D alloy of the temperature is lowered to 380°C. Microstructure with the A and B regions of the film results from diffusion coating at 400°C and a γ-solid solution of Mg and $Mg_{17}Al_{12}$ coexist, C region is a Mg solid solution phase. The concentration of γ-$Mg_{17}Al_{12}$ nano-crystalline layer area A is higher than B zone, B zone, the γ-$Mg_{17}Al_{12}$ and Mg solid solution is in lamellar structure, such as perlite to increase wear resistance. In addition to two-stage process, it has been proposed to improve the bonding strength and corrosion resistance of a resistance component and a coating component into the substrate due to interdiffusion after forming the coating layer by spraying heat treatment. These two processes are the same as the development of practical skills is expected to increase in the effective removal of defects, adhesion strength, and corrosion resistance. Sol-gel coating method is a method to form a film by simple operation of heating and drying after dipping the coating in the sol-gel solution. There is no evaporation or decomposition of organic matter of a liquid precursor to form a thin film, even a complicated shape has been used as a coating of the oxide. Magnesium has the effect of various oxide; sol-gel coating the ZrO_2, SiO_2 has been reported. ZrO_2 sol Zr $(NO_3)_4 \cdots 5H_2O$ precursor, improving agent (acetyl acetone), methanol 1:4; which is a method for producing blended in a molar ratio of 8 to see, in the ZrO_2 coating applied to the AZ91D alloy after the sol-gel treatment the annealing temperature the higher the greater the corrosion resistance is to form a dense film. For a SiO_2 coating, the tetraethyl Ortho-silicate (TEOS) as the precursor is used, in combination with ethanol, water, acetic acid, etc., herein to prepare a sol [35, 36].

A silane coupling agent (triethoxy-vinylsilnae; VTEO) is added to ethylene to prevent the aggregation of SiO_2 sol (CH_2-CH-) and it has been reported to be effective in improving the corrosion resistance.

8.4. Surface treatment techniques for metallic luster and coloring

The above-mentioned chemical treatment, anodic oxidation, diffusion coating, etc. are used for surface treatment of corrosion resistance of magnesium for products with an emphasis on aesthetics, such as cameras used in moderate environments gloss or color in the castle design. Magnesium metal gloss surface also has a close relationship with the surface roughness (roughness).

Expression (6) shown below

$$R / R_o = f(\sigma / \lambda, \psi, \Delta\theta, m) \tag{6}$$

measures the angular width of the formula incident angle (R), a light receiver from the σ / λ, and theoretically mirror measurement results of the gloss square average square also (δ) to quantify gloss, root mean square slope (m) by analysis with respect to the control of the surface roughness and thickness it has been developed acidic solution treatment expressing metallic luster. Here, R is the mirror reflectivity of the sample, R_o is the specular reflectance of a complete smooth surface, λ represents the wavelength of incident light.

Eq. (6) controls the thickness of the colored film or microstructure of the surface without the use of a film formation is possible by the addition of a dye in the various color dyes in the anodizing solution to get the desired color.

8.5. Mg alloy properties

8.5.1. The main composition

Mg alloys have liquidity compared with Al alloy, and it also has small and easy to micro-shrinkage riser effect to occur. Therefore, pressure resistance of small pinhole is not generated. Thus, the factors equipped with cast Al alloy is more difficult. The occurrence of pinhole being the most problematic in the Al alloy castings is the only feature that is not a problem for Mg alloys [37].

Mg alloy is not permition the flow of the molten metal also not so easy to good thermal capacity to cool a molten metal is decreased to a high temperature. Therefore, there is a problem with the minimum thickness of the casting, the minimum thickness must be 4 mm, and the optimal thickness must be 8 mm.

8.5.2. Dimensional accuracy

Mg alloy die-castings and the addition of antioxidants, and the like for molding the molding sand, if the CO_2 detective obtains a desired dimensional accuracy of the casting because it uses a relatively coarse sand or increase applied to the mold can be improved relatively in dimensional accuracy. The degree of finish applied to the casting is shown in **Table 1**.

Casting size	150 mm	150–600 mm			600 mm	
Degree of roughness	▽ ▽▽ ▽▽▽	▽	▽▽	▽▽▽	▽	▽▽ ▽▽▽
Separate values of roughness	2.0 2.5–3.5	2.5–3.5	3.0–4.0	3.0–4.0	3.0–4.0	5.0 below

Table 1. Surface roughness values in casting.

Table 2 displays the tolerance to be considered in the casting. However, the tolerance may be improved by examining the methods of casting required.

8.5.3. Good appearance

According to the foregoing, the molding sand is rough, the strength of a molding sand of the casting surface is lower than the casting for use, and Al is not necessarily satisfactory in the kitchen state. Also, because often use the fill, and also in many cases establishing the riser directly on top of the product, the casting surface is often the finishing touches according to the degree required. After finishing because irradiation of the surface with sand blast, short blast, may generally be obtained a surface roughness of 50–30 μm.

Dimensional space	Parallel in spilt plane	Right angel in spilt plane
50 below	±0.8	±1.3
50–100	±1.0	±1.8
100–200	±1.5	±2.5
200–300	±2.0	±3.3
300–500	±2.5	±4.0
500 below	±3.5	-
Difference in thickness		
5 below	±0.5	
5–10	±1.0	
10–20	±1.5	
Slope	2℃	

Table 2. Dimensional tolerances of the Mg alloy die-castings (unit: mm).

9. Conclusion

Magnesium is a metal element belonging to the second tribe, the alkaline earth metal of the periodic table and is the lightest structural metal. Magnesium is known originally by the compound, such as magnesium sulfate, magnesia, magnesium carbonate alkali, etc. The application of Mg alloy is expected, in particular involving motion or portability of the component, to increase in the future because of the material's high strength-to-weight ratio and a relatively high stiffness. Magnesium and its alloys are increasingly implemented into a number of components where weight reduction is of great concern. However, the poor corrosion resistance of Mg and its alloys has limited their applications in corrosive environments. Recently, the creep behavior of Mg alloy has attracted much attention due to its useful value in practice. However, the conventional magnesium alloy has a relatively low resistance to creep at elevated temperatures in contrast to Al alloy, which has restricted the development of Mg alloy. Also, magnesium alloys have many good properties which make them ideal materials in a number of applications. However, poor corrosion resistance and high chemical reactivity have hindered the wide applications of magnesium alloys.

Author details

Jae-In Yu

Address all correspondence to: yji5945@yumail.ac.kr

DHU Non-Ferrous Surface Treatment Center, Gyeongsan-si, South Korea

References

[1] M. Parco, L. Zhao, J. Zwick, K. Bobzin, E. Lugscheider, Investigation of particle flattening behaviour and bonding mechanisms of APS sprayed coatings on magnesium alloys, Surf. Coat. Technol. 201 (14) (2007) 6290–6296.

[2] M. Qian, D. Li, S.B. Liu, S.L. Gong, Corrosion performance of laser-remelted Al-Si coating on magnesium alloy AZ91D, Corros. Sci. 52 (10) (2010) 3554–3560.

[3] P. Chraska, M. Hrabovsky, An overview of water stabilized plasma guns and their applications, Proceedings of the International Thermal Spray Conference (1992) 81–85.

[4] J. Ilavsky, J. Forman, P. Chraska, Plasma-sprayed aluminium coating, J. Mater. Sci. Lett. 11 (9) (1992) 573–574.

[5] K. Máthis, J. Gubicza, N.H. Nam, Microstructure and mechanical behavior of AZ91 Mg alloy processed by equal channel angular pressing, J. Alloys Compd. 394 (1–2) (2005) 194–199.

[6] L.B. Mccusker, R.B. Von Dreele, D.E. Cox, D. Louër, P. Scardi, Rietveld refinement guidelines, J. Appl. Crystallogr. 32 (1) (1999) 36–50.

[7] S.H. Leigh, C.C. Berndt, A test for coating adhesion on flat substrates—a technical note, J. Therm. Spray Technol. 3 (2) (1994) 184–190.

[8] R. Mušálek, V. Pejchal, M. Vilémová, J. Matějíček, Multiple-approach evaluation of WSP coatings adhesion/cohesion strength, J. Therm. Spray Technol. 22 (2–3) (2013) 221–232.

[9] E. Cooper, M. J. (Crystallographic Laboratory, Cavendish Laboratory, Cambridge), The structure of the intermetallic phase theta (Cr–Al), Acta Crystallogr. 13 (3) (1960) 257–263.

[10] E. Cooper, M.J. (Crystallographic Laboratory, Cavendish Laboratory, Cambridge), E. Robinson, K. (Department of Physics, University of Reading, Whitcknights Park, Reading), The crystal structure of the ternary alloy [alpha](AlMnSi), Acta Crystallogr. 20 (5) (1965) 614–617.

[11] K. Alvarez, H. Nakajima, Materials 2 (2009) 790–832.

[12] J.E. Gray, B. Luan, J. Alloys Compd. 336 (2002) 88–113.

[13] B. Kasemo, Surf. Sci. 500 (2002) 656–677.

[14] T.J. Webster, J.U. Ejiofor, Biomaterials 25 (2004) 4731–4739.

[15] M.M. Stevens, J.H. George, Science 310 (2005) 1135–1138.

[16] M.P. Lutolf, J.A. Hubbell, Nat. Biotechnol. 23 (2005) 47–55.

[17] S. Moon, Y. Nam, Corros.Sci.65 (2012)494–501.

[18] E. Matykina, R. Arrabal, A. Mohamed, P. Skeldon, G.E. Thompson, Corros. Sci. 51 (2009) 2897–2905.

[19] S. Moon, Y. Jeong, Corros. Sci. 51 (2009) 1506–1512.

[20] R. Arrabal, E. Matykina, F. Viejo, P. Skeldon, G.E. Thompson, Corros. Sci. 50 (2008) 1744–1752.

[21] E.K. Tillous, T. Toll-Duchanoy, E. Bauer-Grosse, Surf. Coat. Technol. 203 (2009) 1850–1855.

[22] L. Moore, J.M. Sykes, P.S. Grant, Corros. Sci. 50 (2008) 3233–3240.

[23] S.S. Scherrer, J.B. Quinn, G.D. Quinn, H.W.A. Wiskott, Dent. Mater. 23 (2007) 1397–1404.

[24] F. Croce, L. Persi, B. Scrosati, F. Serraino-Fiory, E. Plichta, M.A. Hendrickson, Electrochim. Acta 46 (2001) 2457–2461.

[25] M.A.K.L. Dissanayake, P.A.R.D. Jayathilaka, R.S.P. Bokalawala, I. Albinsson, B.E. Mellander, J. Power Sources 119–121 (2003) 409–414.

[26] F. Croce, R. Curini, A. Martinelli, F. Ronci, B. Scrosati, R. Caminiti, J. Phy. Chem. B 103 (1999) 10632–10638.

[27] N. Munichandraiah, K. Binod, R. Stanley, J. Power Sources 111 (2002) 165–172.

[28] R.J. Klein, S. Zhang, S. Dou, B.H. Jones, R.H. Colby, J. Chem. Phy. 124 (144903) (2006) 1–8.

[29] J. Liang, P.B. Srinivasan, C. Blawert, W. Dietzel, Comparison of electrochemical corrosion behaviour of MgO and ZrO_2 coatings on AM50 magnesium alloy formed by plasma electrolytic oxidation, Corros. Sci. 51 (2009) 2483–2492.

[30] J. Liang, P.B. Srinivasan, C. Blawert, W. Dietzel, Influence of chloride ion concentration on the electrochemical corrosion behaviour of plasma electrolytic oxidation coated AM50 magnesium alloy, Electrochim. Acta. 55 (2010) 6802–6811.

[31] P.J. Blau, M. Walukas, Sliging friction and wear of magnesium alloys AZ91D produced by two different methods, Tribol. Intl. 33 (2000) 573–579.

[32] J. Piątkowski, F. Binczyk, Właściwości i zastosowanie odlewniczych stopów Mg-Al, Arch. Foundry Eng. 2 (4) (2002) 421–433.

[33] K.N. Braszczyńska-Malik, Odlewnicze tworzywa magnezowe. Postępy Teorii i Praktyki Odlewniczej, Monografia pod redakcją Jana Szajnara, Poland, 2009.

[34] R. Xin, B. Li, L. Li, Q. Liu, Influence of texture on corrosion rate of AZ31 Mg alloy in 3.5 wt.% NaCl, Mater. Design. 32 (2011) 4548–4552.

[35] S.R. Paital, N.B. Dahotre, Biomed. Mater. 2 (2007) 274–281.

[36] C. Xu, C. Reece, M. Kelley, Appl. Surf. Sci. 274 (2013)15–21.

[37] H. Assender, V. Bliznyuk, K. Porfyrakis, Science 297 (2002) 973–976.

Chapter 4

Preparation of Corrosion-Resistant Films on Magnesium Alloys by Steam Coating

Takahiro Ishizaki, Mika Tsunakawa, Ryota Shiratori,
Kae Nakamura and Ai Serizawa

Additional information is available at the end of the chapter

http://dx.doi.org/10.5772/66365

Abstract

This chapter introduces a novel, chemical-free "steam coating" method for preparing films on magnesium (Mg) alloys and assesses their effectiveness in improving the corrosion resistance of two different Mg alloys. A film composed of crystalline $Mg(OH)_2$ and Mg-Al layered double hydroxide (LDH) was successfully formed on AZ31 Mg alloy, and its corrosion resistance was evaluated through electrochemical measurements and immersion tests in an aqueous solution containing 5 wt.% NaCl. An anticorrosive film was also formed on Ca-added flame-resistant AM60 (AMCa602) Mg alloy via the same steam coating method and found to be composed of crystalline $Mg(OH)_2$ and Mg-Al layered double hydroxide (LDH). Its corrosion resistance was also investigated, and the effectiveness of the steam coating method for improving the corrosion resistance of Mg was fully explored.

Keywords: magnesium alloy, steam coating, corrosion resistance, surface treatment, composite film

1. Introduction

Magnesium (Mg) alloys have excellent physical and mechanical properties, including a high specific strength, excellent formability, good vibration adsorption, and high damping capacity [1–5]. However, the greatest advantage of Mg alloys is their lightweight, which has made it possible to achieve energy savings in automobiles, ships, trains, and airplanes through the use of steel-based hybrid materials. In such applications, Mg is superior to aluminum and CFRP (carbon fiber reinforced plastics) in terms of its density and specific strength but offers a much lower corrosion resistance. Various surface treatments have therefore been developed to produce protective coatings on Mg alloys through anodizing, chemical conversion, plasma electrolytic oxidation, or polymer films [6–12].

Chemical conversion has already been used with automotive Mg components because of its ease of operation and low cost, with chromate-based systems the most popular and effective option for producing a protective layer on various metal surfaces [13–15]. The use of chromate, however, is becoming increasingly more regulated due to the high toxicity of hexavalent chromium compounds [16]. Other chemical conversion methods for imparting Mg alloys with a high corrosion resistance have been developed around using tin, rare earth salts, phosphates/permanganates [17–22], vanadium, or composite films of molybdenum-lanthanum or zinc-phosphate-calcium [23, 24].

Films produced by anodizing have also been used to improve the corrosion resistance of Mg alloys, with good results being achieved through the use of micro-arc oxidation (MAO) [25–27]. Using this method, a high anodic voltage (>200 V) is used to locally melt the surface of the Mg alloy and create a ceramic coating [27]. Alkaline electrolytic solutions containing aluminate [28–31], silicate [32–38], or phosphate [39–45] are often used during the MAO of Mg alloys, but the need to treat the liquid waste before disposal creates a risk of environmental pollution due to heavy metal ions. Furthermore, their use may inhibit the recycling of post-use Mg product scraps into Mg ingot. There is therefore a need for an environmentally friendly surface treatment method that is capable of improving the low corrosion resistance of Mg alloys.

A chemical conversion method that creates a protective film of Mg-Al layered double hydroxide (LDH) has recently been used to significantly improve the corrosion resistance of Mg alloys [46–50]. The layered structure of the LDH consists of positively and negatively charged substances with a general molecular formula of $[M_{1-x}^{2+}M_x^{3+}(OH)_2]^{x+}[A_{x/n}^{n-}] \cdot mH_2O$ [51], where M^{2+} and M^{3+} represent divalent and trivalent metal cations, respectively; A^{n-} is an anion such as CO_3^{2-}, SO_4^{2-}, or OH^-; and x has a value of between 0.2 and 0.33 [51]. This structure serves to store and release corrosion inhibitors on demand, thereby creating anion exchange between these inhibitors and chloride ions. This, in turn, results in a self-healing and corrosion-resistant film [52–54]. Tedim et al. reported that LDH nanocontainers loaded with different corrosion inhibitors such as vanadate, phosphate, and 2-mercaptobenzothiazolate (MBT) can improve active corrosion protection [55], with the release of MBT from LDH nanocontainers in a sol-gel film preventing degradation at the metal/film interface. They have also since developed a novel self-healing protective coating using LDH nanocontainers with corrosion inhibitors [56] formed directly on 2024 aluminum alloy via a conversion reaction. Montemor et al. demonstrated that using LDHs loaded with mercaptobenzothiazole in combination with cerium molybdate hollow nanospheres filled with 2-mercaptobenzothiazole results in a synergistic corrosion inhibition effect that has self-healing potential [57].

Though it is clear that LDHs loaded with corrosion inhibitors are effective in substantially improving corrosion resistance, it has also been reported that LDHs devoid of any corrosion inhibitor could also be effective in delaying corrosion by trapping anions such as chloride ions in LDHs [58, 59]. Lin et al., for example, have shown that the CO_3^{2-} in carbonate Mg-Al LDH has Cl^- anion-exchangeability in a corrosive environment and so can protect Mg alloy against corrosion [58]. Tedim et al. have also reported that Cl^- ions in a corrosive environment can be trapped in the interlayer of Zn-Al LDH intercalated with nitrate anions and that the addition of this LDH to a polymer layer drastically reduces the permeability of corrosive Cl^- ions through the protective coating. This confirms that LDHs are a promising material for improving corrosion resistance [59], but anticorrosive films containing LDHs are usually prepared by a multistep process in which LDHs are first synthesized in powder form and then incorporated into a film for corrosion protection [53]. For industrial applications, it is important to develop an environmentally friendly surface treatment method for achieving an anticorrosive film containing LDHs in a single step.

This chapter presents experimental results pertaining to the preparation and corrosion resistance of magnesium hydroxide films containing Mg-Al LDH on two types of Mg alloy: AZ31 and a flame-resistant Ca-added AM60 (AMCa602). A comparison is also made of the corrosion potentials and corrosion current densities obtained by polarization curve measurements in 3.5 mass% NaCl aqueous solution of films prepared on AZ31 alloy by various methods such as physical vapor deposition, chemical vapor deposition, anodization, chemical conversion, plasma electrolytic oxidization (PEO), spin coating, and a newly developed steam coating method for producing a hydroxide film containing Mg-Al LDH [10–12, 23, 24, 60–65]. As shown in **Table 1**, steam coating produces a relatively positive corrosion potential and the lowest corrosion current density of all the films tested, which indicates that this method imparts superior corrosion resistance. Furthermore, the direct growth of such films from the Mg alloy itself can greatly improve their adherence and mechanical stability when compared to other methods such as spin coating and dip coating [66].

Treatment method	Main component in each film	Corrosion potential (V vs. Ag/AgCl)	Corrosion current density (A/cm²)	Ref.
PVD	Al_2O_3	−1.51	9.20×10^{-5}	[60]
PVD	TiO_2	−1.50	1.29×10^{-4}	[61]
CVD	SiOx	−1.37	1.58×10^{-7}	[62]
CVD	SiOx	−1.59	4.3×10^{-7}	[63]
Anodization	Mg-Al LDH + MgO	−1.19	3.82×10^{-7}	[10]
Chemical conversion	$MgO + Mg_2SiO_4 + Al_2O_3$	−0.06	9.47×10^{-6}	[12]
Chemical conversion	Vanadium	−1.05	8.60×10^{-6}	[23]
Chemical conversion	Mo/La	−1.54	1.15×10^{-7}	[24]
PEO	$Zn_3(PO_4)_2$	−1.76	1.04×10^{-7}	[60]
PEO	$MgO + Mg_3(PO_4)_2$	−1.50	4.00×10^{-8}	[64]
PEO	$Mg_2SiO_4 + CeO_2$	−1.20	2.00×10^{-8}	[65]
Spin coating	$TiO_2 + MgAl_2O_4$	−1.26	4.90×10^{-6}	[74]

PVD, plasma vapor deposition; CVD, chemical vapor deposition; PEO, plasma electrolytic oxidation.

Table 1. Corrosion potentials and corrosion current densities obtained by polarization curve measurements in 3.5 mass% NaCl aqueous solution for films prepared by various methods.

2. Preparation of Mg-Al layered double hydroxide films on AZ31 Mg alloy

In this section, we introduce a method for preparing a protective magnesium hydroxide film containing Mg-Al LDH on specimens of AZ31Mg alloy (composition: 2.98% Al, 0.88% Zn, 0.38% Mn, 0.0135% Si, 0.001% Cu, 0.002% Ni, 0.0027% Fe, remainder Mg) measuring 20 × 20 × 1.5 mm. These AZ31 substrates were prepared by ultrasonically cleaning them in absolute ethanol for 10 min and then drying with inert Ar gas. Once clean, the substrates were set on a substrate stage in a 100 ml-capacity, Teflon-lined autoclave, which had 20 mL of ultrapure (resistance of 18.2 MΩ cm) located at the bottom to produce steam. The distance between the water surface and substrate stage was c.a. 3 cm. The autoclave was heated to a temperature of 423–453 K and then held at this temperature for 1–8 h before being allowed to cool naturally to room temperature. After this steam coating treatment, the samples were ultrasonically cleaned in ethanol for 10 min and dried with Ar gas.

The appearance of the samples after steam coating (**Figure 1**) reveals that all the substrates treated at 423–443 K were uniformly covered with film, regardless of the treatment time. However, as partial detachment of the film was observed in the case of the substrates treated at 433 K for 8 h and at 453 K for 3 and 4 h, these treatment conditions are not considered suitable for preparing an anticorrosive film. The detachment of the film was also found to become more prominent with an increase in the treatment time.

Figure 1. Digital photographs showing the surface of 2 × 2 cm-sized samples prepared under different conditions of temperature and duration [reproduced by permission of the Royal Society of Chemistry, Journal of Materials Chemistry A, 2013, 1, 8968–8977].

The GAXRD patterns of the films formed on AZ31 with different treatment times (423, 433, 443, and 453 K) shown in **Figure 2** reveal clear peaks attributable to the Mg alloy substrate at treatment times less than 2 h with all treatment temperatures. An additional peak at around

2θ = 18° assigned to the 0 0 1 diffraction peak of brucite-type $Mg(OH)_2$ is present in samples treated at 423 and 433 K for 2 h and at 443 and 453 K for 1 h. A diffraction peak consistent with [1 0 1] reflection at around 2θ = 38°, which has the highest intensity according to the JCPDS file number 44-1482, becomes stronger with increasing treatment time at all treatment temperatures, which suggests the formation of hexagonal $Mg(OH)_2$. With an increase in treatment time, several reflection lines at 2θ angles of approximately 18, 33, 38, 51, 58, 62, 68, and 72° are clearly observed in all the GAXRD patterns, and these can be assigned to the 0 0 1, 1 0 0, 1 0 1, 1 0 2, 1 1 0, 1 1 1, 2 0 0, and 2 0 1 reflections of brucite-type $Mg(OH)_2$, respectively. When the treatment time is increased to more than 3 h at any of the treatment temperatures, then one or two peaks attributable to a hydrotalcite (HT)-like structure are also observed. These are attributed to the formation of $Mg_{1-x}Al_x(OH)_2(CO_3)_{x/2}{\cdot}nH_2O$ (Mg-Al LDH) intercalated with carbonate anions. The two peaks at 2θ angles of around 11° and 22° were assigned to the [0 0 3] and [0 0 6] reflections of $Mg_{1-x}Al_x(OH)_2(CO_3)_{x/2}{\cdot}nH_2O$, respectively (JCPDS No. 41-1428). Based on these XRD results, it is concluded that the most suitable treatment conditions for preparing a film composed of $Mg(OH)_2$ and Mg-Al LDH are 423 K for 6–8 h, 433 K for 4–6 h, 443 K for 3–4 h, and 453 K for 2 h.

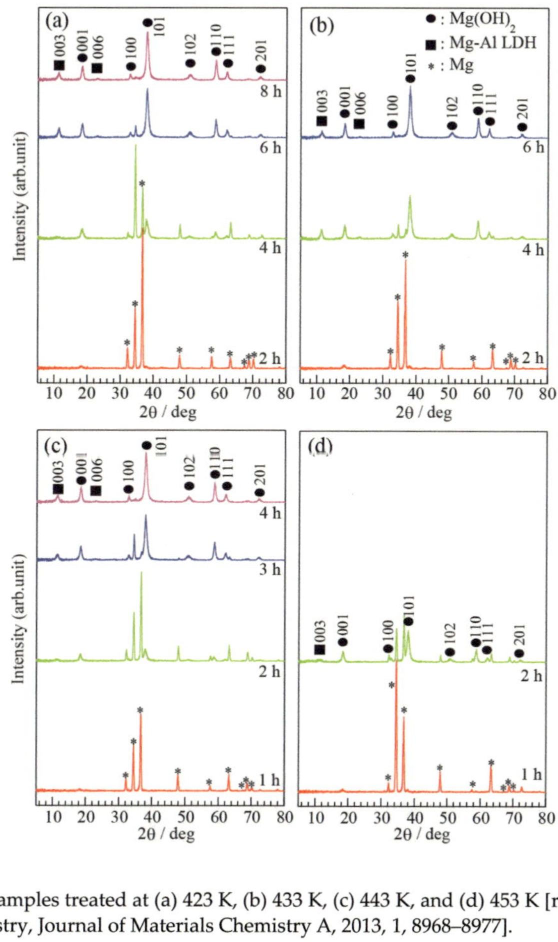

Figure 2. XRD patterns of samples treated at (a) 423 K, (b) 433 K, (c) 443 K, and (d) 453 K [reproduced by permission of the Royal Society of Chemistry, Journal of Materials Chemistry A, 2013, 1, 8968–8977].

The most dominant feature of the FT-IR spectra of the samples treated at 423, 433, 443, and 453 K in **Figure 3** is an intense sharp peak at 3696.9 cm^{-1}. It has been reported that pure $Mg(OH)_2$ exhibits a single band at 3698 cm^{-1} due to the high basicity of its O—H groups [67], and this peak was clearly presented in those samples treated at 423 K for more than 4 h, 433 K for 2–6 h, 443 K for over 2 h, and 453 K for 2 h. These results are in agreement well with their respective XRD profiles. The peak and shoulder bands observed at 950, 781, and 558 cm^{-1} can be attributed to Al—OH translation modes [68], while the bands at 1370–1520 cm^{-1} are attributed to symmetric and asymmetric stretching modes of CO_3^{2-} in the interlayer [68, 69]. These peaks and bands could only be detected in the spectra of samples which exhibited peaks attributable to Mg-Al LDH in their XRD pattern. An additional band at 1630–1650 cm^{-1} was assigned to the bending mode of water molecules in the interlayer. A shoulder band at 3080 cm^{-1} revealed the presence of hydrogen bonding between water and CO_3^{2-} in the interlayer [68]. These results indicate that the films treated at 423 K for more than 4 h; at 433 K for 2, 4, and 6 h; at 443 K for more than 2 h; and at 453 K for 2 h were composed of crystalline $Mg(OH)_2$ and carbonate Mg-Al LDH.

The SEM images in **Figure 4** show the surfaces of samples treated at 433 K, in which we see that steam coating for 2 h led to a rougher and more nodular surface than the untreated AZ31. Several pores of up to several hundred nanometers in diameter were also observed on the surface. When the preparation time was prolonged to 4 h, granular structures were formed and the film was found to be denser than after 2 h. However, a number of pores measuring several hundred nanometers in diameter were still observed on the surface. With the sample treated for 6 h, the film was found to be denser than after 4 h of treatment, with no pores evident on the surface.

Figure 5 presents SEM images showing the sample surfaces treated at 443 K. Note that when the deposition time was less than 1 h, there was only a very low localized surface coverage of the film. However, the surface became fully covered with film when the treatment time was extended to more than 2 h, and this coverage became denser with increasing treatment time. As such, the sample treated for 4 h had a denser film than the sample treated for 3 h, which means that surface topography could be effective way of improving the corrosion resistance.

Figure 6 shows a cross-sectional SEM image and EDX spectrum of the sample treated at 443 K for 4 h, as well as elemental mapping images for Mg, O, and Al. As shown in **Figure 6(a)**, the film produced on this sample had a thickness of about 52 μm, whereas treatment at 423 K for 4 and 6 h, at 433 K for 4 and 6 h, at 443 K for 3 h, and at 453 K for 2 h produced film thicknesses of about 5, 35, 20, 68, 19, and 7 μm, respectively, based on their cross-sectional SEM images. The elemental mapping images in **Figure 6(b)** revealed that the film mainly consisted of Mg and O, which indicates that it was comprised of magnesium hydroxide. This is in agreement well with the XRD results and FT-IR spectrum for the sample. A trace amount of Al was also detected, which provides evidence that Mg-Al LDH exists slightly and locally in the film. The XRD, FT-IR, and SEM-EDX results reveal that steam coating method can produce Mg-Al LDH in the resulting surface film.

Based on these results presented thus far, a model for the formation of the film is proposed as follows. In the closed reaction vessel, steam is produced from water by thermal energy until the vessel becomes saturated. The high pressure and temperature mean that this water vapor has a high kinetic energy and reactivity, which leads to the dissolution of Mg and Al metal into Mg^{2+} and Al^{3+} ions. It is considered likely that $Mg(OH)_2$ crystallites grow via a dissolution-precipitation mechanism, as Nordlien et al. have suggested that amorphous platelets of $Mg(OH)_2$ are initially formed by the precipitation of Mg^{2+} or other soluble Mg species [70]. In addition, Mg^{2+} and Al^{3+} ions might react with the steam to form carbonate Mg-Al LDH, with carbonate ions being sourced from CO_2 present in the enclosed autoclave atmosphere prior to steam treatment. The water vapor can also penetrate the composite film due to the difference in pressure and its high kinetic energy, where it can then react with the Mg alloy substrate.

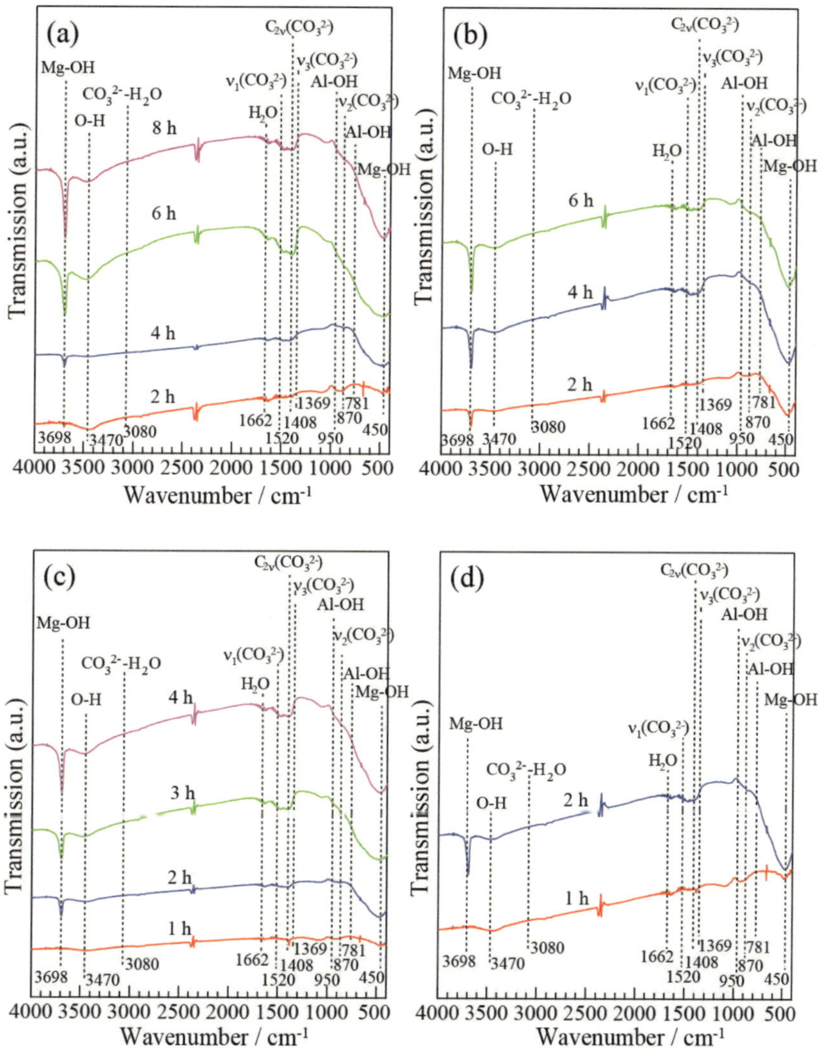

Figure 3. FT-IR spectra of samples treated at (a) 423 K, (b) 433 K, (c) 443 K, and (d) 453 K [reproduced by permission of the Royal Society of Chemistry, Journal of Materials Chemistry A, 2013, 1, 8968–8977].

As a result, a film composed of magnesium oxide/hydroxide and carbonate Mg-Al LDH is prepared directly on the AZ31 substrate. This direct growth of a film from a substrate is a strong advantage over other methods, as this can greatly improve the mechanical stability and adherence of the film. It has also been reported that the existence of Mg-Al LDH in a film prevents metals such as Al and Mg from occurring the corrosion reaction, and so the films produced here are expected to provide a high corrosion resistance.

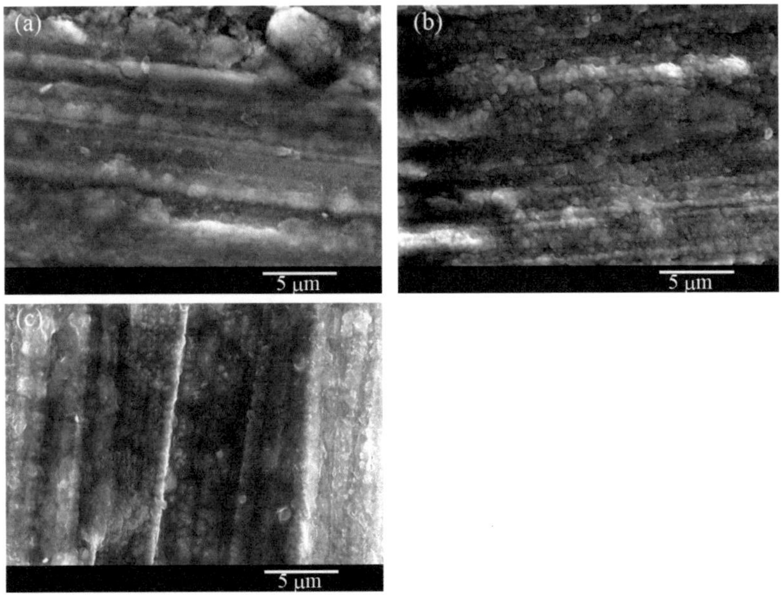

Figure 4. SEM images of samples treated at 433 K for (a) 2, (b) 4, and (c) 6 h [reproduced by permission of the Royal Society of Chemistry, Journal of Materials Chemistry A, 2013, 1, 8968–8977].

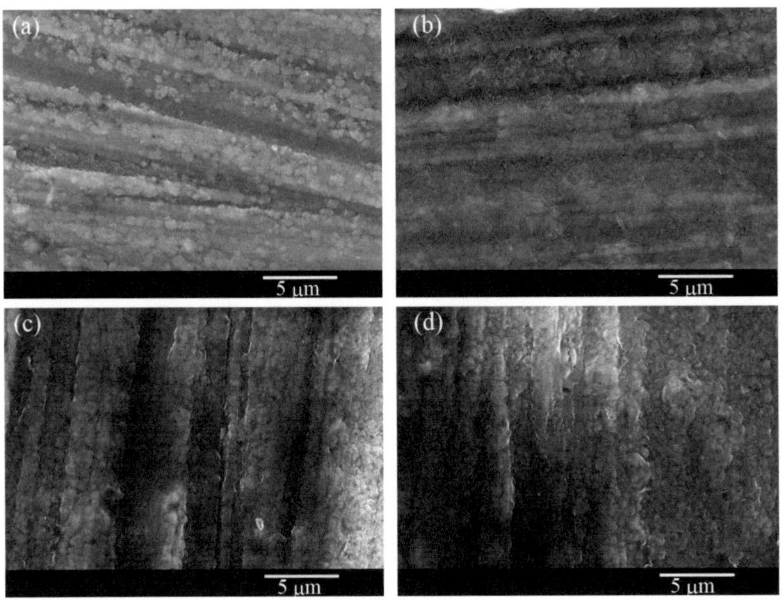

Figure 5. SEM images of samples treated at 443 K for (a) 1, (b) 2, (c) 3, and (d) 4 h [reproduced by permission of the Royal Society of Chemistry, Journal of Materials Chemistry A, 2013, 1, 8968–8977].

The corrosion resistance of the films formed on AZ31 Mg alloy was investigated via potentiodynamic polarization curve measurements, i.e., the lower the polarization current, the better the corrosion resistance. **Figure 7** shows the potentiodynamic polarization curves for the samples treated at 423, 433, 443, and 453 K and untreated AZ31. The corrosion potentials (E_{corr}) and corrosion current densities (i_{corr}) for samples treated under different conditions are listed in **Table 2**. The E_{corr} and i_{corr} values for the bare AZ31 were estimated to be ca. −1.43 V vs. Ag/AgCl and 5.39×10^{-5} A/cm², respectively. The anodic region of the curve for the bare AZ31 can be divided into three fields: (i) first region beginning from E_{corr} (low anodic over potential) that exhibits a linear increase in current density with potential; (ii) a range of abrupt increase in current density from −1.4 V, where the dissolution of Mg to Mg^{+} or Mg^{2+} ions predominantly occurred [71]; and (iii) a current plateau at more positive potentials, where the electrode surface is covered with a film identified as $Mg(OH)_2$ [72]. In the cathodic branch, hydrogen evolution is more dominant at negative potentials than E_{corr}, resulting in an increase in the cathodic current density. The current densities of the anodic and cathodic branches of all samples decreased with film formation, which indicates that the films are effective in improving the corrosion resistance of AZ31 alloy.

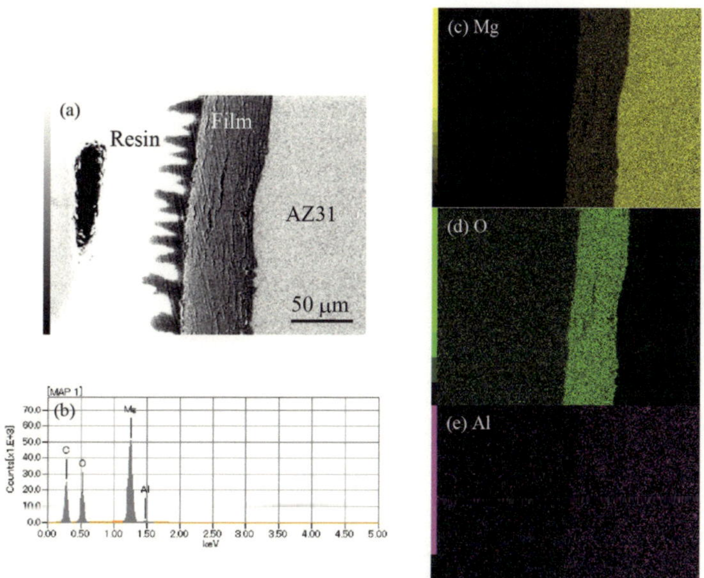

Figure 6. (a) Cross-sectional SEM image of the sample treated at 443 K for 4 h and (b) the corresponding EDX spectra. (c) Elemental maps for (c) Mg, (d) O, and (e) Al [reproduced by permission of the Royal Society of Chemistry, Journal of Materials Chemistry A, 2013, 1, 8968–8977].

At all treatment temperatures, the E_{corr} tends to shift in a positive direction with increasing process time. Furthermore, at potentials more positive than E_{corr}, clear passive ranges can be presented in the curves of the samples treated at 423 K for over 6 h, at 433 K for more than 4 h, at 443 K for more than 2 h, and at 453 K for 2 h. This existence of a long passive range means that the film coatings on AZ31 exhibit highly corrosive-resistant properties in solutions containing Cl^{-} ions. An abrupt increase in current density at more positive potentials than E_{corr}

could be seen in the polarization curves of the samples prepared at 423 K for more than 6 h, at 433 K for 4 h, at 443 K for more than 2 h, and at 453 K for 2 h. For example, the samples treated at 423 K for more than 6 h show an increase in current density at approximately −0.7 and 0.3 V, respectively. This increase of current density could be ascribed to pitting corrosion, which would imply that the electrolytic solution permeated through the film due to the presence of cracks or pores. Similar behavior has been reported before, even in aluminum alloys [73]. However, there was no evidence of pitting corrosion in the curve of the sample treated at 433 K for 6 h, which had E_{corr} and i_{corr} values estimated to be −0.161 V and 4.824 × 10^{-11} A/cm^2, respectively (**Table 2**). These values represent the most positive potential and lowest current density of all the samples, indicating that these conditions produced the best corrosion resistance. We therefore subsequently investigated the durability of the corrosion resistance of a sample treated at 433 K for 6 h when immersed in a 5 wt.% aqueous NaCl solution.

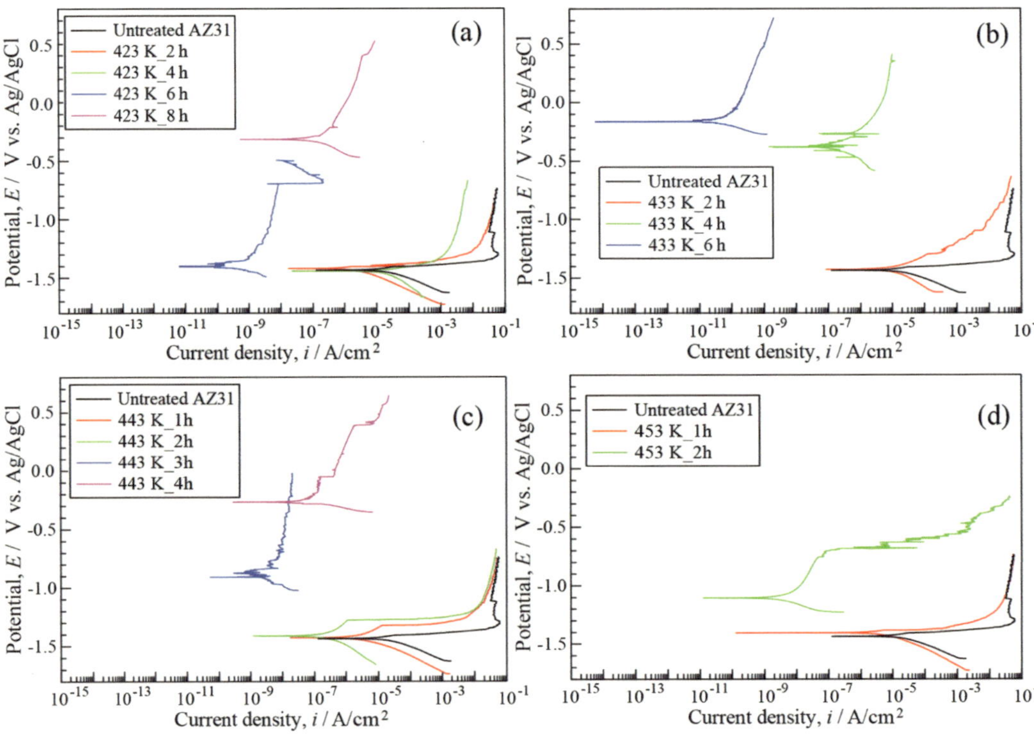

Figure 7. Polarization curves of samples prepared at (a) 423 K, (b) 433 K, (c) 443 K, and (d) 453 K. Polarization curve of bare Mg alloy is also exhibited in respective figure. The scanning rate for all measurements was 0.5 mV/s [reproduced by permission of the Royal Society of Chemistry, Journal of Materials Chemistry A, 2013, 1, 8968–8977].

Figure 8(a) shows the XRD patterns of samples treated at 433 K for 6 h before (black line) and after (dashed red line) immersion in 5 wt.% aqueous NaCl solution for 240 h. The only clear difference in peak intensity is attributable to 0 0 1 reflection of the $Mg(OH)_2$ crystallite, which is much stronger after immersion. This indicates that immersion in a 5 wt.% NaCl solution changes the orientation of the film. The SEM images in **Figure 8(b and c)** show this change in the orientation of the $Mg(OH)_2$ film, with immersion making the film aligned at fairly steeply inclined angles with respect to the surface.

Temp.		1 h	2 h	3 h	4 h	6 h	8 h
423 K	E_{corr} V vs. Ag/AgCl	–	−1.427	–	−1.446	−1.389	−0.316
	i_{corr} A/cm^2	–	1.23×10^{-5}	–	3.22×10^{-5}	5.29×10^{-5}	1.29×10^{-7}
433 K	E_{corr} V vs. Ag/AgCl	–	−1.425	–	−0.376	−0.161	–
	i_{corr} A/cm^2	–	1.28×10^{-5}	–	1.79×10^{-7}	4.82×10^{-11}	–
443 K	E_{corr} V vs. Ag/AgCl	−1.427	−1.421	−0.899	−0.265	–	–
	i_{corr} A/cm^2	3.33×10^{-6}	2.48×10^{-7}	1.90×10^{-9}	9.84×10^{-8}	–	–
453 K	E_{corr} V vs. Ag/AgCl	−1.405	−1.104	–	–	–	–
	i_{corr} A/cm^2	1.21×10^{-5}	7.71×10^{-9}	–	–	–	–

Table 2. Fitting results for potentiodynamic curves in 5.0 wt.% NaCl solution of films formed by steam coating at different temperatures.

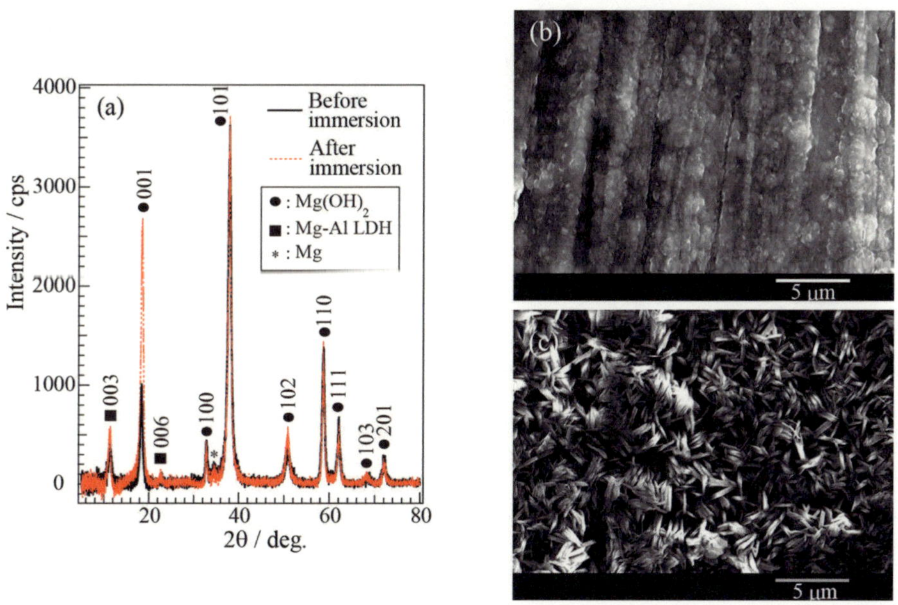

Figure 8. (a) XRD profile of the sample prepared at 433 K for 6 h before and after immersion in 5 wt.% NaCl aqueous solution for 240 h. SEM images of film-coated AZ31 (b) before and (c) after immersion in 5 wt.% NaCl aqueous solution for 240 h [reproduced by permission of the Royal Society of Chemistry, Journal of Materials Chemistry A, 2013, 1, 8968–8977].

The SEM images in **Figure 9** show the sample treated at 433 K for 6 h before and after its immersion in 5 wt.% NaCl solution for 240 h. The visible appearance of these samples is shown in the inset images and reveals that although no physical damage occurred as a result of immersion, there was a slight change in the color of the film. The SEM images show that there was a slight microscale change following immersion, with submicron pores being formed on the surface. A morphological change in the film surface was also observed, with small amounts of nanosheets becoming aligned at fairly inclined angles relative to the surface. This nanosheet formation is in agreement with the XRD patterns and is likely the result of dissolution caused by the film being in contact with the NaCl solution. **Figure 9(c)** presents potentiodynamic polarization curves for the sample treated at 433 K for 6 h before and after immersion in 5 wt.% NaCl solution for 240 h. From this, we see that the E_{corr} after immersion was approximately −0.65 V, which represents a considerable negative shift due to the change in the physicochemical properties of the film. The i_{corr} value, on the other hand, was almost the same before and after immersion. A clear passive region at c.a. 650 mV was observed after immersion, and though there was a slight variation in the current density, this indicates that the film exhibits high corrosion resistance even when immersed in a 5 wt.% NaCl solution for 240 h, as shown by the SEM image in **Figure 10(a)**.

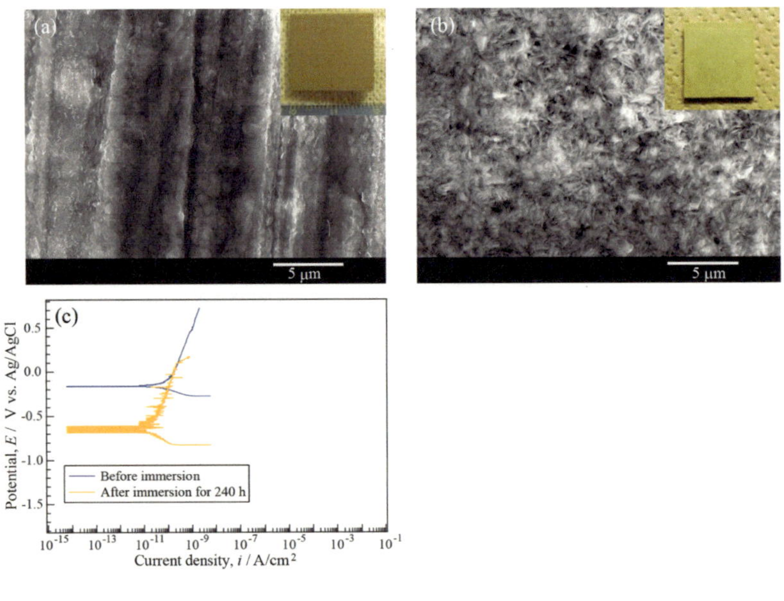

Figure 9. SEM images of the sample prepared at 433 K for 6 h (a) before and (b) after immersion in 5 wt.% NaCl aqueous solution for 240 h. (a) and (b) are photographs showing the appearance of the sample surface. (c) Polarization curves of the sample prepared at 433 K for 6 h before and after immersion in 5 wt.% NaCl aqueous solution for 240 h [reproduced by permission of the Royal Society of Chemistry, Journal of Materials Chemistry A, 2013, 1, 8968–8977].

Figure 10(b–e) shows the EDX spectra obtained at spots (i), (ii), (iii), and (iv) in **Figure 10(a)**. At spots (i) and (ii), there are two peaks attributable to Cl and Al but no peak for Na, whereas no Cl was detected at spots (iii) and (iv). The presence of Al at spots (iii) and (iv) indicates that the Mg-Al LDH layer was still present, which suggests that CO_3^{2-} ions in the Mg-Al LDH layer were partially exchanged for Cl^- anions. From the XRD patterns and EDX spectra, it is evident that Mg-Al LDH is able to trap some Cl^- ions within its structure, which has been reported to delay the degradation of protective coatings [59]. It is also apparent that carbonate Mg-Al LDH has a weak anion exchangeability. This is significant given that Williams et al. have found LDHs loaded with carbonate and nitrates are more effective than a blank coating in preventing the corrosion of a metallic substrate during filiform corrosion tests [74].

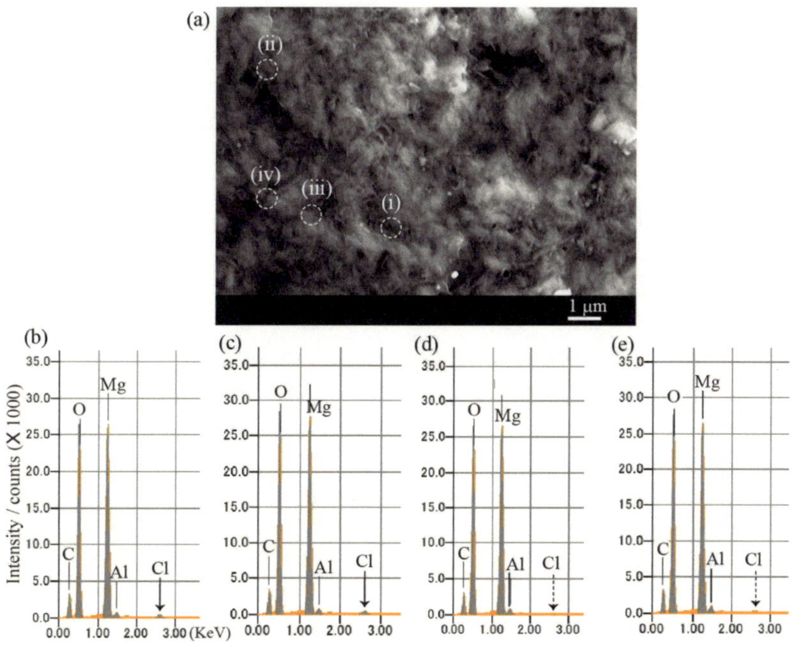

Figure 10. SEM image of the sample treated at 433 K for 6 h after immersion in 5 wt.% NaCl aqueous solution for 240 h. (b), (c), (d), and (e) show EDX spectra obtained at spots (i), (ii), (iii), and (iv) in **Figure 9** (a), respectively [reproduced by permission of the Royal Society of Chemistry, Journal of Materials Chemistry A, 2013, 1, 8968–8977].

This section has shown that a protective magnesium hydroxide film containing Mg-Al layered double hydroxide (LDH) can be produced on Mg alloy via a simple, environmentally friendly, inexpensive, and chemical-free steam coating method. Subsequent XRD and FT-IR studies revealed that these films are composed of $Mg(OH)_2$ and carbonate Mg-Al LDH, and a formation mechanism was proposed based on the results of XRD, FT-IR, SEM, and EDX measurements. Potentiodynamic measurements and immersion tests in 5 wt.% NaCl aqueous solution revealed that a film prepared on AZ31 alloy at 433 K for 6 h provides the greatest corrosion resistance. Increasing the immersion time induced a change in the orientation of the crystalline $Mg(OH)_2$ structure, but no change was observed in the appearance of the sample. With a growing demand for Mg alloys and an increasing focus on environmental issues, it is believed that this steam coating method can provide an effective means of improving the corrosion performance of large, complex-shape Mg alloy components.

3. Formation of Mg-Al layered double hydroxide films on flame-resistant AM60 Mg alloy

In the previous section, we characterized the properties of an anticorrosive film composed of $Mg(OH)_2$ and Mg-Al LDH that is formed on commercial AZ31 Mg alloy by steam coating. In this section, we report on the formation and properties of the anticorrosive film that is produced in a similar manner on Ca-added AM60 Mg alloy (AMCa602), which is one of the most important lightweight materials for transportation components because of its flame-resistant nature. This unique trait is the result of the fact that Ca will oxide before Mg ignites [75], which increases the ignition temperature from 200 to 300 K and greatly restrains the inherently combustible nature of Mg alloys. However, as this addition of Ca does not improve the low corrosion resistance of Mg alloy, there is still a need to develop a suitable surface treatment technology.

Although anodized, chemical conversion and polymer films have all been applied to Mg alloys [6–9, 76–81], there have been few reports pertaining to the surface treatment method of flame-resistant Mg alloys. Xia et al. did find that the corrosion resistance of Mg-4.0Zn-0.2Ca alloy is improved by micro-arc oxidization (MAO) treatment in basic aqueous solutions containing Na_4SiO_4, NaF, and NaOH due to the formation of a film composed mainly of MgO and MgF_2 [82]; however, the presence of fluorine atoms in such coatings creates a risk of environmental pollution. Mori et al. reported on the preparation and corrosion resistance of a plasma electrolytic oxidation layer on nonflammable Ma-Al-Mn-Ca Mg alloy (AMCa602), which was mainly composed of Mg_2SiO_4 and MgO [83]. The liquid waste produced during this anodization process, however, needs to be treated prior to disposal. There is therefore still a need for an environmentally friendly surface treatment capable of improving the low corrosion resistance of flame-resistant Mg alloys.

Specimens of AMCa602 Mg alloy (composition in wt.%: 5.92% Al, 0.36% Mn, 1.99% Ca, <0.01% Si, <0.01% Cu, <0.005% Ni, <0.005% Fe, balance Mg) measuring 20 × 20 × 1.5 mm in size were supplied by Fuji Light Metal Co., Ltd. These Mg alloy substrates were ultrasonically cleaned in absolute ethanol for 10 min and then dried with inert Ar gas. Once clean, they were placed on a substrate stage in a Teflon-lined, 100 ml-capacity autoclave, into which 20 mL of ultrapure water containing 0, 2, 20, or 200 mM $Al(NO_3)_3 \cdot 9H_2O$ was added to the bottom to produce steam. The distance between the surface of the water and the substrate stage was around 3 cm. The autoclave was heated to a temperature of 433 K, where it was held for 6 h before being allowed to cool naturally to room temperature. Following this steam treatment, the samples were ultrasonically cleaned in ethanol for 10 min and dried with Ar gas.

Figure 11 shows the GAXRD patterns of the films formed on AMCa602 at 433 K over a 6 h period from aqueous solutions containing 0, 2, 20, or 200 mM of $Al(NO_3)_3 \cdot 9H_2O$ in 20 mL of ultrapure water. A weak peak attributable to the Mg alloy substrate is visible at a 2θ angle of approximately 34° in all the patterns. A diffraction peak corresponding to 1 0 1 reflection at approximately $2\theta = 38°$, which has the highest intensity according to JCPDS No. 44-1482, appears slightly weaker with an increase in the $Al(NO_3)_3$ concentration. In addition, several peaks were observed in all the GAXRD patterns at 2θ angles of approximately 18, 33, 51, 58, 62, 68, and 72°

that can be attributed to the 0 0 1, 1 0 0, 1 0 2, 1 1 0, 1 1 1, 2 0 0, and 2 0 1 diffraction peaks of brucite-type $Mg(OH)_2$, indicating the formation of hexagonal $Mg(OH)_2$. In addition to these peaks, one or two peaks attributable to a hydrotalcite (HT)-like structure can be distinguished that are assigned to $Mg_{1-x}Al_x(OH)_2(NO_3)_x{\cdot}nH_2O$ intercalated with nitrate anions [84]. These diffraction peaks are indexed according to 3*R* symmetry, and so the two peaks at around $2\theta = 11°$ and 22° are attributed to the 0 0 3 and 0 0 6 reflections, respectively, of $Mg_{1-x}Al_x(OH)_2(NO_3)_x{\cdot}nH_2O$ [51, 85, 86]. This suggests that the treated surfaces were crystallized with a preferential hexagonal orientation along the (0 0 3) plane, which is a characteristic feature of the spontaneous texture axis of $Mg_{1-x}Al_x(OH)_2(NO_3)_x{\cdot}nH_2O$. Furthermore, these XRD profiles reveal that all films were composed of crystalline Mg-Al LDH and $Mg(OH)_2$, regardless of the $Al(NO_3)_3$ concentration used.

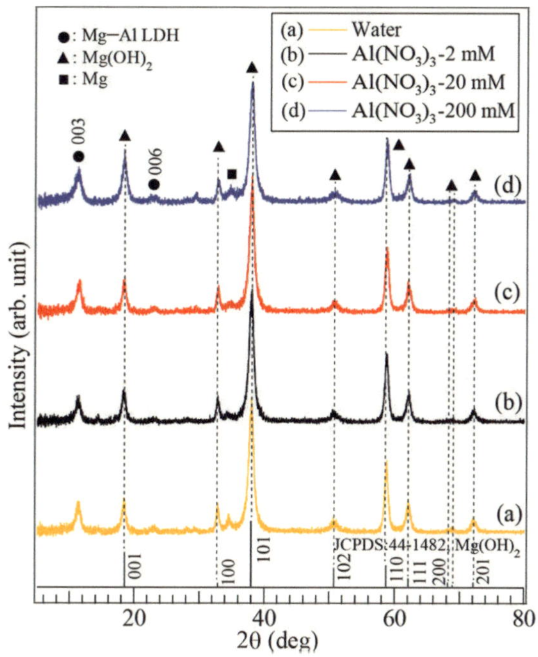

Figure 11. GAXRD patterns of films formed on AMCa602 at 433 K for 6 h from aqueous solutions with $Al(NO_3)_3$ concentrations of (a) 0, (b) 2, (c) 20, and (d) 200 mM [Corrosion Science, 2015, 92, 76–84, doi:10.1016/j.corsci.2014.11.031. Copyright @ELSEVIER (2015)].

The appearances of the samples after steam coating at 433 K for 6 h using aqueous solutions containing $Al(NO_3)_3$ at concentrations of 2, 20, and 200 mM are shown in **Figure 12(a–c)**. Note that in all cases the substrate were uniformly covered with film, but the color of this film did change with $Al(NO_3)_3$ concentration. The SEM images in **Figure 12(d–f)** show the same samples, from which we see that an aqueous solution containing 2 mM $Al(NO_3)_3$ produces a relatively smooth surface with only a few minute pores measuring several hundred nanometers in diameter. The surface of the sample prepared using an aqueous solution containing 20 mM $Al(NO_3)_3$, on the other hand, was quite rough and had a small number of nanosheets that were slightly inclined with respect to the surface.

As shown in **Figure 13**, elemental mapping by EDX revealed traces of elemental Mg, Al, and O primarily in areas where nanosheets existed, which is consistent with the presence of Mg-Al LDH. Aggregated particles are also clearly discernible in these images, which likely formed during steam treatment and condensed on the substrate, and it is on these particles that the nanosheets formed. The sample prepared from an aqueous solution containing 200 mM

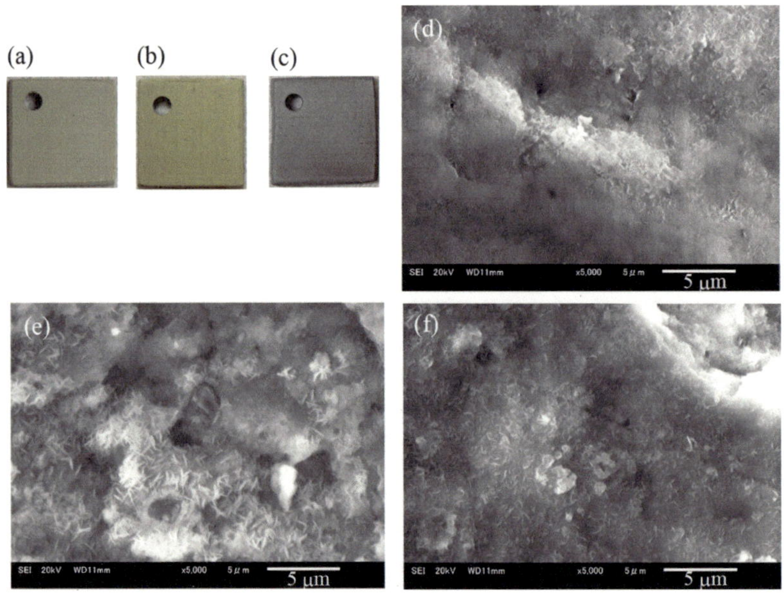

Figure 12. Appearance of samples after steam coating at 433 K for 6 h using aqueous solutions containing $Al(NO_3)_3$ at concentrations of (a) 2, (b) 20, and (c) 200 mM. (d–f) SEM images of the samples shown in (a–c) [Corrosion Science, 2015, 92, 76–84, doi:10.1016/j.corsci.2014.11.031. Copyright @ELSEVIER (2015)].

$Al(NO_3)_3$ was smooth and dense compared to that produced from a 20 mM $Al(NO_3)_3$ solution, and although small nanosheets were formed that were slightly inclined with respect to the surface, no cracks were found in the film.

Figure 14 shows the FT-IR spectra of samples steam coated at 433 K for 6 h using aqueous solutions containing 2, 20, or 200 mM of $Al(NO_3)_3$. The most dominant spectral feature is the intense sharp peak observed at 3696.9 cm^{-1}, which is in good agreement with the XRD data in that pure $Mg(OH)_2$ has been reported to exhibit a single band at 3698 cm^{-1} due to the high basicity of its O—H groups [67]. The shoulder band at 781 cm^{-1} can be ascribed to the Al-OH translational mode [68], while the bands at 1370–1520 cm^{-1} are attributed to the symmetric and asymmetric stretching modes of CO_3^{2-} in the interlayer [69, 84]. The peak at approximately 1385 cm^{-1} is attributed to the ν_3 vibrational mode of NO_3^- with D_{3h} symmetry in the interlayer [46]. These peaks and bands were present in the spectra of all samples that had peaks attributable to Mg-Al LDH in the irrespective XRD patterns. An additional band at 1630–1650 cm^{-1} is assigned to the bending mode of water molecules in the interlayer, which are confirmed by

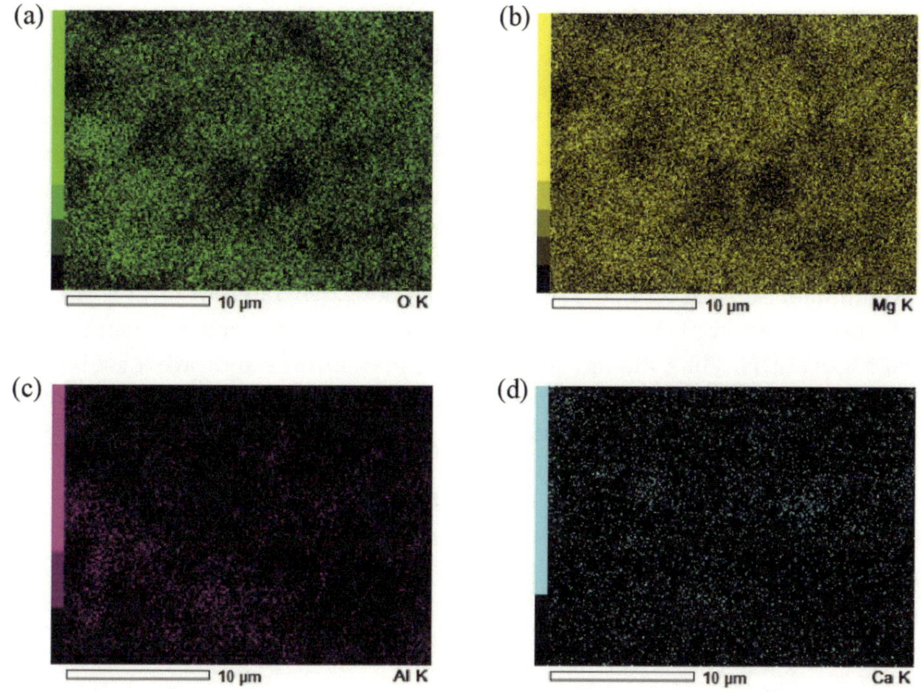

Figure 13. Elemental mapping images of AMCa602 Mg alloy after steam coating at 433 K for 6 h using an aqueous solution containing 20 mM $Al(NO_3)_3$: (a) O, (b) Mg, (c) Al, and (d) Ca [Corrosion Science, 2015, 92, 76–84, doi:10.1016/j.corsci.2014.11.031. Copyright @ELSEVIER (2015)].

the shoulder band at 3080 cm^{-1} and means that hydrogen bonding can occur between H_2O and carbonate ions in the interlayer [84]. These results demonstrate that all films were composed of crystalline $Mg(OH)_2$, nitrate-based Mg-Al LDH ($Mg_{1-x}Al_x(OH)_2(NO_3)_x{\cdot}nH_2O$), and carbonate-based Mg-Al LDH ($Mg_{1-x}Al_x(OH)_2(CO_3)_{x/2}{\cdot}nH_2O$).

Figure 15(a) presents a cross-sectional SEM image of the film prepared at 433 K for 6 h from an aqueous solution containing 200 mM of $Al(NO_3)_3$, which reveals that it had a thickness of about 197 μm. In comparison, the films prepared under the same conditions from 2 and 20 mM $Al(NO_3)_3$ solutions were estimated to be about 308 and 193 μm, respectively. All of the films formed from an aqueous solution containing $Al(NO_3)_3$ were much thicker than the 53 μm achieved with only H_2O, which implies that the existence of $Al(NO_3)_3$ in a closed autoclave accelerates film growth. The elemental mapping images in **Figure 5(b–d)** reveal that the films contained mostly Mg and O, implying that they were composed mainly of $Mg(OH)_2$. This result is in agreement well with the XRD profiles and FT-IR spectra obtained from each of the samples. A small trace of Al was also detected, demonstrating that Mg-Al LDH is presented locally in the film. Based on these XRD, FT-IR, and SEM-EDX results, it is concluded that steam coating is capable of forming $Mg(OH)_2$ and Mg-Al LDH on both AMCa602 and AZ31.

From the results presented in this section, the proposed mechanism by which films are formed by steam treatment can be described as follows. First, an aqueous solution containing

$Al(NO_3)_3$ is heated in the electric oven to generate steam containing Al^{3+} and NO_3^- ions that then saturates a closed reaction vessel. When this steam makes contact with the AMCa602 Mg alloy, the Mg metal it contains is readily dissolved into Mg^{2+} ions due to the high temperature and pressure. It is likely that $Mg(OH)_2$ crystallites are subsequently formed via a dissolution-precipitation mechanism, resulting in the formation of $Mg(OH)_2$ on the surface of the AMCa602 substrate. This is predicated on a notion by Nordlien et al. that amorphous platelets of $Mg(OH)_2$ are initially formed by the precipitation of Mg^{2+} or other soluble Mg species [70]. In addition, Mg^{2+}, Al^{3+}, and NO_3^- ions have the potential to react with the steam and form nitrate-based Mg-Al LDH. Any carbonate ions generated from CO_2 present in the enclosed autoclave atmosphere before steam treatment can result in the formation of carbonate-based Mg-Al LDH. Thus, along with $Mg(OH)_2$, both nitrate- and carbonate-based Mg-Al LDHs can form on AMCa602. Even if these reactions only occur at the surface, they can still progress at the interface between AMCa602, and the overlying film as the steam is able to penetrate thanks to the difference in pressure and its high kinetic energy. The end result is that a film composed of $Mg(OH)_2$ and Mg-Al LDH is directly formed on AMCa602. A porous film formed during the early stage of the process will allow more steam to pass through, resulting

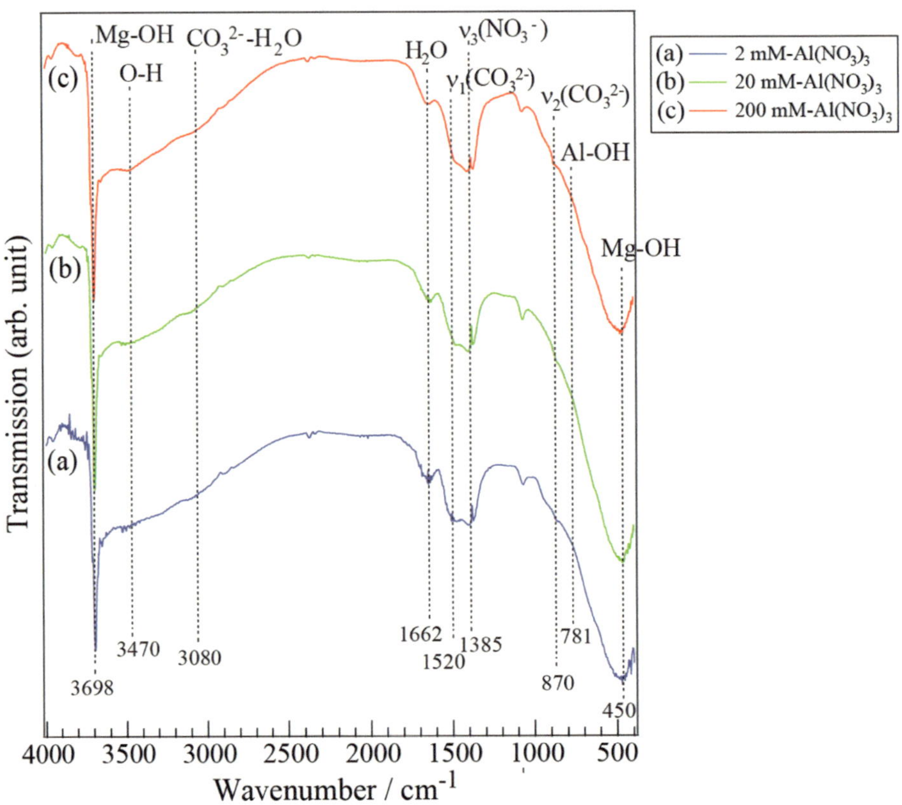

Figure 14. FT-IR spectra of AMCa602 Mg alloy after steam coating at 433 K for 6 h using aqueous solutions containing (a) 2, (b) 20, and (c) 200 mM of $Al(NO_3)_3$ [Corrosion Science, 2015, 92, 76–84, doi:10.1016/j.corsci.2014.11.031. Copyright @ELSEVIER (2015)].

in a thicker final film. In contrast, it is difficult for steam to easily penetrate a dense initial film. Thus, the film thickness is considered to be dependent on the preparation conditions.

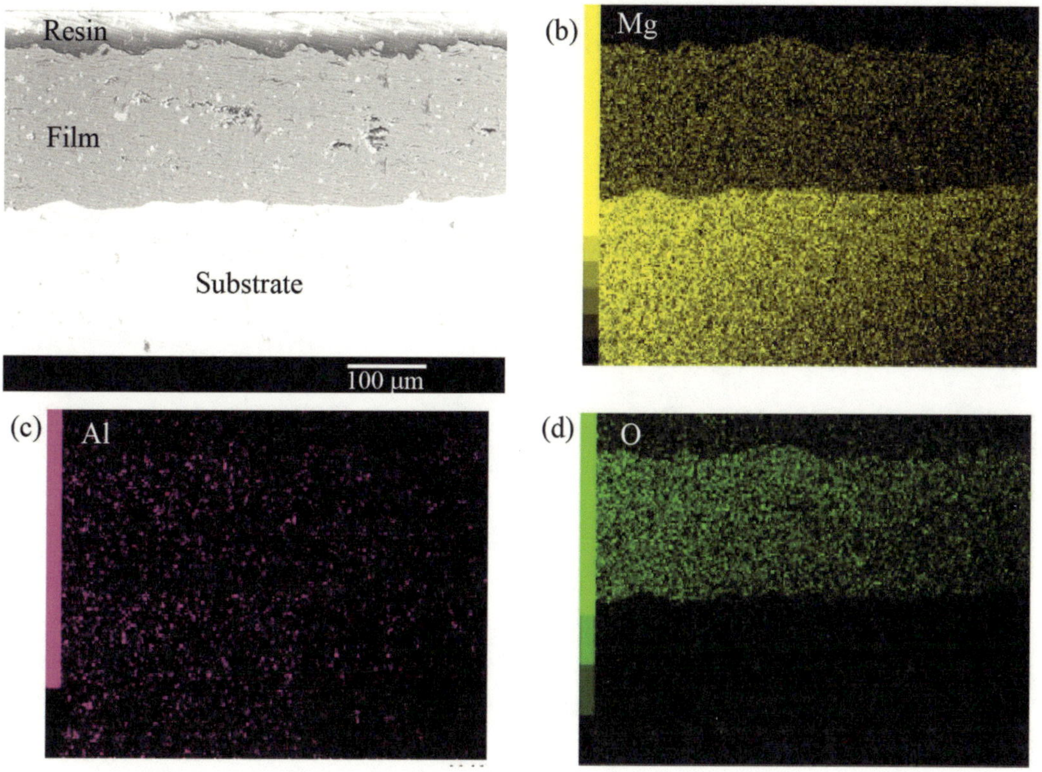

Figure 15. (a) Cross-sectional SEM image of the film prepared at 433 K for 6 h from an aqueous solution containing 200 mM $Al(NO_3)_3$ and elemental mapping images for (b) Mg, (c) Al, and (d) O [Corrosion Science, 2015, 92, 76–84, doi:10.1016/j.corsci.2014.11.031. Copyright @ELSEVIER (2015)].

The corrosion resistance of the film formed on the AMCa602 Mg alloy at 433 K for 6 h from aqueous solutions containing 0, 2, 20, and 200 mM $Al(NO_3)_3$ was estimated from the potentiodynamic polarization curves shown in **Figure 16**. For reference, the potentiodynamic polarization curve of untreated AMCa602 is shown in **Figure 6**. The corrosion potentials (E_{corr}) and corrosion current densities (j_{corr}) of the untreated AMCa602 and samples prepared under various conditions are listed in **Table 3** along with the Tafel slopes (β_a, anodic; β_c, cathodic) obtained from the polarization curve for each sample. The E_{corr} and j_{corr} values for the untreated AMCa602 were estimated to be −1.524 V vs. Ag/AgCl/sat.-KCl and 4.214×10^{-5} A cm^{-2}, respectively. In the cathodic region of the bare AMCa602, hydrogen evolution dominated at potentials that were more negative than E_{corr} (i.e., negative potentials with magnitudes above 1.524 V), resulting in an increase in the cathodic current density. The anodic region of the curve for the bare AMCa602 can be distinguished into two ranges. The first of these starts from E_{corr} (low anodic overpotential, −1.524 V) and is defined by an abrupt linear increase in the logarithm of the current density, with a shift toward a positive voltage in the potential. It is in this region that dissolution reaction of Mg to Mg^+ or Mg^{2+} ions is most likely to have occurred [71]. As the potential moved toward values more positive than −1.4 V vs. Ag/AgCl/sat.-KCl (i.e., negative potentials with magnitudes below 1.4 V and positive potentials), a current plateau occurred as a result of the electrode surface being covered with $Mg(OH)_2$ [72]. It has been reported that the anodic region of the polarization curves for AZ31 can be distinguished into three ranges [72],

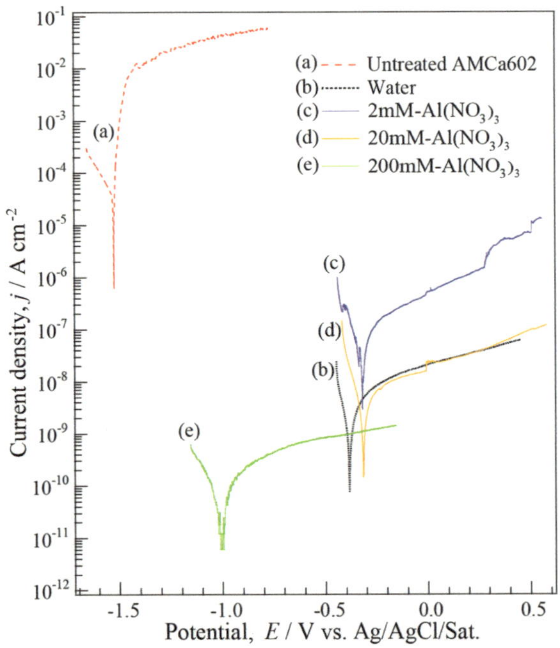

Figure 16. Potentiodynamic polarization curves of samples treated at 433 K for 6 h using aqueous solutions containing 0, 2, 20, and 200 mM of $Al(NO_3)_3$. The scanning rate was 0.5 mV s^{-1} [Corrosion Science, 2015, 92, 76–84, doi:10.1016/j.corsci.2014.11.031. Copyright @ELSEVIER (2015)].

which means that the overpotential for the dissolution of metallic magnesium to magnesium ions (Mg^+ and/or Mg^{2+}) for AMCa602 is lower than that of AZ31. This improved dissolution might have resulted from the addition of Ca, as this is more reactive than Mg.

Figure 17 shows the potentiodynamic polarization curves for the bare AMCa602 and the samples that were treated at 433 K for 6 h from aqueous solutions containing 0, 2, 20, and 200 mM $Al(NO_3)_3$. Note that the current densities in the anodic and cathodic regions for all film-coated samples were lowered compared to the bare AMCa602 (**Figure 6**) and reproducibly low (**Figure 7**), demonstrating that all films improved the corrosion resistance of AMCa602. The E_{corr} also shifted toward a more positive value independent of the $Al(NO_3)_3$ concentration. At potentials more positive than E_{corr}, extended passive regions were clearly observed in the curves of the samples prepared from aqueous solutions containing 0 and 200 mM $Al(NO_3)_3$, as shown in **Figure 6**. These perfect passive regions imply that the film effectively protected AMCa602 in a solution containing Cl^- ions. The j_{corr} values decreased with an increase in the $Al(NO_3)_3$ concentration, and this was independent of the film thickness. Any increase in the thickness and/or density of a film is generally beneficial to improving corrosion resistance by inhibiting contact between the metal surface and a corrosive solution. In this case, the film prepared from an aqueous solution containing 2 mM $Al(NO_3)_3$ was the thickest of all of the samples, and so this is expected to provide the best corrosion resistance. However, the

	E_{corr} (V vs. Ag/AgCl)	j_{corr} (A/cm²)	Tafel slope (mV/decade)
Untreated AMCa602	−1.524	$4.21 \times 10^{-5} \pm 4.82 \times 10^{-5}$	b_a: 21.2 b_c: −178.7
Film coated AMCa602 prepared from ultrapure water	−0.385	$1.85 \times 10^{-9} \pm 9.85 \times 10^{-10}$	b_a: 198.7 b_c: −66.9
Film coated AMCa602 prepared from 2 mM $Al(NO_3)_3$	−0.322	$7.86 \times 10^{-8} \pm 7.68 \times 10^{-7}$	b_a: 278.1 b_c: −164.0
Film coated AMCa602 prepared from 20 mM $Al(NO_3)_3$	−0.316	$1.16 \times 10^{-9} \pm 2.39 \times 10^{-9}$	b_a: 235.4 b_c: −56.2
Film coated AMCa602 prepared from 200 mM $Al(NO_3)_3$	−1.007	$9.56 \times 10^{-11} \pm 2.17 \times 10^{-10}$	b_a: 288.6 b_c: −198.0

Table 3. Fitting results for potentiodynamic polarization curves in 5.0 wt.% NaCl solution for films formed using different concentrations of $Al(NO_3)_3$.

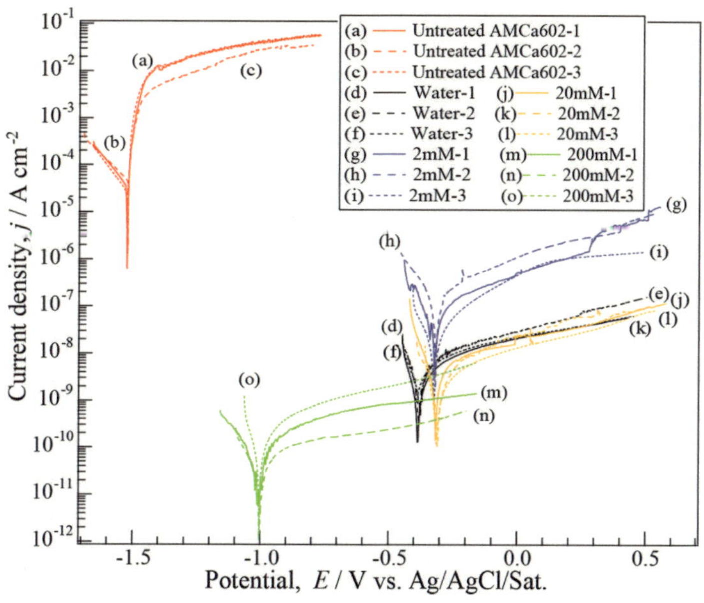

Figure 17. Potentiodynamic polarization curves for AMCa602 Mg alloy treated at 433 K for 6 h using aqueous solutions containing 0, 2, 20, and 200 mM of $Al(NO_3)_3$. Each system represents three repeat polarization curves (labeled 1, 2, and 3) measured under identical conditions to demonstrate reproducibility [Corrosion Science, 2015, 92, 76–84, doi:10.1016/j.corsci.2014.11.031. Copyright @ELSEVIER (2015)].

low density of this film produced the highest j_{corr} value, as explained earlier in the discussion regarding the formation mechanism. Thus, although the films prepared from aqueous solutions containing 20 and 200 mM $Al(NO_3)_3$ were of comparable thickness, their j_{corr} values suggest that they had very different densities.

An increase in current density at approximately 0.3 and 0 V vs. Ag/AgCl/sat.-KCl was observed in the polarization curves of the samples prepared from aqueous solutions containing 2 and 20 mM $Al(NO_3)_3$, which suggests that the corrosive solution permeated through cracks or pores in the film and initiated pitting corrosion. Similar behavior has been reported even in aluminum alloys [73], but in this case, the increase in current density was suppressed to produce a passive-like behavior with a shift in potential toward more positive values. This behavior suggests that the film might have some degree of self-healing functionality, which has been reported with Zn-Al LDH films formed on Al substrates that exhibited an increase in passive current density to initial values after immersion in a corrosive electrolyte [87]. In contrast, there is no evidence of pitting corrosion in the curve of the sample prepared from an aqueous solution containing 200 mM $Al(NO_3)_3$, which had E_{corr} and j_{corr} values estimated to be −1.007 V vs. Ag/AgCl/sat.-KCl and 9.560×10^{-11} A cm^{-2}, respectively, as shown in **Table 3**. This j_{corr} value was the lowest of all the samples, and as this is closely related to the rate of corrosion, this would suggest that a film prepared from 200 mM $Al(NO_3)_3$ offers the best corrosion resistance.

The correlations between the Mg-Al LDH content of the films, $Al(NO_3)_3$ concentration in solution, and corrosion current density are shown in **Figure 18**. The Mg-Al LDH content of the films was roughly estimated from the XRD peak intensity ratio of the (0 0 3) plane of Mg-Al LDH, $I_{LDH\text{-}003}$, to that of the (1 0 1) plane of $Mg(OH)_2$, $I_{Mg(OH)2\text{-}101}$, i.e., $I_{LDH\ content} = I_{LDH\text{-}003}/I_{Mg(OH)2\text{-}101}$. The peak intensities for these planes correspond to the strongest intensities in the standard powder diffraction data for $Mg(OH)_2$ and Mg-Al LDH. As the $Al(NO_3)_3$ concentration in the aqueous solution increased, the I_{LDH} content increased and the corrosion current density decreased, suggesting a strong relationship between the two. It has been reported that carbonate-based Mg-Al LDH shows anion exchangeability, whereby Cl^- ions are incorporated from NaCl solution and CO_3^{2-} ions are exhausted [51]. Moreover, nitrate-based Mg-Al LDH can abstract Cl^- by ejecting NO_3^- and so has an anion exchangeability that is higher than that of carbonate-based Mg-Al LDH. The Mg-Al LDH content of the films is therefore considered to have a significant effect on the corrosion current density, which is closely related to the rate of corrosion.

The durability of the corrosion-resistance performance of the sample prepared from an aqueous solutions containing 200 mM $Al(NO_3)_3$ was investigated by immersing it in a 5 wt.% NaCl solution. **Figure 19** shows the surface of the sample before immersion and after intervals of 5, 12, 24, 48, and 72 h. Note that the beige color of the film surface appeared to change slightly upon immersion, but no corrosion product or physical change in the appearance of the film was observed for up to 12 h. After 24 h of immersion, however, some spotty contrasts emerged that are likely the result of the film being damaged by the corrosive solution. An increase in the immersion time from 48 to 72 h led to a slight dissolution of the AMCa602 substrate at the corners of the sample, which is believed to be the progression of corrosion initiated by corrosive medium coming into contact with the substrate through a pit or minute pore in the film. Additionally, as corrosion resistance is known to be closely related to the film thickness, it is possible that the film was simply thinner at the corner than at other points on the sample. This nonuniformity in film thickness could have occurred through differences in crystal growth at the edge of the substrate.

The SEM images in **Figure 20** show the sample surface before and after immersion in 5 wt.% NaCl solution for 24 and 72 h. It is evident from this that with an increase in immersion time, there was a change in the surface morphology and nanosheets that were slightly inclined with respect to the surface were more likely to be noticed on the surface. The EDX spectrum in **Figure 20(d)** revealed that there were two peaks attributable to Cl and Al at spot (i) in **Figure 20(c)**, but no evidence of a Na peak, which supports the earlier assertion that the Mg-Al LDH layer remained in the film and exhausted CO_3^{2-} and NO_3^- in exchange for Cl^- anions. Similarly, the absence of Cl and the presence of Al at spot (ii) also indicate that the Mg-Al LDH layer remained, which means that the nitrate or carbonate anions in the Mg-Al LDH layer were partially replaced with Cl^- anions.

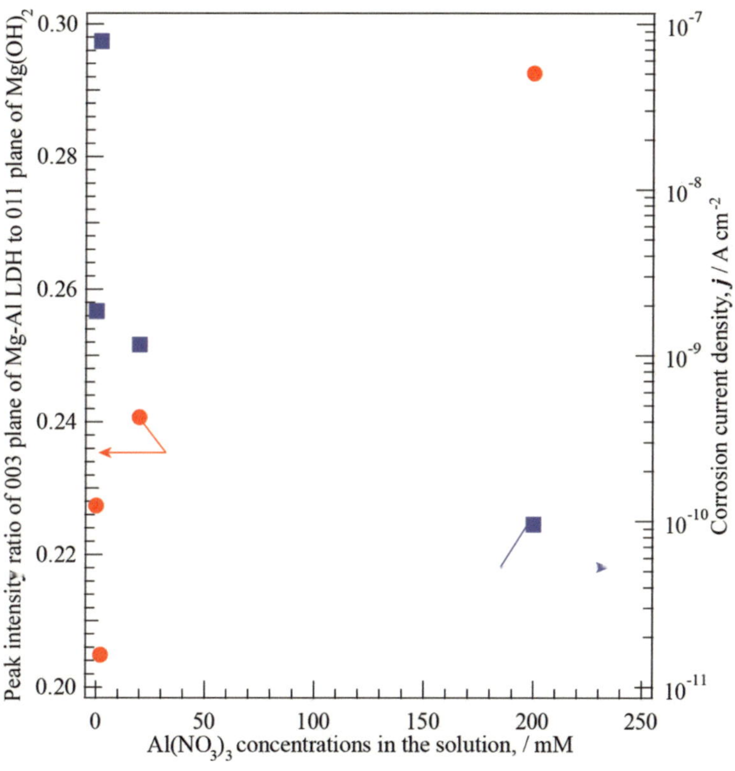

Figure 18. Correlation between Mg-Al LDH content of the film, $Al(NO_3)_3$ concentration in the aqueous solution, and corrosion current densities. The standard deviation of j_{corr} data for the samples obtained from aqueous solution containing 0, 2, 20, and 200 of mM $Al(NO_3)_3$ was 9.85×10^{-10}, 7.68×10^{-7}, 2.39×10^{-9}, and 2.17×10^{-10} A cm^{-2}, respectively [Corrosion Science, 2015, 92, 76–84, doi:10.1016/j.corsci.2014.11.031. Copyright @ELSEVIER (2015)].

Figure 21(a) presents XRD patterns obtained from the film-coated AMCa602 before and after immersion in 5 wt.% NaCl solution for 24 to 72 h. This shows that the peak attributable to the

(1 0 1) plane of the $Mg(OH)_2$ crystallite decreased slightly in intensity with an increase in immersion time, whereas that attributable to the (0 0 3) plane of the Mg-Al LDH crystallite increased in intensity. This indicates that even though the $Mg(OH)_2$ film was slightly dissolved in the corrosive solution, the Mg-Al LDH crystallite remained intact. The potentiodynamic polarization curves for the film-coated AMCa602 before and after immersion are shown in **Figure 21(b)**, from which the E_{corr} and j_{corr} values were determined to be −0.348 V vs. Ag/AgCl/sat.-KCl and 1.821×10^{-9} A cm^{-2}, respectively, after 24 h immersion, and −0.065 V vs. Ag/AgCl/sat.-KCl and 1.326×10^{-9} A cm^{-2} after 72 h. The E_{corr} shifted toward a more positive value with an increase in immersion time, which is possibly a result of the change in the ratio of anodic to cathodic sites on the surface that is caused by the dissolution of $Mg(OH)_2$ in the film. The j_{corr} value of the film-coated AMCa602 increased after immersion, indicating an increase in the corrosion rate that agrees well with the results of the immersion tests. This increase, however, was independent of the immersion time, with the j_{corr} value almost comparable after 24 and 72 h of immersion.

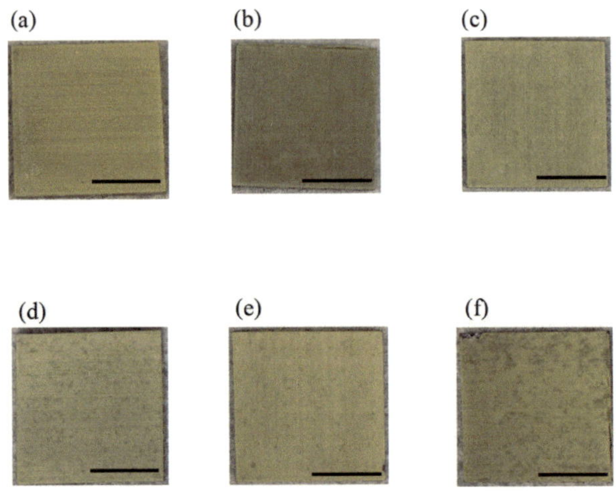

Figure 19. (a) Surface of AMCa602 Mg alloy after steam coating at 433 K for 6 h using an aqueous solution containing 200 mM $Al(NO_3)_3$ before immersion in a 5 wt.% NaCl aqueous solution. Same surface after immersion in a 5 wt.% NaCl aqueous solution for (b) 5, (c) 12, (d) 24, (e) 48, and (f) 72 h. All scale bars are 1 cm [Corrosion Science, 2015, 92, 76–84, doi:10.1016/j.corsci.2014.11.031. Copyright @ELSEVIER (2015)].

From the XRD patterns, SEM observations, EDX spectra, and polarization curve measurements, the mechanism by which a film produced by steam coating provides corrosion resistance to AMCa602 alloy can be explained as follows. At first, $Mg(OH)_2$ from the $Mg(OH)_2$/Mg-Al LDH composite film formed earlier on AMCa602 is dissolved to some extent through contact with a corrosive solution. This dissolution of $Mg(OH)_2$ causes the amount of Mg-Al LDH with anion exchangeability present on the film surface to increase gradually, leading to an increase in the intensity of the 0 0 3 reflection of Mg-Al LDH, as shown in **Figure 21(a)**. Due to the presence of Mg-Al LDH on the film surface, Cl^- ions can become trapped in the Mg-Al LDH, as shown in the EDS analysis, resulting in delaying corrosion. These results support the hypothesis that the Mg-Al content of a $Mg(OH)_2$/Mg-Al LDH composite film is an key factor in improving the durability and corrosion resistance of the film. Moreover, steam coating produces a film that is thick and has a high density due to the presence of two phases of different sizes, namely, $Mg(OH)_2$ and Mg-Al LDH, which further contributes to improving the corrosion resistance of the film.

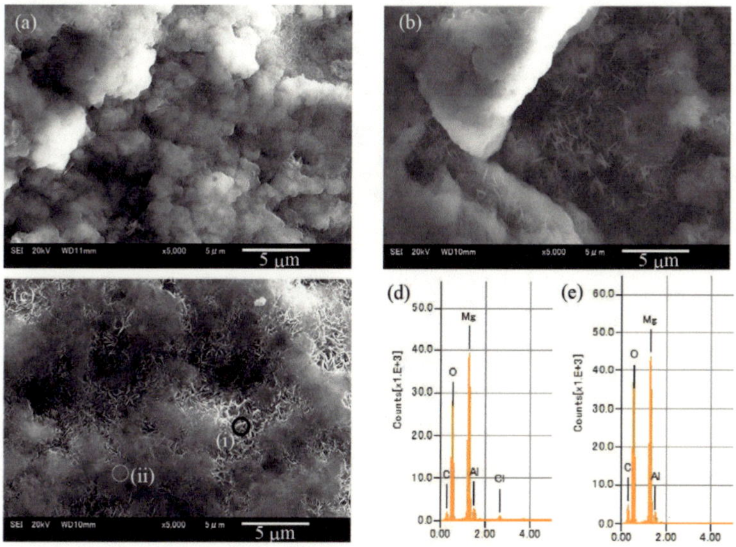

Figure 20. SEM images of sample surface (a) before and after immersion in 5 wt.% NaCl solution for (b) 24 and (c) 72 h. (d, e) EDX spectra obtained at spots (i) and (ii) in (c) [Corrosion Science, 2015, 92, 76–84, doi:10.1016/j.corsci.2014.11.031. Copyright @ELSEVIER (2015)].

In conclusion for this section, a film composed of $Mg(OH)_2$ and Mg-Al LDH was successfully formed on combustion-resistant AMCa602 Mg alloy by using an environmentally friendly steam coating method. Subsequent XRD and FT-IR studies confirmed that the film was composed of $Mg(OH)_2$ in combination with nitrate- and carbonate-based Mg-Al LDHs. The film thickness varied from 52 to 308 μm depending on the $Al(NO_3)_3$ concentration. A mechanism for the formation of such films was proposed based on the results of XRD, FT-IR, SEM, and EDX measurements, and the corrosion resistance was estimated by potentiodynamic measurements and immersion tests in 5 wt.% NaCl solution. A film prepared by steam coating at 433 K for 6 h using an aqueous solution of 200 mM $Al(NO_3)_3$ was found to have the highest corrosion resistance of all samples tested. The correlation between the Mg-Al LDH content of the film, $Al(NO_3)_3$, its corrosion current density, and the concentration of the aqueous solution used to create it was investigated, which revealed a strong correlation between the Mg-Al LDH content and corrosion current density. Based on the results achieved here with flame-resistant Mg alloy, we believe that our steam coating technique is an effective method for improving the corrosion resistance of both existing and future-generation Mg alloys.

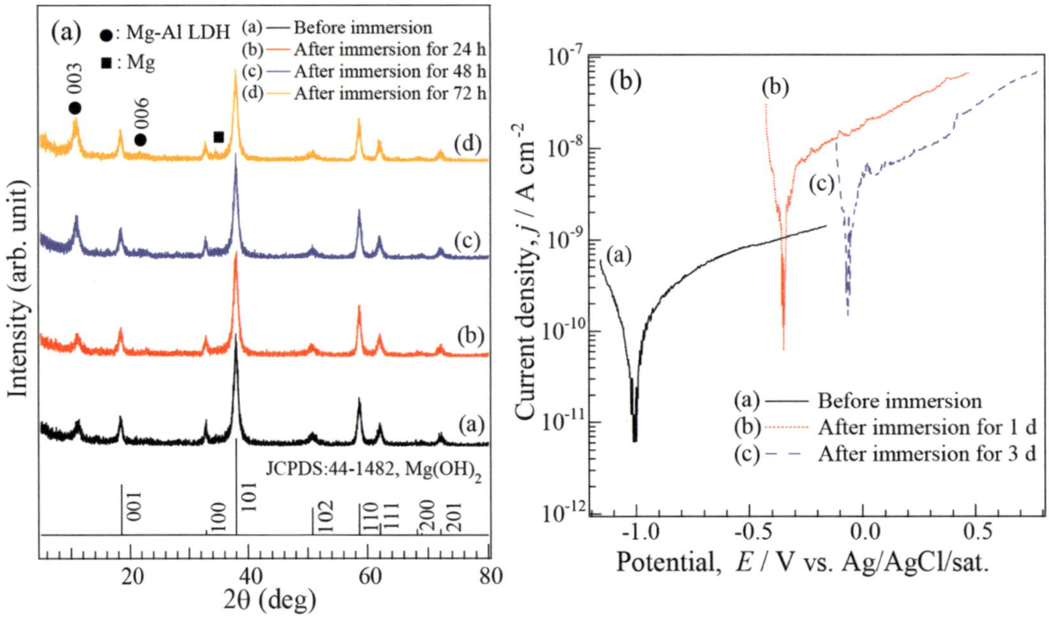

Figure 21. (a) GAXRD patterns and (b) potentiodynamic polarization curves of film-coated AMCa602 Mg alloy before and after immersion in 5 wt.% NaCl solution for 24 to 72 h [Corrosion Science, 2015, 92, 76–84, doi:10.1016/j.corsci.2014.11.031. Copyright @ELSEVIER (2015)].

4. Conclusion and outlook

This chapter has presented the results of experimental studies into the formation and corrosion resistance of an anticorrosive film composed of $Mg(OH)_2$ and Mg-Al LDH produced on AZ31 and AMCa602 Mg alloys by steam treatment. The physicochemical properties of the films were investigated using XRD, SEM, EDX, and FT-IR, and a mechanism for their formation was proposed based on these results. The corrosion resistance of the films produced on AZ31 and AMCa602 was estimated based on potentiodynamic measurements and immersion tests in 5 wt.% NaCl solution, which revealed that corrosion resistance of both alloys was improved by steam treatment. The Mg-Al LDH content of the films was also found to be strongly related to the corrosion current density. As Mg alloys are one of the more promising materials for reducing vehicle weight, thereby lowering fuel consumption and reducing CO_2 emission, any improvement in their inherently low corrosion resistance is of great value to increasing their wider-scale use. Thus, although there has already been a great deal of academic work and commercial interest in surface treatments for Mg alloys, we believe future work should focus on developing the steam coating process described here to greatly improve the corrosion resistance of a range of Mg alloys.

http://dx.doi.org/10.5772/66365

Acknowledgements

This research was partially supported by a grant for Advanced Industrial Technology Development from the New Energy and Industrial Technology Development Organization (NEDO) of Japan (no. 11B06024d), a grant-in-aid for young scientists (B) (no. 16 K18249) from the Japan Society for the Promotion of Science, and the Japan Science and Technology Agency (JST), Adaptable and Seamless Technology Transfer Program through Target-driven R&D (A-step: no. AS251Z02997K and AS2815047S).

Author details

Takahiro Ishizaki*[,1], Mika Tsunakawa[2], Ryota Shiratori[2], Kae Nakamura[2] and Ai Serizawa[1]

*Address all correspondence to: ishizaki@shibaura-it.ac.jp

1 Department of Materials Science and Engineering, College of Engineering, Shibaura Institute of Technology, Koto-ku, Japan

2 Department of Materials Science and Engineering, Graduate School of Engineering and Science, Shibaura Institute of Technology, Koto-ku, Japan

References

[1] Song G L, Atrens A, Quondam MX. Corrosion mechanisms of magnesium alloys. Advanced Engineering Materials. 1999; 1:1:11–33. DOI: 10.1002/(SICI)1527-2648(199909)1:1<11::AID-ADEM11>3.0.CO;2-N

[2] Shreir L L, editor. Corrosion Volume 1. London: Newnes-Butterworths; 1965. 86 p. ISBN: 0408001097

[3] Pokhmurska H, Wielage B, Lampke T, Grund T, Student M, Chervinska N, Quondam MX. Post-treatment of thermal spray coatings on magnesium. Surface and Coatings Technology. 2008; 202:18:4515–4524. DOI: 10.1016/j.surfcoat.2008.04.036

[4] Ghali E. Magnesium and magnesium alloys. In: Revie R W, editor. Uhlig's Corrosion Handbook. 3rd ed. New York: John Wiley & Sons; 2000, 793 p. DOI: 10.1002/9780470872864

[5] Pardo A, Merino M C, Coy A E, Arrabal R, Viejo F, Matykina E, Quondam MX. Corrosion behaviour of magnesium/aluminium alloys in 3.5 wt.% NaCl. Corrosion Science. 2008; 50:823–834. DOI: 10.1016/j.corsci.2007.11.005

[6] Liu Z, Gao W, Quondam MX. Electroless nickel plating on AZ91 Mg alloy substrate. Surface and Coatings Technology. 2006; 200:5087–5093. DOI: 10.1016/j.surfcoat.2005.05.023

[7] Lian J S, Li G Y, Niu L Y, Gu C D, Jiang Z H, Jiang Q, Quondam MX. Electroless Ni–P deposition plus zinc phosphate coating on AZ91D magnesium alloy. Surface and Coatings Technology. 2006; 200:5956–5962. DOI: 10.1016/j.surfcoat.2005.09.007

[8] Barchiche C E, Rocca E, Juers C, Hazan J, Steinmetz J. Corrosion resistance of plasma-anodized AZ91D magnesium alloy by electrochemical methods. Electrochimica Acta. 2007; 53:417–425. DOI: 10.1016/j.electacta.2007.04.030

[9] Montemor M F, Ferreira M G S, Quondam MX. Electrochemical study of modified bis-[triethoxysilylpropyl] tetrasulfide silane films applied on the AZ31 Mg alloy. Electrochimica Acta. 2007; 52:7486–7495. DOI: 10.1016/j.electacta.2006.12.086

[10] Zhang D, Gou Y, Liu Y, Guo X, Quondam MX. A composite anodizing coating containing superfine Al_2O_3 particles on AZ31 magnesium alloy. Surface and Coatings Technology. 2013; 236:52–57. DOI: 10.1016/j.surfcoat.2013.04.059

[11] Guo X, Du K, Guo Q, Wang Y, Wang F, Quondam MX. Experimental study of corrosion protection of a three-layer film on AZ31B Mg alloy. Corrosion Science. 2012; 65:367–375. DOI: 10.1016/j.corsci.2012.08.055

[12] Hamdy A S, Doench I, Mohwald H, Quondam MX. Smart self-healing anti-corrosion vanadia coating for magnesium alloys. Progress in Organic Coatings. 2011; 72:387–393. DOI: 10.1016/j.porgcoat.2011.05.011

[13] Gigandet M P, Faucheu J, Tachez M, Quondam MX. Formation of black chromate conversion coatings on pure and zinc alloy electrolytic deposits: role of the main constituents. Surface and Coatings Technology. 1997; 89:3:285–291. DOI: 10.1016/S0257-8972(96)03013-7

[14] Anandan C, William Grips V K, Rajam K S, Jayaram V, Bera P, Quondam MX. Investigation of surface composition of electrodeposited black chrome coatings by X-ray photoelectron spectroscopy. Applied Surface Science. 2002; 191:254–260. DOI: 10.1016/S0169-4332(02)00192-7

[15] Bayati M R, Shariat M H, Janghorban K, Quondam MX. Design of chemical composition and optimum working conditions for trivalent black chromium electroplating bath used for solar thermal collectors. Renewable Energy. 2005; 30:2163–2178. DOI: 10.1016/j.renene.2005.02.003

[16] Budavari S, O'Neil M, Smith A, Heckelman P, Obenchain J. The Merck Index: An Encyclopedia of a Chemical, Drugs and Biologicals. 12th ed. Boca Raton: CRC Press; 1996. 1475 p. ISBN: 0911910123

[17] Elsentriecy H H, Azumi K, Konno H, Quondam MX. Effect of surface pretreatment by acid pickling on the density of stannate conversion coatings formed on AZ91 D magnesium alloy. Surface and Coatings Technology. 2007; 202:532–537. DOI: 10.1016/j.surfcoat.2007.06.033

[18] Chong K Z, Shih T S, Quondam MX. Conversion-coating treatment for magnesium alloys by a permanganate–phosphate solution. Materials Chemistry and Physics. 2003; 80:191–200. DOI: 10.1016/S0254-0584(02)00481-9

[19] Rudd A L, Breslin C B, Mansfeld F, Quondam MX. The corrosion protection afforded by rare earth conversion coatings applied to magnesium. Corrosion Science. 2000; 42:275–288. DOI: 10.1016/S0010-938X(99)00076-1

[20] Takenaka T, Ono T, Narazaki Y, Naka Y, Kawakami M, Quondam MX. Improvement of corrosion resistance of magnesium metal by rare earth elements. Electrochimica Acta. 2007; 53: 117–121. DOI: 10.1016/j.electacta.2007.03.027

[21] Zucchi F, Frignani A, Grassi V, Trabanelli G, Monticelli C, Quondam MX. Stannate and permanganate conversion coatings on AZ31 magnesium alloy. Corrosion Science. 2007; 49:4542–4552. DOI: 10.1016/j.corsci.2007.04.011

[22] Scholes F H, Soste C, Hughes A E, Hardin S G, Curtis P R, Quondam MX. The role of hydrogen peroxide in the deposition of cerium-based conversion coatings. Applied Surface Science. 2006; 253:1770–1780. DOI: 10.1016/j.apsusc.2006.03.010

[23] Yang L, Li J, Lin C, Zhang M, Wu J, Quondam MX. Study of molybdenum/lanthanum-based composite conversion coatings on AZ31 magnesium alloy. Applied Surface Science. 2011; 257:2838–2842. DOI: 10.1016/j.apsusc.2010.10.077

[24] Zeng R, Lan Z, Kong L, Huang Y, Cui H, Quondam MX. Characterization of calcium-modified zinc phosphate conversion coatings and their influences on corrosion resistance of AZ31 alloy. Surface and Coatings Technology. 2011; 205:3347–3355. DOI: 10.1016/j.surfcoat.2010.11.027

[25] Khaselev O, Yahalom J, Quondam MX. Constant voltage anodizing of Mg-Al Alloys in KOH-$Al(OH)_3$ solutions. Journal of The Electrochemical Society. 1998; 145:190–193. DOI: 10.1149/1.1838234

[26] Yerokhin A L, Nie X, Matthews L A, Dowey S J, Quondam MX. Plasma electrolysis for surface engineering. Surface and Coatings Technology. 1999; 122:73–93. DOI: 10.1016/S0257-8972(99)00441-7

[27] Kuhn A, Quondam MX. Plasma anodizing of magnesium alloys. Metal Finishing. 2003; 101:44–50. DOI: 10.1016/S0026-0576(03)90261-3

[28] Ma Y, Nie X, Northwood D O, Hu H, Quondam MX. Systematic study of the electrolytic plasma oxidation process on a Mg alloy for corrosion protection. Thin Solid Films. 2006; 494:296–301. DOI: 10.1016/j.tsf.2005.08.156

[29] Luo H, Cai Q, He J, Wei B, Quondam MX. Preparation and properties of composite ceramic coating containing Al_2O_3-ZrO_2-Y_2O_3 on AZ91D magnesium alloy by plasma electrolytic oxidation. Current Applied Physics. 2009; 9:1341–1346. DOI: 10.1016/j.cap.2009.02.017

[30] Li X, Liu X Y, Luan B L, Quondam MX. Corrosion and wear properties of PEO coatings formed on AM60B alloy in $NaAlO_2$ electrolytes. Applied Surface Science. 2011; 257:9135–9141. DOI: 10.1016/j.apsusc.2011.05.115

[31] Hussein R O, Zhang P, Nie X, Xia T, Northwood D O, Quondam MX. The effect of current mode and discharge type on the corrosion resistance of plasma electrolytic oxidation (PEO) coated magnesium alloy AJ62. Surface and Coatings Technology. 2011; 206:1990–1997. DOI: 10.1016/j.surfcoat.2011.08.060

[32] Birss V, Xia S, Yue R, Richard G R Jr, Quondam MX. Characterization of oxide films formed on Mg-based WE43 alloy using AC/DC anodization in silicate solutions. Journal of the Electrochemical Society. 2004; 151:B1–10. DOI: 10.1149/1.1629095

[33] Wang Y, Wang J, Zhang J, Zhang Z, Quondam MX. Characteristics of anodic coatings oxidized to different voltage on AZ91D Mg alloy by micro-arc oxidization technique. Materials and Corrosion. 2005; 56:88–92. DOI: 10.1002/maco.200403822

[34] Hsiao H Y, Tsung H C, Tsai W T, Quondam MX. Anodization of AZ91D magnesium alloy in silicate-containing electrolytes. Surface and Coatings Technology. 2005; 199:127–134. DOI: 10.1016/j.surfcoat.2004.12.010

[35] Blawert C, Heitmann V, Dietzel W, Nykyforchyn H M, Klapkiv M D, Quondam MX. Influence of process parameters on the corrosion properties of electrolytic conversion plasma coated magnesium alloys. Surface and Coatings Technology. 2005; 200:68–72. DOI: 10.1016/j.surfcoat.2005.02.035

[36] Srinivasan P B, Liang J, Blawert C, Störmer M, Dietzel W, Quondam MX. Effect of current density on the microstructure and corrosion behaviour of plasma electrolytic oxidation treated AM50 magnesium alloy. Applied Surface Science. 2009; 255:4212–4218. DOI: 10.1016/j.apsusc.2008.11.008

[37] Sreekanth D, Rameshbabu N, Venkateswarlu K, Quondam MX. Effect of various additives on morphology and corrosion behavior of ceramic coatings developed on AZ31 magnesium alloy by plasma electrolytic oxidation. Ceramics International. 2012; 38:4607–4615. DOI: 10.1016/j.ceramint.2012.02.040

[38] Stojadinović S, Vasilić R, Petković M, Belča I, Kasalica B, Perić M, Zeković L J, Quondam MX. Luminescence during anodization of magnesium alloy AZ31. Electrochimica Acta. 2012; 59:354–359. DOI: 10.1016/j.electacta.2011.10.084

[39] Khaseleva O, Weiss D, Yahalom J, Quondam MX. Anodizing of pure magnesium in KOH-aluminate solutions under sparking. Journal of the Electrochemical Society. 1999; 146:1757–1761. DOI: 10.1149/1.1391838

[40] Bonilla F A, Berkani A, Liu Y, Skeldon P, Thompson G E, Habazaki H, Shimizu K, John C, Stevens K, Quondam MX. Formation of anodic films on magnesium alloys in an alka-

line phosphate electrolyte. Journal of The Electrochemical Society. 2002; 149:B4–B13. DOI: 10.1149/1.1424896

[41] Ghasemi A, Raja V S, Blawert C, Dietzel W, Kainer K U, Quondam MX. Study of the structure and corrosion behavior of PEO coatings on AM50 magnesium alloy by electrochemical impedance spectroscopy. Surface and Coatings Technology. 2008; 202:3513–3518. DOI: 10.1016/j.surfcoat.2007.12.033

[42] Arrbal R, Matykina E, Viejo F, Skeldon P, Thompson G E, Quondam MX. Corrosion resistance of WE43 and AZ91D magnesium alloys with phosphate PEO coatings. Corrosion Science. 2008; 50:1744–1752. DOI: 10.1016/j.corsci.2008.03.002

[43] Liang J, Srinivasan P B, Blawert C, Dietzel W, Quondam MX. Influence of pH on the deterioration of plasma electrolytic oxidation coated AM50 magnesium alloy in NaCl solutions. Corrosion Science. 2010; 52:540–547. DOI: 10.1016/j.corsci.2009.10.011

[44] Srinivasan P B, Liang J, Balajeee R G, Blawert C, Stormer M, Dietzel W, Quondam MX. Effect of pulse frequency on the microstructure, phase composition and corrosion performance of a phosphate-based plasma electrolytic oxidation coated AM50 magnesium alloy. Applied Surface Science. 2010; 256:3928–3935. DOI: 10.1016/j.apsusc.2010.01.052

[45] Yagi S, Sengoku A, Kubota K, Matsubara E, Quondam MX. Surface modification of ACM522 magnesium alloy by plasma electrolytic oxidation in phosphate electrolyte. Corrosion Science. 2012; 57:74–80. DOI: 10.1016/j.corsci.2011.12.032

[46] Collazo A, Hernández M, Nóvoa X R, Pérez C, Quondam MX. Effect of the addition of thermally activated hydrotalcite on the protective features of sol–gel coatings applied on AA2024 aluminium alloys. Electrochimica Acta. 2011; 56:7805–7814. DOI: 10.1016/j.electacta.2011.03.067

[47] Lin J K, Uan J Y, Quondam MX. Formation of Mg,Al-hydrotalcite conversion coating on Mg alloy in aqueous HCO_3^-/CO_3^{2-} and corresponding protection against corrosion by the coating. Corrosion Science. 2009; 51:1181–1188. DOI: 10.1016/j.corsci.2009.02.007

[48] Chen J, Song Y, Shan D, Han E H, Quondam MX. In situ growth of Mg–Al hydrotalcite conversion film on AZ31 magnesium alloy. Corrosion Science. 2011; 10:3281–3288. DOI: 10.1016/j.corsci.2011.06.0033

[49] Lin J K, Jeng K L, Uan J Y, Quondam MX. Crystallization of a chemical conversion layer that forms on AZ91D magnesium alloy in carbonic acid. Corrosion Science. 2011; 53:3832–3839. DOI: 10.1016/j.corsci.2011.07.035

[50] Chen J, Song Y, Shan D, Han E H, Quondam MX. Study of the in situ growth mechanism of Mg–Al hydrotalcite conversion film on AZ31 magnesium alloy. Corrosion Science. 2012; 33:148–158. DOI: 10.1016/j.corsci.2012.05.022

[51] Cavani F, Trifirò F, Vaccari A, Quondam MX. Hydrotalcite-type anionic clays: preparation, properties and applications. Catalysis Today. 1991; 11:2:173–301. DOI: 10.1016/0920-5861(91)80068-K

[52] Poznyak S K, Tedim J, Rodrigues L M, Salak A N, Zheludkevich M L, Dick L F P, Ferreira M G S, Quondam MX. Novel inorganic host layered double hydroxides intercalated with guest organic inhibitors for anticorrosion applications. ACS Applied Materials & Interfaces. 2009; 1:10:2353–2362. DOI: 10.1021/am900495r

[53] Zheludkevich M L, Poznyak S K, Rodrigues L M, Raps D, Hack T, Dick L F, Nunes T, Ferreira M G S, Quondam MX. Active protection coatings with layered double hydroxide nanocontainers of corrosion inhibitor. Corrosion Science. 2010; 52:2:602–611. DOI: 10.1016/j.corsci.2009.10.020

[54] Buchheit R G, Guan H, Mahajanam S, Wong F, Quondam MX. Active corrosion protection and corrosion sensing in chromate-free organic coatings. Progress in Organic Coatings. 2003; 47:3–4:174–182. DOI: 10.1016/j.porgcoat.2003.08.003

[55] Tedim J, Poznyak S K, Kuznetsova A, Raps D, Hack T, Zheludkevich M L, Ferreira M G S, Quondam MX. Enhancement of active corrosion protection via combination of inhibitor-loaded nanocontainers. ACS Applied Materials & Interfaces. 2010; 2:5:1528–1535. DOI: 10.1021/am100174teira

[56] Tedim J, Zheludkevich M L, Salak A N, Lisenkov A, Ferreira M G S, Quondam MX. Nanostructured LDH-container layer with active protection functionality. Journal of Materials Chemistry. 2011; 21:15464–15470. DOI: 10.1039/C1JM12463C

[57] Montemor M F, Snihirova D V, Taryba M G, Lamaka S V, Kartsonakis I A, Balaskas A C, Kordas G C, Tedim J, Kuznetsova A, Zheludkevich M L, Ferreira M G S, Quondam MX. Evaluation of self-healing ability in protective coatings modified with combinations of layered double hydroxides and cerium molibdate nanocontainers filled with corrosion inhibitors. Electrochimica Acta. 2012; 60: 31–40. DOI: 10.1016/j.electacta.2011.10.078

[58] Lin J K, Hsia C L, Uan J Y, Quondam MX. Characterization of Mg, Al-hydrotalcite conversion film on Mg alloy and Cl^- and CO_3^{2-} anion-exchangeability of the film in a corrosive environment. Scripta Materialia. 2007; 56:11:927–930. DOI: 10.1016/j.scriptamat.2007.02.020

[59] Tedim J, Kuznetsova A, Salak A N, Montemor M F, Snihirova D, Pilz M, Heludkevich M L, Ferreira M G S, Quondam MX. Zn-Al layered double hydroxides as chloride nanotraps in active protective coatings. Corrosion Science. 2012; 55:1–4. DOI: 10.1016/j.corsci.2011.10.003

[60] Wu G, Zeng X, Ding W, Guo X, Yao S, Quondam MX. Characterization of ceramic PVD thin films on AZ31 magnesium alloys. Applied Surface Science 2006; 252:7422–7429. DOI: 10.1016/j.apsusc.2005.08.095

[61] Kuo Y L, Chang K H, Quondam MX. Atmospheric pressure plasma enhanced chemical vapor deposition of SiO_x films for improved corrosion resistant properties of AZ31 magnesium alloys. Surface and Coatings Technology. 2015; 283:194–200. DOI: 10.1016/j.surfcoat.2015.11.004

[62] Wang J, Li D, Liu Q, Yin X, Zhang Y, Jing X, Zhang M, Quondam MX. Fabrication of hydrophobic surface with hierarchical structure on Mg alloy and its corrosion resistance. Electrochimica Acta. 2010; 55:6897–6906. DOI: 10.1016/j.electacta.2010.05.070

[63] Lom T S, Ryu H S, Hong S H, Quondam MX. Electrochemical corrosion properties of CeO_2-containing coatings on AZ31 magnesium alloys prepared by plasma electrolytic oxidation. Corrosion Science. 2012; 62:104–111. DOI: 10.1016/j.corsci.2012.04.043

[64] Ryu H S, Park D S, Hong S H, Quondam MX. Improved corrosion protection of AZ31 magnesium alloy through plasma electrolytic oxidation and aerosol deposition duplex treatment. Surface and Coatings Technology. 2013; 219:82–87. DOI: 10.1016/j.surfcoat.2013.01.008

[65] Zhang F, Sun M, Xu S, Zhao L, Zhang B, Quondam MX. Fabrication of oriented layered double hydroxide films by spin coating and their use in corrosion protection. Chemical Engineering Journal. 2008; 141:362–367. DOI: 10.1016/j.cej.2008.03.016

[66] Ishizaki T, Cho S P, Saito N, Quondam MX. Morphological control of vertically self-aligned nanosheets formed on magnesium alloy by surfactant-free hydrothermal synthesis. CrystEngComm. 2009; 11:2338–2343. DOI: 10.1039/B907490B

[67] Xu Z P, Zeng H C, Quondam MX. In-situ generation of maximum trivalent cobalt in synthesis of hydrotalcite-like compounds $Mg_xCoII_{1-x-y}CoIII_y(OH)_2(NO_3)_y{\cdot}nH_2O$. Chemistry of Materials. 2000; 12:9:2597–2603. DOI: 10.1021/cm000053f

[68] Millange F, Walton R I, O'Hare D, Quondam MX. Time-resolved in situ X-ray diffraction study of the liquid-phase reconstruction of Mg–Al–carbonate hydrotalcite-like compounds. Journal of Materials Chemistry. 2000; 10:1713–1720. DOI: 10.1039/D002827O

[69] Regnier P, Lasaga A C, Berner R A, Han K W, Zilm O H, Quondam MX. Mechanism of CO_3^{2-} substitution in carbonate-fluorapatite: evidence from FTIR spectroscopy, 13C NMR, and quantum mechanical calculations. American Mineralogist. 1994; 79:9–10:809–818.

[70] Nordlien J S, Ono S, Masuko N, Nisancioglu K, Quondam MX. A TEM investigation of naturally formed oxide films on pure magnesium. Corrosion Science. 1997; 39:8:1397–1414. DOI: 10.1016/S0010-938X(97)00037-1

[71] Tanaka D K, Long G G, Kruger J. Structure of the passive films on cast and rapidly solidified Mg alloys. In: Proceedings of 11th International Corrosion Congress; 2-9, April, 1990; Florence. London: The International Corrosion Council. 1990. 605–610.

[72] Baril G, Galicia G, Deslouis C, Pébère N, Tribollet B, Vivier V, Quondam MX. An impedance investigation of the mechanism of pure magnesium corrosion in sodium sulfate

solutions corrosion, passivation, and anodic films. Journal of the Electrochemical Society. 2007; 154:2:C108–C113. DOI: 10.1149/1.2401056

[73] Bobby Kannan M, Raja V S, Quondam MX. Influence of heat treatment and scandium addition on the electrochemical corrosion behavior of Al-Zn-Mg-Cu-Zr (7010) alloy. Metallurgical and Materials Transactions A. 2007; 38A:2843–2852. DOI: 10.1007/s11661-007-9303-6.

[74] Williams G, McMurray H N, Quondam MX. Anion-exchange inhibition of filiform corrosion on organic coated AA2024-T3 aluminum alloy by hydrotalcite-like pigments. Electrochemical and Solid-State Letters. 2003; 6:B9–B11. DOI:10.1149/1.1539771

[75] Masaki K, Ochi Y, Kakiuchi T, Kurata K, Hirasawa T, Matsumura T, Takigawa Y, Higashi K, Quondam MX. Thermodynamic properties of selenium in Ag-Pb alloy and lead oxide phases at 1273 K. Materials Transactions. 2009; 50:5:1148–1151. DOI: 10.2320/matertrans.M2009021

[76] Zeng R C, Zhang F, Lan Z D, Cui H Z, Han E H, Quondam MX. Corrosion resistance of calcium-modified zinc phosphate conversion coatings on magnesium–aluminium alloys. Corrosion Science. 2014; 88:452–459. DOI: 10.1016/j.corsci.2014.08.007

[77] Yang H Y, Guo X W, Chen X B, Birbilis N, Quondam MX. A homogenisation pre-treatment for adherent and corrosion-resistant Ni electroplated coatings on Mg-alloy AZ91D. Corrosion Science. 2014; 79:41–49. DOI: 10.1016/j.corsci.2013.10.024

[78] Cui X J, Liu C H, Yang R S, Fu Q S, Lin X Z, Gong M, Quondam MX. Duplex-layered manganese phosphate conversion coating on AZ31 Mg alloy and its initial formation mechanism. Corrosion Science. 2013; 76:474–485. DOI: 10.1016/j.corsci.2013.07.024

[79] Wan T T, Liu Z X, Bu M Z, Wang P C, Quondam MX. Effect of surface pretreatment on corrosion resistance and bond strength of magnesium AZ31 alloy. Corrosion Science. 2013; 66:33–42. DOI: 10.1016/j.corsci.2012.08.051

[80] Zhang J, Kang Z, Quondam MX. Effect of different liquid–solid contact models on the corrosion resistance of superhydrophobic magnesium surfaces. Corrosion Science. 2014; 87:452–459. DOI: 10.1016/j.corsci.2014.07.010

[81] Ivanou D K, Starykevich M, Lisenkov A D, Zheludkevich M L, Xue H B, Lamaka S V, Ferreira M G S, Quondam MX. Plasma anodized ZE41 magnesium alloy sealed with hybrid epoxy-silane coating. Corrosion Science. 2013; 73:300–308. DOI: 10.1016/j.corsci.2013.04.019

[82] Xia Y H, Zhang B P, Lu C X, Geng L, Quondam MX. Improving the corrosion resistance of Mg–4.0Zn–0.2Ca alloy by micro-arc oxidation. Materials Science and Engineering C, 2013; 33:5044–5050. DOI: 10.1016/j.msec.2013.08.033

[83] Mori Y, Koshi A, Liao J, Quondam MX. Corrosion resistance of plasma electrolytic oxidation layer of a non-ignitable Mg–Al–Mn–Ca magnesium alloy. Corrosion Science. 2016; 104:207–216. DOI: 10.1016/j.corsci.2015.12.013

[84] Xu Z P, Zeng H C, Quondam MX. Abrupt structural transformation in hydrotalcite-like compounds $Mg_{1-x}Al_x(OH)_2(NO_3)_x \cdot nH_2O$ as a continuous function of nitrate anions. The Journal of Physical Chemistry B. 2001; 105:9:1743–1749. DOI: 10.1021/jp0029257

[85] Rives V, Ulibarri M A, Quondam MX. Layered double hydroxides (LDH) intercalated with metal coordination compounds and oxometalates. Coordination Chemistry Reviews. 1999; 181:1:61–120. DOI: 10.1016/S0010-8545(98)00216-1

[86] Kuma K, Paplawsky W, Gedulin B, Arrhenius G, Quondam MX. Mixed-valence hydroxides as bioorganic host minerals. Origins of Life and Evolution of the Biosphere. 1989; 19:6:573–602. DOI: 10.1007/BF01808119

[87] Yan T, Xu S, Peng Q, Zhao L, Lei X, Zhang F, Quondam MX. Self-healing of layered double hydroxide film by dissolution/recrystallization for corrosion protection of aluminum. Journal of the Electrochemical Society. 2013; 160:10:C480–C486. DOI: 10.1149/2.053310jes

Section 3

Corrosion

Chapter 5

A Measuring Approach to Assess the Corrosion Rate of Magnesium Alloys Using Electrochemical Impedance Spectroscopy

Maria C. Delgado, Federico R. García-Galvan,
Violeta Barranco and Sebastian Feliu Batlle

Additional information is available at the end of the chapter

http://dx.doi.org/10.5772/65018

Abstract

An attempt was made to estimate the corrosion rate of AZ31 and AZ61 magnesium alloys immersed in 0.6 M NaCl during long-term exposure using electrochemical impedance spectroscopy (EIS). The EIS results were compared with the corrosion rate independently assessed by the hydrogen evolution test. A correlation was established between the integration of the polarization resistance (R_p) and charge transfer resistance (R_t) over time, as evaluated by EIS and hydrogen gas measurements. Regardless of the immersion time, a strong link was found between the R_t and R_p values determined by EIS. This relation seems to depend on the composition of the alloy. The influence of immersion time on the estimated corrosion rate reliability was investigated. The typical deviations of the measurement methods are apparently decreasing upon prolonging the immersion time. No significant errors were obtained in the measurement of the corrosion rate when using R_t or R_p determined by EIS with their corresponding "apparent" Stern-Geary coefficient values compared with the real values determined by gravimetric measurements.

Keywords: Magnesium, weight loss, electrochemical impedance spectroscopy, corrosion rate, hydrogen evolution

1. Introduction

Mg-Al alloys have aroused great scientific and technological interest in recent decades. From a practical point of view, their high strength-weight ratio makes them ideal alloys for automotive, aerospace and electronics applications, where weight reduction is of significant concern [1]. The usage of magnesium alloys in automotive applications can significantly contribute to greater fuel savings and environmental conservation [2]. Also, magnesium and its alloys have been investigated recently by many authors as a suitable biodegradable biomaterial [3]. However, their corrosion performance at the current stage of development is still not good enough for increasingly diverse their practical applications [4]. Given the growing use of Mg alloys as structural alloys, an accurate assessment of corrosion is imperative and the unambiguous determination of electrode kinetics in the case of Mg electrodes is essential [5].

Electrochemical impedance spectroscopy (EIS) has been shown in innumerable studies to be an efficient and accurate tool for investigating the nature of surface films and the associated corrosion mechanisms of metals and alloys [6]. The most outstanding and well-known advantages of EIS are that it allows continuous monitoring of the corrosion with measurements of the instantaneous corrosion rate. Such measurements of the corrosion rate require only minor disturbance in the system because a very small AC signal is applied [7]. The use of electrochemical impedance spectroscopy (EIS) to assess electrochemical corrosion behavior of magnesium and its alloys is of great technical relevance. Since stationary values take significant time to become established, it should be noted that short-term measurements may give rise to misleading comparisons of corrodibility of the magnesium alloys [8]. As James et al. [9] noted, EIS is an ideal method for monitoring the long-term corrosion behavior and ranking the corrosion protective ability of Mg and its alloys. The term "long term" is used to distinguish it from the standard electrochemical corrosion studies, which are performed within the first 1 or 2 h [9].

Many Mg corrosion studies assign the first capacitive semicircle, appearing at high frequencies, to the charge transfer reaction of Mg-Al alloys with a partial protective oxide/hydroxide layer formed as the corrosion product in the test solution [6, 9–14]. The charge transfer resistance R_t is determined by the diameter of this capacitive loops [15–20]. The corrosion rate of freely corroding metal in the absence of coverage effects is inversely proportional to polarization resistance as described by the Stern-Geary relationship [21, 22]:

$$i_{corr} = \frac{B}{R_t} \tag{1}$$

where B is the Stern-Geary constant and can be determined using anodic (β_a) and cathodic (β_c) Tafel slopes [23] ($B = (1/2.303) \times (\beta_a \cdot \beta_c/(\beta_a+\beta_c))$.

The determination of the anodic Tafel slope is not trivial because the magnesium electrode can sustain very high anodic current with minimal anodic polarization, accompanied by substantial hydrogen evolution and local alkalinization, producing a large Ohmic drop in the proximity of the electrode surface [24]. Anodic potentiodynamic polarization of Mg and Mg alloys initially exhibits a very low apparent Tafel slope, representative of an ideally non-polarizable electrode behavior, followed by the bending of the curve [22, 24–26].

In our previous studies [15, 19, 20], due to the uncertainty in the β_a values from the polarization curves commented above [22, 24–26], it was decided to use the constant B estimated empirically from the correlation between electrochemical and gravimetric measurements [15]. Previous investigation [27] yielded B values of around 65 mV for the AZ31 alloy and of around 120 mV for the AZ61 alloy. Cumulative weight loss after an extended period of time was determined by the integration of the weight changes for individual exposure periods. In this way, the average corrosion rates of the specimens after 14 days of exposure to the 0.6 M NaCl solution were obtained from EIS, which were similar to that determined from hydrogen evolution and gravimetric measurements [28].

Recently, King et al. [22] and Bland et al. [26] have shown that the determination of polarization resistance, R_p, at the zero frequency limit as evaluated from analysis of impedance data, provided excellent correlation between the volume of hydrogen collected and mass loss, for both commercially pure Mg and the AZ31 alloy. King et al. [22] critically address previous studies which have utilized the impedance behavior of magnesium and which claim electrochemical tests may underestimate the corrosion rate of Mg; in most cases by a factor of 2 or more, while attempting to use a charge transfer resistance at intermediate frequencies and do not account for the electrochemically justified inductive behavior. Later, Curioni et al. [24] have revealed that the reciprocal of the values of resistance estimated by EIS is linearly correlated with the corrosion current measured by hydrogen collection. The experimentally measured proportionality constant provides an "apparent" Stern-Geary coefficient that can be used to evaluate the corrosion current from EIS data [24].

In these studies [22, 24, 25], impedance measurements were taken during free corrosion while monitoring in real time the amount of hydrogen evolved from the corroding electrodes to obtain instantaneous values of corrosion current of pure magnesium and the AZ31 alloy in NaCl solutions. As Curioni et al. [24] commented, the lack of reproducibility of the corrosion current is due to statistical variation in the corrosion process, but the combined measurement of the corrosion rate by impedance spectroscopy and hydrogen evolution is reliable. From a practical viewpoint, however, the possibilities of obtaining "apparent" Stern-Geary coefficients from the relationship between independent hydrogen evolution and electrochemical impedance spectroscopy measurements are of interest since the experimentation is easy to replicate and to implement; although long exposure times may be involved, the relatively simple procedure reduces the propensity to introduce systematic errors.

The effect of exposure time to the corrosive solution is important since there are frequent cases in which the corrosion rate of Mg tends to change significantly with the immersion time [8, 29–34]. Typically, the initial corrosion rate is less, and then the corrosion rate accelerates to a steady state [35–38]. Due to the high protective nature of the pre-existing native oxide film formed on the polished surfaces of some magnesium alloys, significant fractions of uncorroded silvery regions can be found even after 24 h immersion in 3.5% NaCl aqueous solution [29, 32]. Bland et al. [26] noted that because of the rapid corrosion rates normally observed for the Mg system, the Tafel characteristics can change dramatically between a short- and long-term test. Hsieh et al. [39] examined the relationship between weight loss (WL) measurements and polarization resistance measurements of corrosion in metals and metal alloys (mild steel, aluminum, copper, and cupronickel) commonly used in industrial cooling systems. The experimentally measured proportionality constant *B*', based on the correlation between electrochemically and gravimetric measured corrosion rates, changes with the exposure time until steady state is reached, which may take up to 3 days. At longer immersion times (>3 days), the *B*′ value begins to stabilize and changes little with the corrosivity of water. For this reason, long-term exposure to saline aqueous solutions was selected (AZ31 specimen after 4 days and AZ61 specimen after 14 days of testing) for the measurement of an "apparent" Stern-Geary coefficient for these alloys. The purpose of this study is to report new results within a broader study carried out on the factors affecting the estimation of the corrosion rate in magnesium alloys by EIS [31]. The ultimate aim is to expand our overall understanding of such phenomena and improve the reliability of the experimental results by devising better study designs.

The summary of main results of corrosion rates of pure Mg and magnesium alloys evaluated from electrochemical measurements [5, 16, 17, 22, 24, 26, 32, 38, 40–46] can be seen in **Table 1**.

Authors	Work title	Main aspects	Reference
Makar L, Kruger J	Corrosion Studies of Rapidly Solidified Magnesium Alloys	The charge transfer resistance could be used to accurately measure the corrosion rates of magnesium alloys.	[17]
Kirkland NT, Birbilis N, Staiger MP	Assessing the corrosion of biodegradable magnesium implants: A critical review of current methodologies and their limitations	EIS also does not directly yield a corrosion rate of magnesium implants.	[40]
Pebere N, Riera C, Dabosi F	Investigation of magnesium corrosion in aerated sodium sulfate solution by electro-chemical impedance spectroscopy	The electrochemical determination of the corrosion rate are in agreement with those obtained by atomic absorption spectroscopy.	[16]
Liu M, Schmutz P, Uggowitzer PJ, Song GL, Atrens A	The influence of yttrium (Y) on the corrosion of Mg-Y binary alloys	The corrosion rate evaluated by electrochemical impedance spectroscopy was somewhat smaller than that evaluated from H evolution.	[41]
King AD, Birbilis N, Scully JR	Accurate electrochemical measurement of magnesium corrosion rates; a combined impedance, mass-loss and hydrogen collection study	The determination corrosion rate from EIS polarization resistance correlated to the mass loss and volume of hydrogen collected.	[22]
Qiao ZX, Shi ZM, Hort N, Abidin NIZ, Atrens A	Corrosion behaviour of a nominally high purity Mg ingot produced by permanent mould direct chill casting	The corrosion rate measured using electrochemical techniques was consistently lower than that measured independently by weight loss or hydrogen evolution.	[42]
Bland LG, King AD, Birbilis N, Scully JR.	Assessing the corrosion of commercially pure magnesium	The integration of the polarization resistance over time, as evaluated from EIS provided	[26]

http://dx.doi.org/10.5772/65018

Authors	Work title	Main aspects	Reference
	and commercial AZ31B by electrochemical impedance, mass-loss, hydrogen collection, and inductively coupled plasma optical emission spectrometry solution analysis	correlation to the cumulative mass loss, ICP-OES solution analysis, and volume of hydrogen collected for commercially pure Mg and for AZ31.	
Shkirskiy V, King AD, Gharbi O, Volovitch P, Scully JR, Ogle K, Birbilis N.	Revisiting the electrochemical impedance spectroscopy of magnesium with online inductively coupled plasma atomic emission spectroscopy	In real time, electrochemical impedance spectroscopy (EIS) was coupled with online atomic emission spectroelectrochemistry (AESEC) to simultaneously measure Mg corrosion.	[5]
Curioni M, Scenini F, Monetta T, Bellucci F	Correlation between electrochemical impedance measurements and corrosion rate of magnesium investigated by real-time hydrogen measurement and optical imaging	A very good correlation between values of resistances estimated by EIS and corrosion currents obtained from real-time hydrogen measurement was found.	[24]
Shi ZM, Cao FY, Song GL, Atrens A	Low apparent valence of Mg during corrosion	The values of the corrosion evaluated from EIS data were typically not good measurements of the corrosion rate of Mg as measured by weight loss.	[43]
Atrens A, Song GL, Cao F, Shi Z, Bowen PK.	Advances in Mg corrosion and research suggestions	The reliability of corrosion rate obtained by EIS is questionable.	[38]
Shi ZM, Liu M, Atrens A	Measurement of the corrosion rate of magnesium alloys using Tafel extrapolation	In principle, the electrochemical measurement hypothesis for Mg can be disproved.	[44]
Cao FY, Shi ZM, Hofstetter J, Uggowitzer PJ, Song GL, Liu M, Atrens A.	Corrosion of ultra-high-purity Mg in 3.5% NaCl solution saturated with Mg(OH)(2)	The amount of dissolution under electrochemical control was a small amount of the total dissolution.	[45]
Atrens A, Song GL, Liu M, Shi Z, Cao F, Dargusch MS	Review of recent developments in the field of magnesium corrosion	Good values of the Tafel slopes are needed to EIS measurements of the corrosion rate.	[32]
Frankel GS, Samaniego A, Birbilis N	Evolution of hydrogen at dissolving magnesium surfaces	Cathodic reaction of magnesium is enhanced during dissolution	[46]

Table 1. Summary of previous studies about the estimation the corrosion rate using electrochemical techniques.

2. Materials and methods

AZ31 and AZ61 magnesium alloys manufactured under wrought conditions with the chemical composition indicated in **Table 2** (as detailed by the manufacturer, Magnesium Elektron Company) were used in the present study. Samples were cut into test strips, 2 ×2 cm squares with a thickness of 0.3 cm and then mechanically dry grounded successively with finer and finer abrasive paper up to P2000 SiC grit papers and polished to produce a uniformly flat surface with mirror-like reflectivity. All samples were rinsed in distilled water and dried in air stream as the final step.

Alloy	Al	Zn	Mn	Si	Fe	Ca	Mg
AZ31	3.1	0.73	0.25	0.02	0.005	0.0014	Bal.
AZ61	6.2	0.74	0.23	0.04	0.004	0.0013	Bal.

Table 2. Chemical composition of AZ31 and AZ61 alloys (wt. %).

The corrosion rate of magnesium alloys was calculated from the volume of collected hydrogen during the corrosion test. Specimens were immersed in a beaker containing 700 ml of 0.6 M NaCl. The beaker was covered with an inverted funnel, and the released gas was collected in a graduated burette. All experiments were conducted at room temperature. The hydrogen evolution was measured for 4 days in the case of the AZ31 specimen and for 14 days in the case of the AZ61 specimen. Three replicated specimens were tested to confirm reproducibility of the results. Also, the corrosion rates were determined from the weight loss of the tested specimens, after removing all the corrosion products from the sample surface using chromic acid (200 g L^{-1} CrO_3 + 10 g L^{-1} $AgNO_3$).

Electrochemical experiments were performed using an AUTOLAB PGSTAT30, with a frequency response analyzer. EIS scans were acquired from 100 kHz to 0.005 Hz with 5 points per decade and an AC amplitude of ±10 mV. A three electrode configuration was used in which the sample was the working electrode, a graphite rod was the counter electrode, and an Ag/AgCl (Sat.) electrode was the reference electrode. The tested specimens exposed an area of 9.0 cm^2 to the 0.6 M NaCl solution. Measurements were performed in triplicate to ensure the reproducibility of the results. ZView software was used for equivalent circuit fitting to the impedance plots obtained in these experiments [47].

3. Results and discussion

3.1. Influence of the immersion time on the error margin of the hydrogen evolution and electrochemical impedance measurements in 0.6 M NaCl solution

Figure 1 compares the time evolution of the corrosion current measured by the hydrogen collection method of tested materials immersed in 0.6 M NaCl solution. The corrosion current

http://dx.doi.org/10.5772/65018

for the AZ31 alloy initially increased slowly during the first day of immersion, after which it increased approximately linearly with the immersion time. This can be attributed to an incubation period in which the initial surface films were partially protective and break down, leading to the steady state corrosion indicated by the later linear trend [48]. The AZ31 alloy was completely dissolved after 4 days of immersion. In contrast, for the AZ61 alloy (**Figure 1b**), the corrosion current measured by the hydrogen evolution was relatively small for the whole immersion duration (14 days). This difference may be due to the strong protective effect of the oxide film that forms spontaneously on the polished AZ61 alloy surface, much more perfect and protective than the film on the AZ31 alloy [29, 31, 49]. The absence of significant variations in the corrosion current measured by the hydrogen evolution for the AZ61 alloy (**Figure 1b**) indicates the presence of a very protective film in this case.

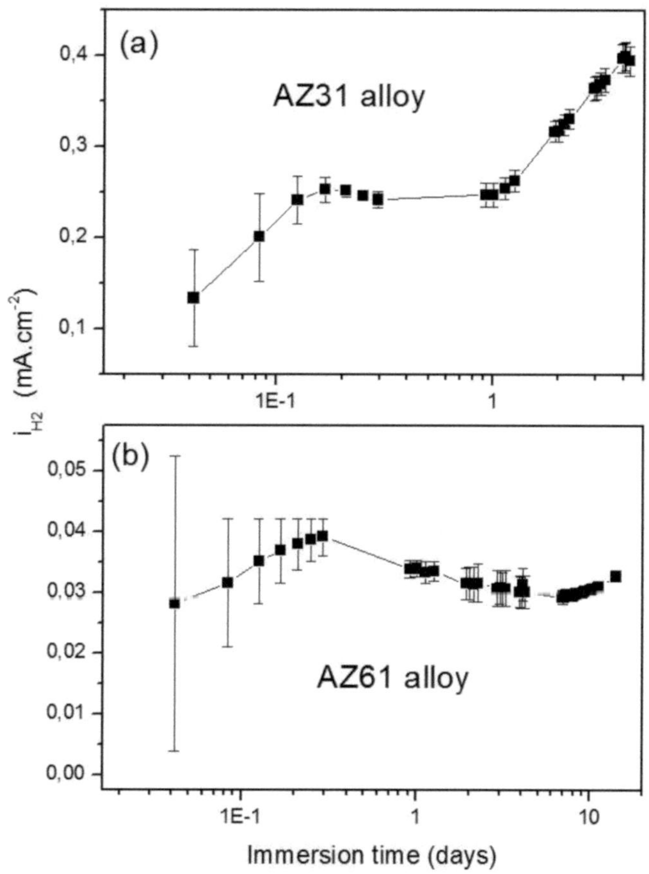

Figure 1. Time evolution of the corrosion current obtained from hydrogen measurement during immersion in 0.6 M NaCl.

Figure 1 also presents the error bars, which indicate the data variability, as deduced from three parallel measurements. It is interesting to note that the error tends to decrease quickly with immersion time. The observed variations in the error bars-time behavior are attributed to a larger quantity of evolved volume of hydrogen with increasing immersion time, which improves the scale-reading precision [50]. On the other hand, hydrogen gas is very soluble in aqueous environments [22, 51], so it is likely that some H_2 was lost by dissolution in the first stages of experimentation [52].

The Nyquist diagrams (**Figure 2**) obtained for the AZ31 and AZ61 specimens immersed in 0.6 M NaCl show the presence of a capacitive loop at high frequencies (HF) and an inductive loop at low frequencies (LF). **Figure 3** shows the representative impedance spectra of the tested specimens in term of Bode plots.

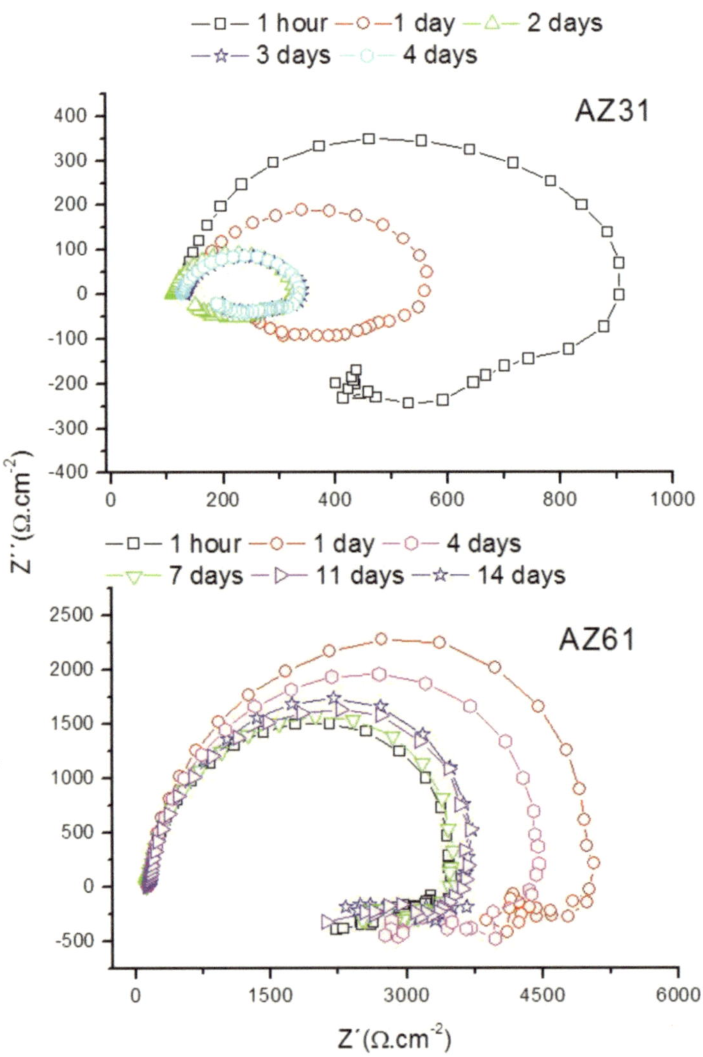

Figure 2. Variation in Nyquist plots for AZ31 and AZ61 samples with immersion time in 0.6 M NaCl solution.

The impedance spectra fitted with the equivalent circuit, as shown in **Figure 4**, were reported by King et al. [22]. R_s represents the electrolyte resistances, C_1 and C_2 are associated with the oxide/hydroxide layers capacitance and the double layer capacitance at the metal/electrolyte interface, respectively. R_1 is related to the initial metallic corrosion step, and R_2 represents the discharge of the intermediate adsorbed species formed in the initial step [53, 54]. Finally, L accounts for the variation in the extension of active anodic regions during the sinusoidal polarization, and R_3 represents the resistances associated with local environmental changes (precipitation of gels, presence of bubbles) nearby the anodic and cathodic regions [24]. In full compliance with the suggestions of King et al. [22], for the analysis of EIS data, in this work, we have included capacitors in the place of constant phase element (CPE) for the representation of the capacitive elements.

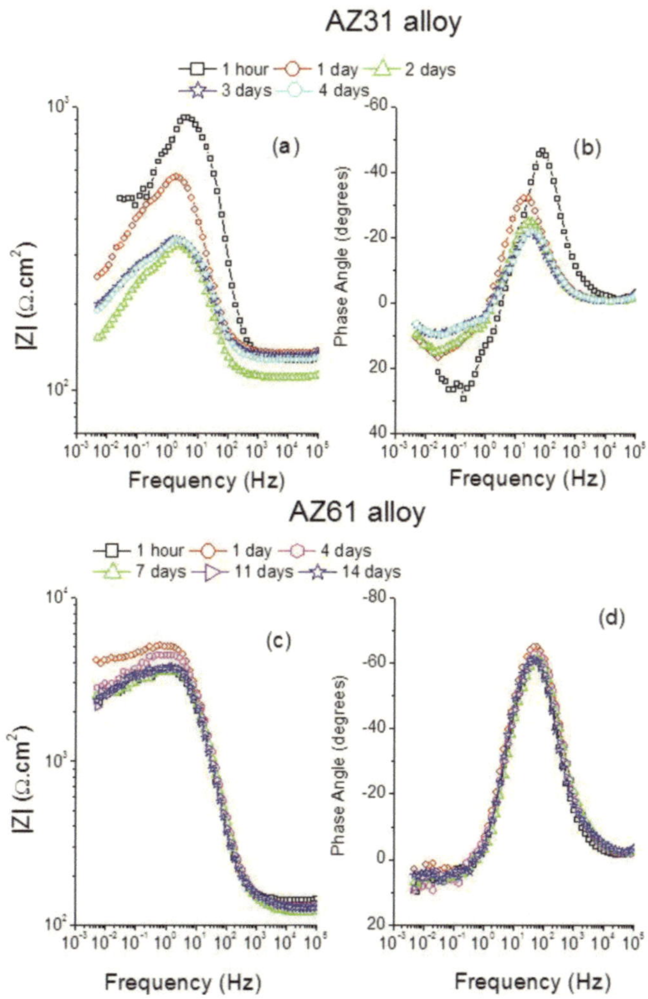

Figure 3. Variation in Bode plots for AZ31 and AZ61 samples with immersion time in 0.6 M NaCl solution.

The Nyquist and Bode plots in **Figure 5** include the measured data along with a fit employing the equivalent circuit proposed and validated by King et al [22]. Generally, the simulated and experimental impedance spectra were closely similar, and the center of the Nyquist plots does not appear depressed, which is a near ideal capacitive behavior [22, 26]. Polarization resistance (R_p) can profile the corrosion rate with time [55]. According to King et al. [22], the R_p can be evaluated as the zero frequency impedance ($f \rightarrow 0$) at $-Z'' = 0$ [56]. On the other hand, R_t is defined as the value corresponding to Z' when $-Z'' = 0$, which is commonly obtained at intermediate frequencies [56]. The R_p [22, 26] and R_t values have been determined using the following equations derived from the equivalent circuit:

$$\frac{1}{R_p} = \frac{1}{R_1 + R_2} + \frac{1}{R_3} \tag{2}$$

$$R_t = R_1 + R_2 \tag{3}$$

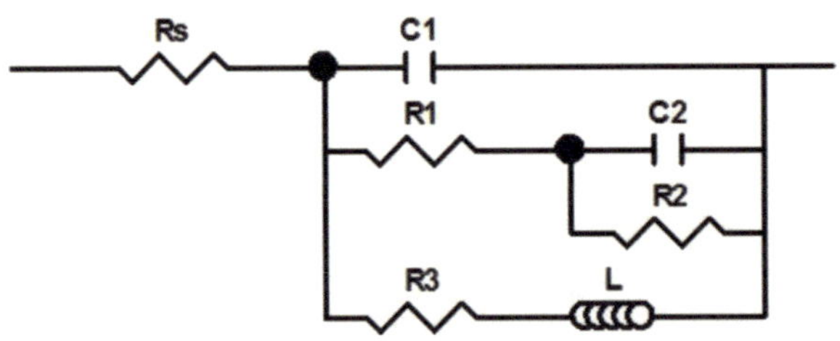

Figure 4. Equivalent circuit used for fitting experimental EIS spectra of the samples.

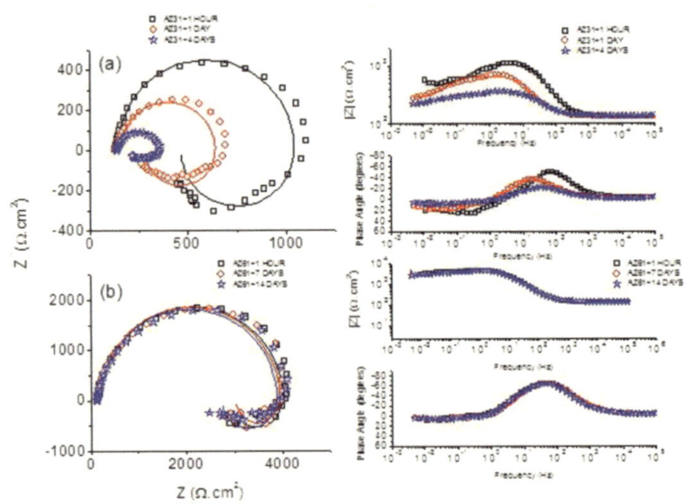

Figure 5. Example of Nyquist and Bode plots with respective fitting for AZ31 and AZ61 alloys with different immersion times in 0.6 M NaCl solution.

http://dx.doi.org/10.5772/65018

The quantitative results from spectra fitting are given in **Tables 3** and **4**, in which the data represent the mean values for the three parallel measurements, and the standard deviation suggests the reproducibility of the EIS measurements.

Immersion time	C_1 (μF cm^{-2})	R_1 (kΩ cm^2)	C_2 (μF cm^{-2})	R_2 (kΩ cm^2)	L (kΩs cm^2)	R_3 (kΩ cm^2)	R_p (kΩ cm^2)	R_t (kΩ cm^2)	I_{H2} (mA cm^{-2})
1 h	5 ± 0.2	0.09 ± 0.03	3 ± 1	0.64 ± 0.15	0.50 ± 0.18	0.41 ± 0.17	0.25 ± 0.09	0.75 ± 0.18	0.13 ± 0.05
4 h	13 ± 6	0.09 ± 0.02	15 ± 12	0.27 ± 0.09	0.26 ± 0.06	0.29 ± 0.12	0.15 ± 0.02	0.37 ± 0.10	0.25 ± 0.01
7 h	16 ± 1	0.12 ± 0.05	20 ± 5	0.31 ± 0.08	0.61 ± 0.19	0.42 ± 0.22	0.21 ± 0.09	0.43 ± 0.13	0.24 ± 0.00
1 day	20 ± 2	0.11 ± 0.02	26 ± 5	0.28 ± 0.10	1.26 ± 0.34	0.19 ± 0.04	0.13 ± 0.03	0.39 ± 0.12	0.25 ± 0.01
2 days	28 ± 2	0.07 ± 0.01	31 ± 6	0.13 ± 0.04	0.61 ± 0.19	0.07 ± 0.01	0.05 ± 0.01	0.20 ± 0.05	0.32 ± 0.01
3 days	25 ± 2	0.08 ± 0.01	45 ± 3	0.11 ± 0.02	0.82 ± 0.10	0.14 ± 0.01	0.08 ± 0.01	0.20 ± 0.03	0.37 ± 0.01
4 days	24 ± 1	0.08 ± 0.01	41 ± 2	0.09 ± 0.01	0.77 ± 0.14	0.14 ± 0.02	0.08 ± 0.01	0.18 ± 0.02	0.40 ± 0.01

Table 3. Fitting results for the EIS measurements performed in 0.6 M NaCl and corrosion current densities obtained from independent measurement of hydrogen evolution for the AZ31 alloy.

Immersion time	C_1 (μF cm^{-2})	R_1 (kΩ cm^2)	C_2 (μF cm^{-2})	R_2 (kΩ cm^2)	L (kΩs cm^2)	R_3 (kΩ cm^2)	R_p (kΩ cm^2)	R_t (kΩ cm^2)	I_{H2} (mA cm^{-2})
1 h	4.18 ± 0.06	1.38 ± 0.02	4.08 ± 0.68	2.16 ± 0.49	43.43 ± 5.63	8.82 ± 1.70	2.53 ± 0.88	3.55 ± 1.82	0.028 ± 0.024
4 h	3.66 ± 0.16	1.25 ± 0.01	2.85 ± 0.08	3.39 ± 0.25	22.58 ± 0.41	16.63 ± 1.94	3.62 ± 0.95	4.64 ± 1.10	0.037 ± 0.005
1 day	3.18 ± 0.05	1.21 ± 0.18	2.25 ± 0.36	4.45 ± 1.32	34.60 ± 5.79	22.72 ± 0.34	4.51 ± 0.72	5.66 ± 1.15	0.034 ± 0.001
2 days	3.20 ± 0.17	1.67 ± 0.18	2.48 ± 0.09	4.31 ± 0.43	61.21 ± 0.54	22.04 ± 1.82	4.70 ± 0.46	5.98 ± 0.61	0.034 ± 0.003
3 days	2.94 ± 0.07	1.02 ± 0.36	2.30 ± 0.60	3.97 ± 0.83	51.80 ± 5.30	18.94 ± 5.40	3.86 ± 0.41	4.92 ± 0.45	0.032 ± 0.003
4 days	3.06 ± 0.04	1.00 ± 0.06	2.20 ± 0.32	3.54 ± 0.59	33.05 ± 2.70	10.66 ± 3.26	3.17 ± 0.55	4.55 ± 0.53	0.030 ± 0.002
7 days	3.03 ± 0.36	0.59 ± 0.21	2.21 ± 0.64	2.96 ± 0.15	31.05 ± 5.96	9.77 ± 0.39	2.60 ± 0.22	3.55 ± 0.36	0.029 ± 0.001
8 days	2.70 ± 0.19	0.33 ± 0.01	2.21 ± 0.17	3.34 ± 0.15	32.45 ± 2.08	8.87 ± 2.39	2.53 ± 0.19	3.68 ± 0.39	0.030 ± 0.000
9 days	2.91 ± 0.31	0.33 ± 0.03	2.34 ± 0.14	3.32 ± 0.19	29.91 ± 3.20	10.16 ± 1.11	2.68 ± 0.04	3.65 ± 0.22	0.030 ± 0.000
10 days	2.94 ± 0.17	0.37 ± 0.16	2.43 ± 0.13	3.53 ± 0.44	25.83 ± 3.79	9.81 ± 0.93	2.65 ± 0.41	3.90 ± 0.25	0.031 ± 0.000
11 days	3.18 ± 0.38	0.42 ± 0.17	2.50 ± 0.08	3.06 ± 0.02	39.67 ± 4.70	7.81 ± 1.29	2.39 ± 0.08	3.47 ± 0.18	0.031 ± 0.000
14 days	3.18 ± 0.35	0.51 ± 0.11	3.09 ± 0.01	3.10 ± 0.30	33.26 ± 2.43	8.43 ± 0.08	2.53 ± 0.11	3.11 ± 0.23	0.033 ± 0.000

Table 4. Fitting results for the EIS measurements performed in 0.6 M NaCl and corrosion current densities obtained from independent measurement of hydrogen evolution for the AZ61 alloy.

The evolution of resistance (R_p and R_t) values as a function of immersion time in 0.6 M NaCl solution is presented in **Figure 6**. After the first day of immersion test, the EIS data for the AZ31 specimen, illustrated in **Figure 6a** and **c** indicate that the resistance (R_p and R_t) values decreased markedly, approximately consistent with the linear increase in the corrosion current measured by the hydrogen collection presented in **Figure 1**, suggesting that the corrosion product layer formed on this alloy in 0.6 M NaCl is not particularly protective. In contrast, it is interesting to note that there is little effect or a moderate increase observed in the case of the AZ61 specimen over the 14 days immersion time (**Figure 6b** and **d**) in comparison to the high resistance (R_p and R_t) values obtained after 1h of immersion. During the first 2 days of immersion, the AZ61 specimens present a slight increase in the resistance (R_p and R_t) values (**Figure 6b** and **d**), which may be attributed to the growth of partially protective corrosion product film on the surface [30, 57]. According to a previous study [29], filiform corrosion was initiated almost immediately on the magnesium specimens after immersion, and the population of filaments expanded in less than 1 day across the entire exposed surface.

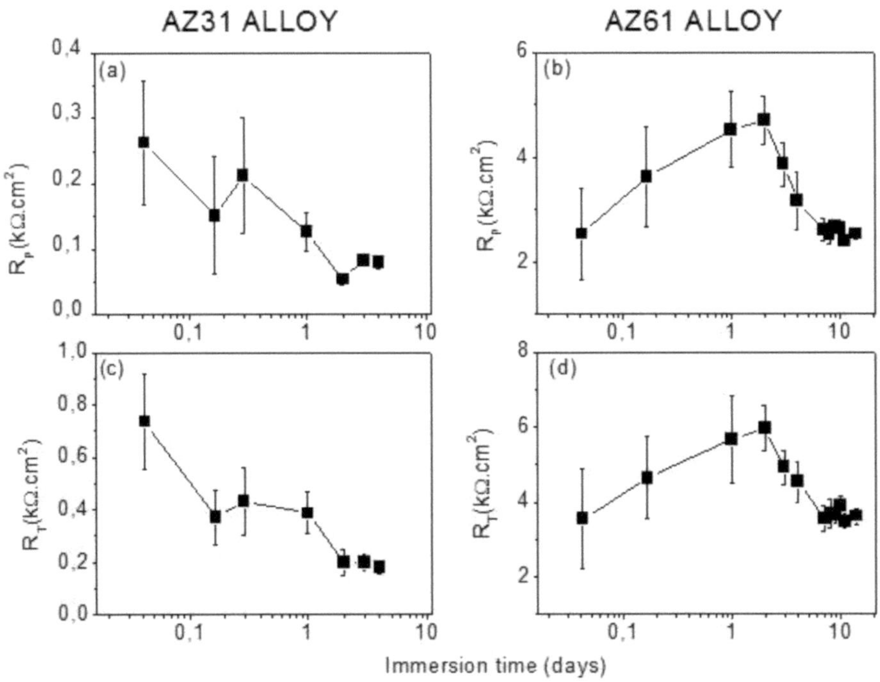

Figure 6. Evolution of the resistance (R_p and R_t) values with immersion time in NaCl 0.6 M solution.

The standard deviations of the average resistance (R_p and R_t) values decrease quickly and strongly with the immersion time indicating the formation of an homogeneous corrosion layer, as it can be seen in **Figure 6**. Curioni et al. [24] observed significant variations in the values of impedance measured under similar conditions attributed to the statistical nature of the corrosion process and to the relatively small electrode size. Because EIS measurements in this study were made with a large (9.0 cm^2) exposed area, it is reasonable to assume that the stochastic nature of the filament formation and growth processes initiated immediately after immersion and expanded in less than 1 day across the entire exposed surface, Samaniego et al. [29] contributed to this initial increase in the deviation of the resistance values. Increasing immersion time showed a decrease in the deviation of the resistance values, probably due to the formation of a uniform corrosion layer on the surface.

3.2. Low-frequency inductive behaviors

As already commented, the impedance diagrams in this study are characterized by a capacitive loop at HF followed by a well-marked inductive loop at LF. Numerous examples of such inductive loops can be found in the literature. An inductive effect may be explained by different reasons. In the corrosion of magnesium and its alloys, the low-frequency inductive loop in the Nyquist plot has been interpreted as a manifestation of localized corrosion damage of the magnesium substrate [58–67], produced by the relaxation of adsorbed species such as Mg(OH)ads+ or $Mg(OH)_2$ [9, 35, 68–73], to the formation and precipitation of a salt film [74, 75], to the incorporation of elements with properties of a semiconductor in the passive films [76] or most recently with accelerated anodic dissolution [22, 77]. However, the species responsible for the inductive response are yet to be determined [22], and the physical interpretation of inductive behavior associated with electrochemical processes is non-trivial, and that is probably the reason why this inductive behavior is often neglected [24].

The notable size of the inductive loops in this work may be seen in **Figure 2**. If the loop size is expressed as the diameter of the semicircle that fits it, and δ is the ratio between the diameters of the inductive and capacitive loops, δ values of the order of 0.6–0.7 have been obtained throughout exposure with the AZ31 alloy but decrease significantly with alloy AZ61 approximately to 0.2–0.3. Our previous EIS data of the specimens corroded under experimental conditions similar to those in this work have produced similar Nyquist plots [8, 15, 27, 28, 29, 49]. Asmussen et al. [77] suggested that the change in EIS behavior seems to be related somehow, with the alloy microstructure and subsequent distribution of the alloying elements, especially Al. According to Song et al. [78], the inductive points could be associated with the corrosion of the α phase while the inert β phase normally shows a capacitive spectrum [79]. **Figure 7** gives the microstructures of the AZ31 and AZ61 magnesium alloys. In the case of AZ61, the electrochemical signals from small amounts of β-phase ($Mg_{17}Al_{12}$), which is formed at the grain boundaries (**Figure 7b**) may modify the α phase formation [78]. In contrast, it can be found that the alloying element Al in the AZ31 alloy completely dissolved into the Mg matrix and no precipitation of the β-phase is observed (**Figure 7a**).

Figure 8 illustrates the relationship between the R_p values and the R_t values obtained from impedance measurements. It is evident that a linear relationship between R_p and R_t values was observed for the two alloys. The absence of significant modifications in the shape of the impedance diagrams throughout the test (**Figure 2**) may explain the high correlation between measured R_p and R_t values. As it has been suggested by Curioni [24], the values of charge transfer resistance (RCT) and corrosion product resistance (RCP) indirectly determine the value of adsorption resistance (RA), and therefore also the zero frequency resistance.

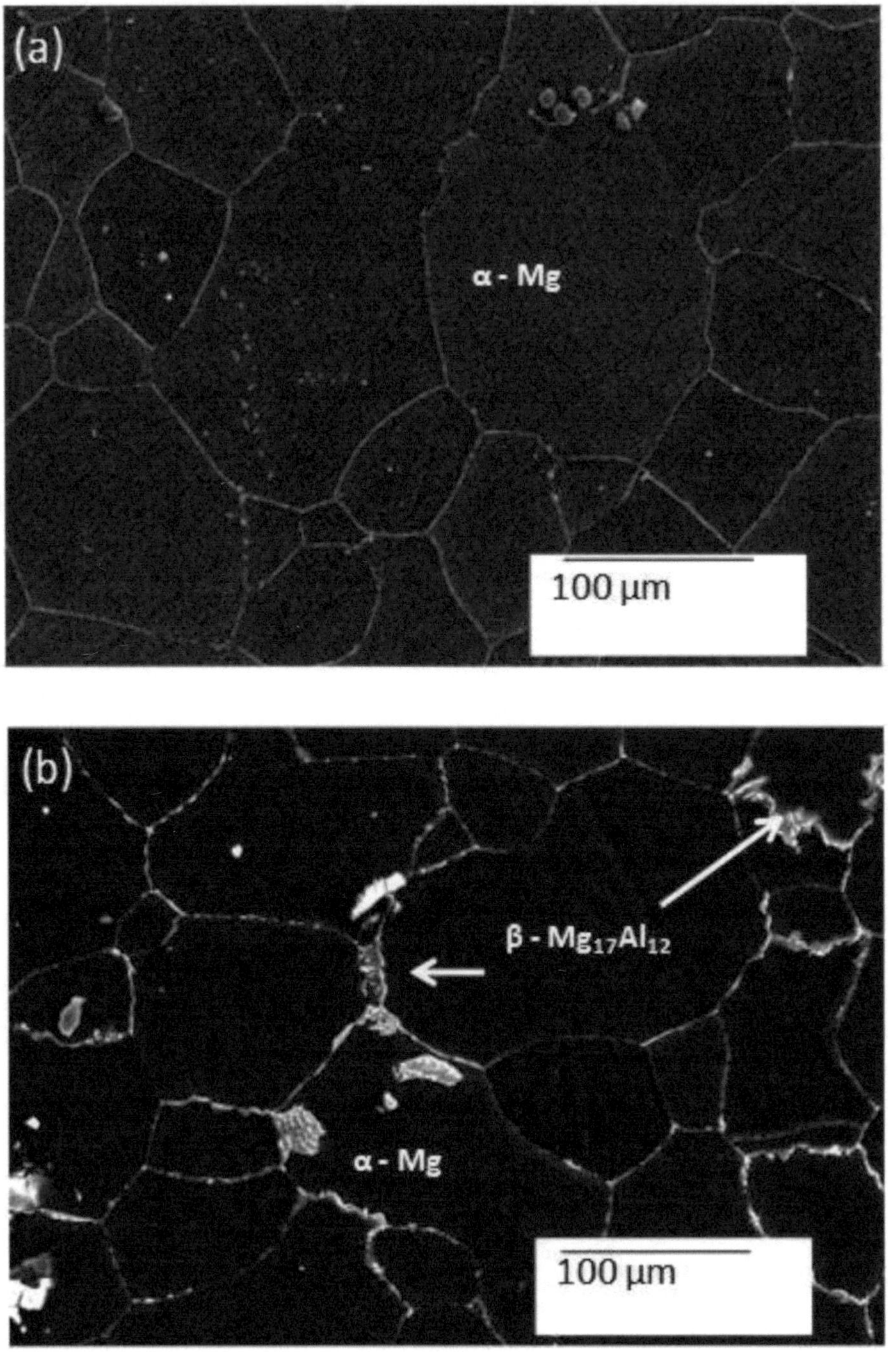

Figure 7. (a) SEM micrographs of AZ31 and (b) AZ61 magnesium alloys.

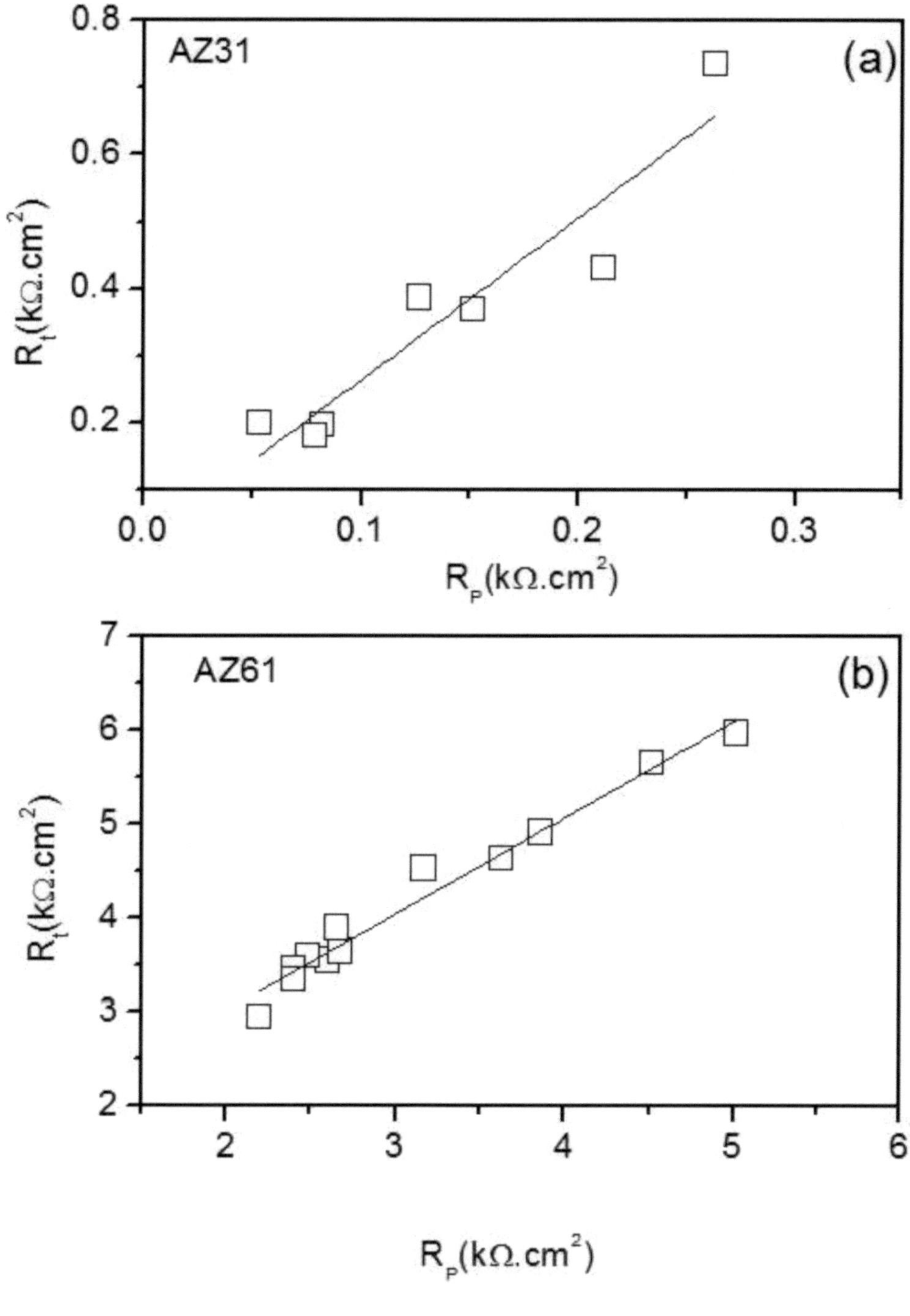

Figure 8. R_t values as a function of R_p values (obtained from **Tables 3** and **4**).

3.3. "Apparent" Stern and Geary values obtained from the correlation between independent EIS and hydrogen evolution measurements

King et al. [22] and Bland et al. [26] have shown that the integration of the corrosion current density over the exposure period *via* Eq. (4), and conversion of the volume of H_2 gas evolved to a corresponding anodic charge ($Q_a^{H_2}$) *via* the ideal gas law and Faraday's Law, all resulted in anodic charge estimations that are similar in value .

$$Q_a^{H_2} = i_{H_2}.t = \int i_{corr}.dt = \int \frac{B}{R}.dt \tag{4}$$

where i_{H_2} is the corrosion current density obtained from hydrogen measurement, i_{corr} is the corrosion current density obtained *via* Eq (1), *B* is the Stern-Geary coefficient, and *R* is the value of resistance obtained from fitting of the EIS spectra. In our previous studies [19] and [28], we have also observed high degree of similarity between the amounts of corroded metal determined by hydrogen evolution, gravimetric, and electrochemical measurements.

Under the assumption that B does not change with time for a metal or a metal alloy in a certain aqueous solution [39], one can obtain

$$B' = \frac{Q_a^{H_2}}{\int \left(\frac{1}{R} \right).dt} \tag{5}$$

where B' is the "apparent" Stern-Geary coefficient.

The averages and standard deviations of charge values obtained from hydrogen evolution measurements as a function of time-integrated reciprocal of the polarization resistance R_p or charge transfer resistance R_t values determined by EIS from 1 h and a maximum of 4 days immersion for the AZ31 alloy and 14 days immersion for the AZ61 alloy were used in this study to determine B' values. The results are shown in **Figures 9** and **10**. These figures show that for each alloy immersed in 0.6 M NaCl, charge values from hydrogen evolution measurements and time-integrated reciprocal of the polarization or charge transfer values were approximately proportional. Consequently, the value of B' for that particular system can be directly estimated from the graphs of **Figures 9** and **10** by considering the slope of the fitting lines [24, 39]. A very good correlation between R_p or R_t values estimated by EIS and corrosion currents obtained from independent hydrogen measurements was found. Due to the direct dependence of both resistance values commented above and shown in **Figure 8**, the points in the plots for the determination of the "apparent" Stern-Geary coefficient appear similarly scattered (**Figures 9** and **10**).

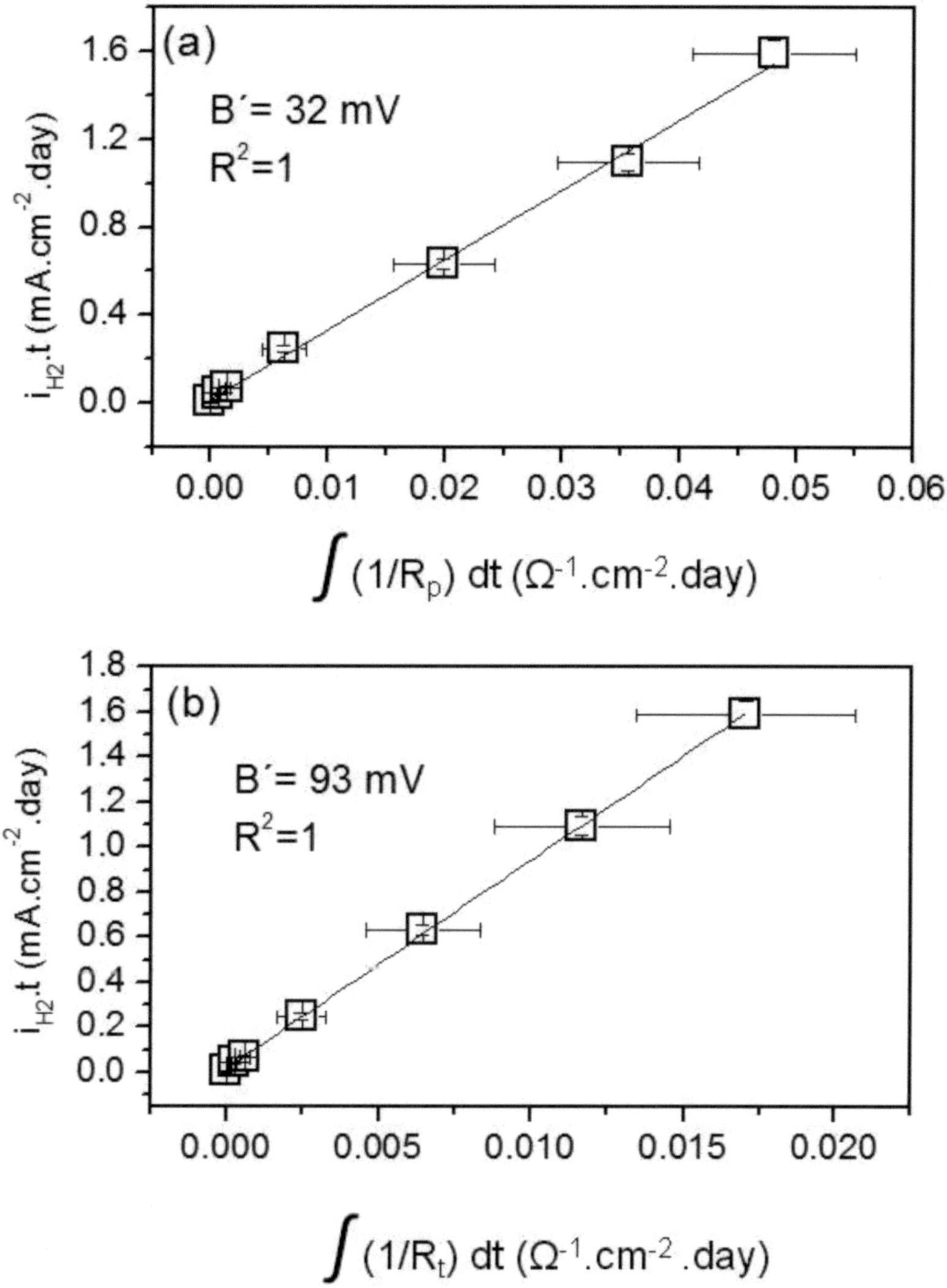

Figure 9. (a) Charge values from hydrogen evolution measurements as a function of time-integrated reciprocal of the polarization resistance R_p or (b) charge transfer resistance R_t values determined by EIS for AZ31 alloy from measurement in 0.6 M NaCl (data from **Table 3**).

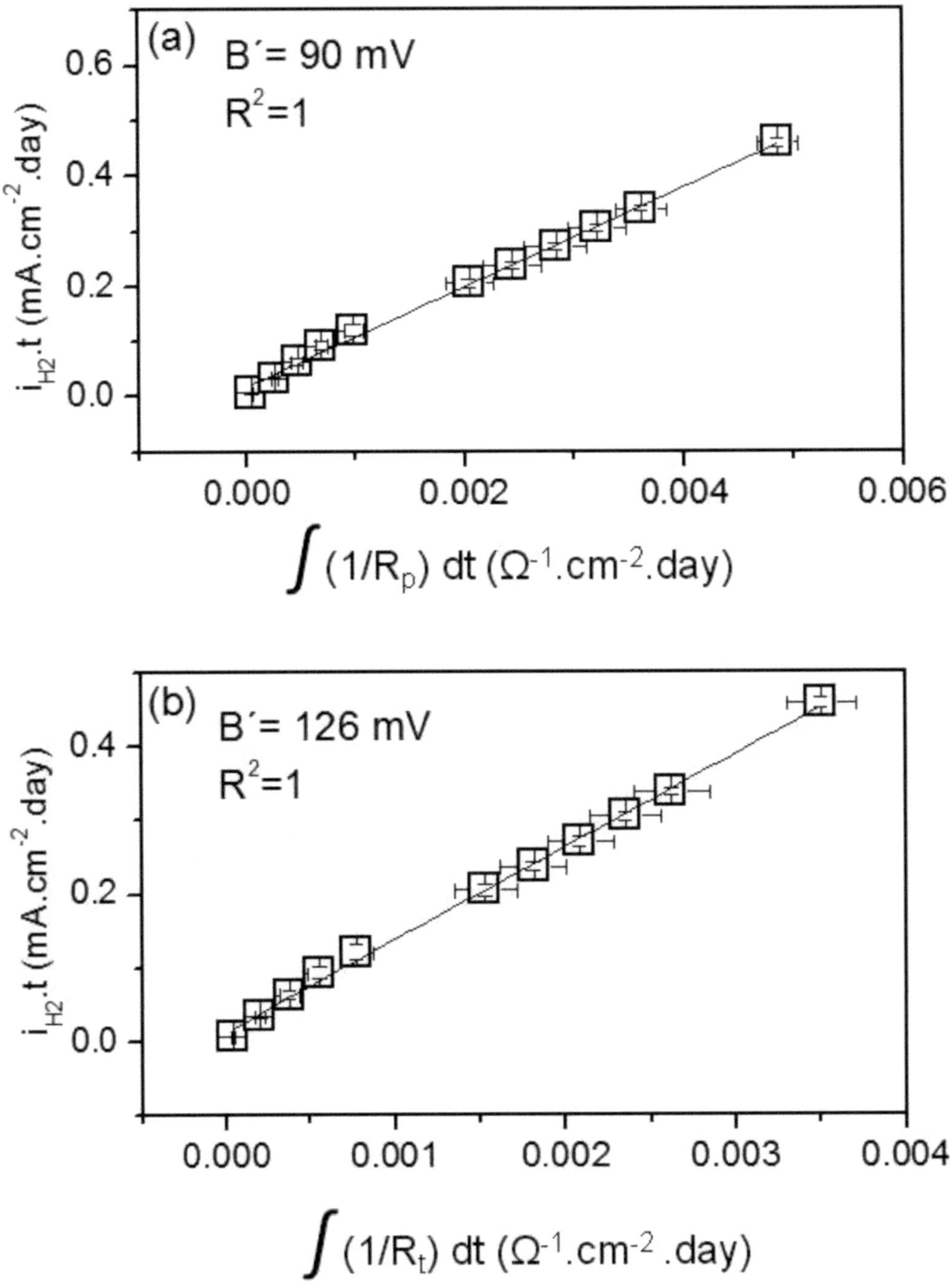

Figure 10. (a) Charge values from hydrogen evolution measurements as a function of time-integrated reciprocal of the polarization resistance R_p or (b) charge transfer resistance R_t values determined by EIS for AZ61 alloy from measurement in 0.6 M NaCl (data from **Table 4**).

Clearly, depending on the resistance and the alloy selected for the calculation, the "apparent" Stern-Geary coefficient is different; 93 or 32 mV for the AZ31 alloy if the charge transfer resistance or the polarization resistance is used and a value of around 126 mV or around 90 mV for the AZ61 alloy if the charge transfer resistance or the polarization resistance is used, respectively. These values are significantly lower compared to the "apparent" Stern-Geary values of 223 mV for the R_t and 116 mV for the R_p previously obtained by Curioni [24] for pure magnesium.

In the case of the AZ31 specimen, the "apparent" Stern-Geary coefficient obtained when the polarization resistance is used is significantly lower (approximately by a factor of 3) compared to when the values of the charge transfer resistance are used (**Figure 8**), while in the case of AZ61 specimen (**Figure 9**), little effect or a moderate decrease in the "apparent" Stern-Geary coefficient is observed. This behavior tends to suggest that the differences between the "apparent" Stern-Geary coefficients obtained from the polarization resistance and charge transfer resistance are dependent on the different sizes of the inductive loops in the impedance diagrams of the studied alloys (**Figure 2**).

A compilation of values of B quoted in the literature of AZ31 and AZ61 alloys arranged by determination method, test solution, time of exposure, author and reference is given in **Table 5**. Quoted B values obtained from Tafel determination method were 5–37 mV for the AZ31 alloy [22, 27, 44, 61, 80–86]. In the literature for AZ61, the *B* values are quoted ranging from 21 to 52 mV [27, 87–91]. The calculated *B′* value of 31 mV for the R_p for the AZ31 alloy is consistent with B values calculated by Bland et al. [26], Cao et al. [44], and Shi et al. [86] (**Table 5**). On the other hand, the *B'* values of around 118 mV for the AZ61 alloy if the charge transfer resistance is similar to that previously obtained by empirical estimation, from electrochemical and gravimetric measurements [27] (**Table 5**).

Alloy	B determination method	Immersion time	B (mV)	NaCl Conc	Author	Ref
AZ31B	Tafel extrapolation method	60 min	7.0	0.6 M	Tekin et al.	[80]
AZ31	Fitting software	48 h	10.1	0.6 M	Lim et al.	[61]
AZ31	Tafel extrapolation method	1800 sg	13.4	0.6 M	Li et al.	[81]
AZ31 Mg alloy	Tafel extrapolation method		16.2	0.6 M	Shen et al.	[82]
Bare AZ31 Mg alloy	Tafel extrapolation method	an initial delay of 5 min	16.5	0.6 M	Einkhah et al.	[83]
AZ31 alloy	Fitting software	30 min of exposure.	17.4	0.6 M	Singh et al.	[84]
AZ31	Fitting software		20.8	0.6 M	Zhang et al	[85]

Alloy	B determination method	Immersion time	B (mV)	NaCl Conc	Author	Ref
AZ31	Tafel assumption	24 h	31.1	0.6 M saturated with Mg(OH)2	Cao et al.	[44]
AZ31	Tafel assumption	24 h	36	0.6 M	King et al.	[22]
AZ31	Tafel assumption	24 h	36.8	0.6 M saturated with Mg(OH)2	Shi et al.	[86]
AZ31	Empirical estimation from electrochemical and gravimetric measurements	10 d	65	0.6 M	Feliu et al.	[27]
Mg6Al	LEV method from polarization curves	for 1 day	21.2	0.6 M saturated with Mg(OH)2	Cao et al.	[87]
Untreated AZ60	Fitting software		21.5	0.6 M	Su et al.	[88]
AZ61 magnesium alloy		30 min immersion	24.2	0.6 M	Rajabalizadeh and Seifzadeh	[91]
AZ61 magnesium alloy		30 min immersion	35.5	0.6 M	Seifzadeh, and Rajabalizadeh	[90]
Heat treated Mg6Al02	LEV method	10 d	38.6	0.6 M saturated with Mg(OH)2	Cao et al.	[91]
Heat treated Mg6Al02	LEV method	10 d	46.1	0.6 M saturated with Mg(OH)2	Cao et al.	[91]
As-cast Mg6Al02	LEV method	10 d	46.7	0.6 M saturated with Mg(OH)2	Cao et al.	[91]
As-cast Mg6Al01	LEV method	10 d	52.4	0.6 M saturated with Mg(OH)2	Cao et al.	[91]
AZ61	Empirical estimation from electrochemical and gravimetric measurements	10 d	120	0.6 M NaCl	Feliu et al.	[27]

Table 5. A selection of B values obtained from the literature.

3.4. Estimation of the corrosion rate from EIS measurements

In **Figure 11**, the variations in EIS-estimated corrosion rate using Eq. (4) and the "apparent" B' values obtained in this study with immersion time are compared with the corrosion rate as determined from hydrogen collection. A decrease in the differences between the corrosion rate values derived from impedance measurements and those derived from the hydrogen evolution rate occurs with increasing exposure time. As commented previously, this might be associated with the transformation of an air formed film to the steady corrosion layer for Mg specimens exposed to saline solutions and the lower hydrogen evolution rates at the beginning of the immersion tests.

Finally, **Figure 12** compares long-term cumulative corrosion rates determined by mass loss, measurement of the H_2 evolved over the entire testing period and integration of the current density (i_{corr}) versus time data determined by the temporal evolution of R_p and R_t over time. All resulted in corrosion rate estimations that are similar in value for each alloy. This corroborating outcome is similar to that from King et al. [22] and Bland et al. [26], indicating that the comparison of unique methods not only provides good comparison for the corrosion rate of magnesium alloys but validates the electrochemical approach as a standalone method.

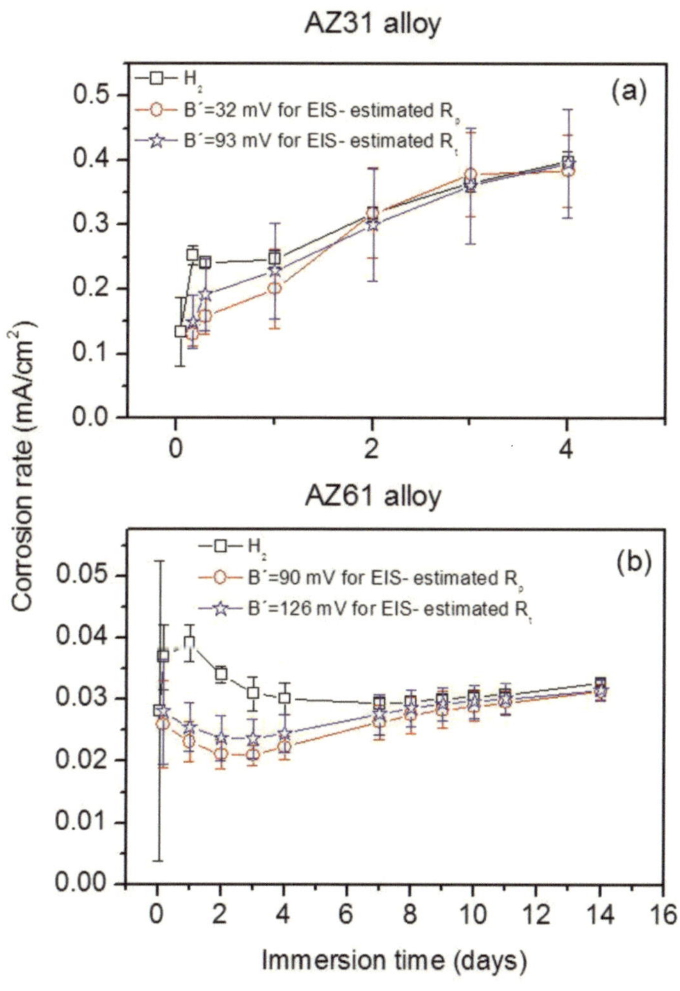

Figure 11. Variation in corrosion rates (mA/cm^2) as a function of immersion time obtained from EIS and hydrogen evolution measurements during 4 days immersion for AZ31 alloy (a) and 14 days of immersion for AZ61 alloy (b) in 0.6 M NaCl.

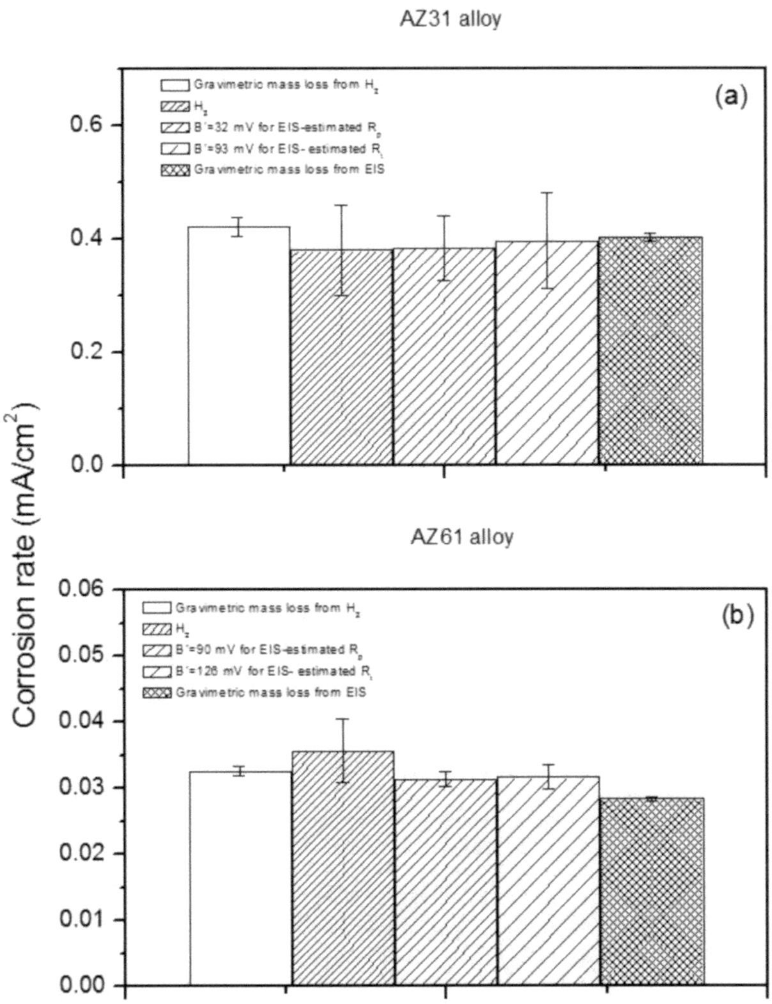

Figure 12. Comparison of corrosion rates (mA/cm²) obtained from EIS with weight loss and hydrogen evolution measurements after 4 days immersion for AZ31 alloy (a) and 14 days of immersion for AZ61 alloy (a) in 0.6 M NaCl.

4. Conclusions

1. In this research, it has been investigated the possible relationships between independent hydrogen evolution and electrochemical impedance spectroscopy measurements during long-term exposure of AZ31 and AZ61 magnesium alloys in 0.6 M NaCl.

2. Linear relationships with high correlation coefficients have been obtained between time-integrated polarization resistance R_p or charge transfer resistance R_t values determined by EIS and hydrogen evolution measurements.

3. The changes of R_t values with the immersion time have shown a direct and close relationship with the changes of R_p values determined by EIS, being the relationship characteristic of the microstructure of the alloy.

4. With immersion time, the system becomes more stable and the reproducibility of the EIS data appears to be improved. Also, the average error for hydrogen measurements after prolonged periods of exposure is usually smaller.

5. In this investigation, an empirical determination of the "apparent" Stern-Geary coefficient from the correlation between electrochemical and hydrogen measurements has yielded values of around 118 mV for the AZ61 alloy if the charge transfer resistance is used and of around 83 mV if the polarization resistance is used and values of around 97 mV for the AZ31 alloy if the charge transfer resistance is used and of around 31 mV if the polarization resistance is used.

6. No significant errors are obtained in the measurement of the corrosion rate when using R_t or R_p with their corresponding "apparent" Stern-Geary coefficient values.

Acknowledgements

The authors gratefully acknowledge the financial support provided by the Ministry of Economy and Competitiveness of Spain (MAT 2009-13530, MAT2012-30854, and MAT2015-65445-C2-1-R) to carry out the present study, and Prof. S. Feliu is duly acknowledged for fruitful discussions in the frame of this work.

Author details

Maria C. Delgado[1,2], Federico R. García-Galvan[1], Violeta Barranco[1] and Sebastian Feliu Batlle[1*]

*Address all correspondence to: sfeliu@cenim.csic.es

1 National Centre for Metallurgical Research (CENIM-CSIC), Madrid, Spain

2 Department of Materials Science, Simon Bolivar University, Baruta, Venezuela

References

[1] Gray JE, Luan B. Protective coatings on magnesium and its alloys – a critical review. Journal of Alloys and Compounds. 2002; 336: 88–113. DOI: 10.1016/S0925-8388(01)01899-0

[2] Kulekci MK. Magnesium and its alloys applications in automotive industry. International Journal of Advanced Manufacturing Technology. 2008; 39: 851–865. DOI: 10.1007/s00170-007-1279-2

[3] Witte F, Hort N, Vogt C, Cohen S, Kainer KU, Willumeit R, Feyerabend F. Degradable biomaterials based on magnesium corrosion. Current Opinion in Solid State and Materials Science. 2008; 12: 63–72. DOI: 10.1016/j.cossms.2009.04.001

[4] Song GL. Recent progress in corrosion and protection of magnesium alloys. Advanced Engineering Materials. 2005; 7: 563–586. DOI: 10.1002/adem.200500013

[5] Shkirskiy V, King AD, Gharbi O, Volovitch P, Scully JR, Ogle K, Birbilis N. Revisiting the electrochemical impedance spectroscopy of magnesium with online inductively coupled plasma atomic emission spectroscopy. ChemPhysChem. 2015; 16: 536–539. DOI: 10.1002/cphc.201402666

[6] Banerjee PC, Raman RKS, Durandet Y, McAdam G. Electrochemical investigation of the influence of laser surface melting on the microstructure and corrosion behaviour of ZE41 magnesium alloy – an EIS based study. Corrosion Science. 2011; 53: 1505–1514. DOI: 10.1016/j.corsci.2011.01.017

[7] Feliu Jr. S, Barajas R, Bastidas JM, Morcillo M, Feliu S. Study of protection mechanisms of zinc-rich paints by electrochemical impedance spectroscopy. In: Scully JR, Silverman DC, Kendig MW, editors. Electrochemical Impedance: Analysis and Interpretation. Vol. 1188. Philadelphia: ASTM; 1993. pp. 438–449. DOI: 10.1520/STP18084S

[8] Feliu Jr. S, Maffiotte C, Samaniego A, Galvan JC, Barranco V. Effect of naturally formed oxide films and other variables in the early stages of Mg-alloy corrosion in NaCl solution. Electrochimica Acta. 2011; 56: 4454–4565. DOI: 10.1016/j.electacta.2011.02.077

[9] Jamesh M, Kumar S, Sankara Narayanan TSNS. Corrosion behavior of commercially pure Mg and ZM21 Mg alloy in Ringer's solution – Long term evaluation by EIS. Corrosion Science. 2011; 53: 645–654. DOI: 10.1016/j.corsci.2010.10.011

[10] Jian SY, Chu YR, Lin CS. Permanganate conversion coating on AZ31 magnesium alloys with enhanced corrosion resistance. Corrosion Science. 2015; 93: 301–309. DOI: 10.1016/j.corsci.2015.01.040

[11] Rocca E, Juers C, Steinmetz J. Corrosion behaviour of chemical conversion treatments on as-cast Mg–Al alloys: Electrochemical and non-electrochemical methods. Corrosion Science. 2010; 52: 2172–2178. DOI: 10.1016/j.corsci.2010.02.036

[12] Ardelean H, Frateur I, Marcus P. Corrosion protection of magnesium alloys by cerium, zirconium and niobium-based conversion coatings. Corrosion Science. 2008; 50: 1907–1918. DOI: 10.1016/j.corsci.2008.03.015

[13] Chu YR, Lin CS. Citrate gel conversion coating on AZ31 magnesium alloys. Corrosion Science. 2014; 87: 288–296. DOI: 10.1016/j.corsci.2014.06.034

[14] Chen J, Song YW, Shan DY, Han EH. Study of the corrosion mechanism of the *in situ* grown Mg–Al–CO32- hydrotalcite film on AZ31 alloy. Corrosion Science. 2012; 65: 268–277. DOI: 10.1016/j.corsci.2012.08.026

[15] Feliu Jr. S, Samaniego A, Barranco V, El-Hadad AA, Llorente I, Adeva P. The effect of low temperature heat treatment on surface chemistry and corrosion resistance of commercial magnesium alloys AZ31 and AZ61 in 0.6 M NaCl solution. Corrosion Science. 2014; 80: 461–472. DOI: 10.1016/j.corsci.2013.11.065

[16] Pebere N, Riera C, Dabosi F. Investigation of magnesium corrosion in aerated sodium-sulfate solution by electrochemical impedance spectroscopy. Electrochimica Acta. 1990; 35: 555–561. DOI: 10.1016/0013-4686(90)87043-2

[17] Makar L, Kruger J. Corrosion studies of rapidly solidified magnesium alloys. Journal of the Electrochemical Society 1990; 137: 414–421. DOI: 10.1149/1.2086455

[18] Mathieu S, Rapin C, Hazan J, Steinmetz P. Corrosion behaviour of high pressure die-cast and semi-solid cast AZ91D alloys. Corrosion Science. 2002; 44: 2737–2756. DOI: 10.1016/S0010-938X(02)00075-6

[19] Feliu Jr. S, Samaniego A, Barranco V, El-Hadad AA, Llorente I, Serra C, Galvan JC. A study on the relationships between corrosion properties and chemistry of thermally oxidised surface films formed on polished commercial magnesium alloys AZ31 and AZ61. Applied Surface Science. 2014; 295: 219–230. DOI: 10.1016/j.apsusc.2014.01.033

[20] Feliu Jr. S, Samaniego A, Bermudez EA, El-Hadad AA, Llorente I, Galvan JC. Effect of native oxide film on commercial magnesium alloys substrates and carbonate conversion coating growth and corrosion resistance. Materials. 2014; 7: 2534–2560. DOI: 10.3390/ma7042534

[21] Stern M, Geary AL. Electrochemical polarization 1. A theoretical analysis of the shape of polarization curves. Journal of the Electrochemical Society. 1957; 104: 56–63. DOI: 10.1149/1.2428496

[22] King AD, Birbilis N, Scully JR. Accurate electrochemical measurement of magnesium corrosion rates; a combined impedance, mass-loss and hydrogen collection study. Electrochimica Acta. 2014; 121: 394–406. DOI: 10.1016/j.electacta.2013.12.124

[23] Poursaee A. Potentiostatic transient technique, a simple approach to estimate the corrosion current density and Stern–Geary constant of reinforcing steel in concrete. Cement and Concrete Research. 2010; 40: 1451–1458. DOI: 10.1016/j.cemconres.2010.04.006

[24] Curioni M, Scenini F, Monetta T, Bellucci F. Correlation between electrochemical impedance measurements and corrosion rate of magnesium investigated by real-time hydrogen measurement and optical imaging. Electrochimica Acta. 2015; 166: 372–384. DOI: 10.1016/j.electacta.2015.03.050

[25] Fajardo S, Frankel GS. Gravimetric method for hydrogen evolution measurements on dissolving magnesium. Journal of the Electrochemical Society. 2015; 162: C693–C701. DOI: 10.1149/2.0241514jes

[26] Bland LG, King AD, Birbilis N, Scully JR. Assessing the corrosion of commercially pure magnesium and commercial AZ31B by electrochemical impedance, mass-loss, hydrogen collection, and inductively coupled plasma optical emission spectrometry solution analysis. Corrosion. 2015; 71: 128–145. DOI: 10.5006/1419

[27] Feliu Jr. S, Maffiotte C, Samaniego A, Galván JC, Barranco V. Effect of the chemistry and structure of the native oxide surface film on the corrosion properties of commercial AZ31 and AZ61 alloys. Applied Surface Science. 2011; 257: 8558–8568. DOI: 10.1016/j.apsusc.2011.05.014

[28] Feliu Jr. S, Llorente I. Corrosion product layers on magnesium alloys AZ31 and AZ61: Surface chemistry and protective ability. Applied Surface Science. 2015; 347: 736–746. DOI: 10.1016/j.apsusc.2015.04.189

[29] Samaniego A, Llorente I, Feliu Jr. S. Combined effect of composition and surface condition on corrosion behavior of magnesium alloys AZ31 and AZ61. Corrosion Science. 2013; 68: 66–71. DOI: 10.1016/j.corsci.2012.10.034

[30] Feliu Jr. S, Maffiotte C, Galvan JC, Barranco V. Atmospheric corrosion of magnesium alloys AZ31 and AZ61 under continuous condensation conditions. Corrosion Science. 2011; 53: 1865–1872. DOI: 10.1016/j.corsci.2011.02.003

[31] Samaniego A, Feliu Jr. S. Different performance of factors affecting the estimation of the corrosion rate in magnesium alloys by implementation of the common methods for electrochemical measurements. In: Monteiro WA, editor. Light Metal Alloys Applications. Croatia: Intech; 2014. p. 145–161. DOI: 10.5772/58370

[32] Atrens A, Song GL, Liu M, Shi Z, Cao F, Dargusch MS. Review of recent developments in the field of magnesium corrosion. Advanced Engineering Materials. 2015; 17: 400–453. DOI:10.1002/adem.201400434

[33] Abidin NIZ, Rolfe B, Owen H, Malisano J, Martin D, Hofstetter J, Uggowitzer PJ, Atrens A. The in vivo and in vitro corrosion of high-purity magnesium and magnesium alloys WZ21 and AZ91. Corrosion Science. 2013; 75: 354–366. DOI: 10.1016/j.corsci.2013.06.019

[34] Abidin NIZ, Atrens AD, Martin D, Atrens A. Corrosion of high purity Mg, $Mg_2Zn_{0.2}Mn$, ZE41 and AZ91 in Hank's solution at 37°C. Corrosion Science. 2011; 53: 3542–3556. DOI: 10.1016/j.corsci.2011.06.030

[35] Song GL, Song SZ. A possible biodegradable magnesium implant material. Advanced Engineering Materials. 2007; 9: 298–302. DOI: 10.1002/adem.200600252

[36] Song GL. Control of biodegradation of biocompatible magnesium alloys. Corrosion Science. 2007; 49: 1696–1701. DOI: 10.1016/j.corsci.2007.01.001

[37] Wang H, Estrin Y, Fu H, Song G, Zuberova Z. The effect of pre-processing and grain structure on the bio-corrosion and fatigue resistance of magnesium alloy AZ31. Advanced Engineering Materials. 2007; 9: 967–972. DOI: 10.1002/adem.200700200

[38] Atrens A, Song GL, Cao F, Shi Z, Bowen PK. Advances in Mg corrosion and research suggestions. Journal of Magnesium and Alloys. 2013; 1: 177–200. DOI: 10.1016/j.jma.2013.09.003

[39] Hsieh MK, Dzombak DA, Vidie RD. Bridging gravimetric and electrochemical approaches to determine the corrosion rate of metals and metal alloys in cooling systems: Bench scale evaluation method. Industrial and Engineering Chemistry Research. 2010; 49: 9117–9123. DOI: 10.1021/ie100217k

[40] Kirkland NT, Birbilis N, Staiger MP. Assessing the corrosion of biodegradable magnesium implants: A critical review of current methodologies and their limitations. Acta Biomaterialia. 2012; 8: 925–936. DOI: 10.1016/j.actbio.2011.11.014

[41] Liu M, Schmutz P, Uggowitzer PJ, Song GL, Atrens A. The influence of yttrium (Y) on the corrosion of Mg-Y binary alloys. Corrosion Science. 2010; 52: 3687–3701. DOI: 10.1016/j.corsci.2010.07.019

[42] Qiao ZX, Shi ZM, Hort N, Abidin NIZ, Atrens A. Corrosion behaviour of a nominally high purity Mg ingot produced by permanent mould direct chill casting. Corrosion Science. 2012; 61: 185–207. DOI: 10.1016/j.corsci.2012.04.030

[43] Shi ZM, Cao FY, Song GL, Atrens A. Low apparent valence of Mg during corrosion. Corrosion Science. 2014; 88: 434–443. DOI: 10.1016/j.corsci.2014.07.060

[44] Shi ZM, Liu M, Atrens A. Measurement of the corrosion rate of magnesium alloys using Tafel extrapolation. Corrosion Science. 2010; 52: 579–588. DOI: 10.1016/j.corsci.2009.10.016

[45] Cao FY, Shi ZM, Hofstetter J, Uggowitzer PJ, Song GL, Liu M, Atrens A. Corrosion of ultra-high purity Mg in 3.5% NaCl solution saturated with Mg(OH)2. Corrosion Science. 2013; 75: 78–99. DOI: 10.1016/j.corsci.2013.05.018

[46] Frankel GS, Samaniego A, Birbilis N. Evolution of hydrogen at dissolving magnesium surfaces. Corrosion Science. 2013; 70: 104–111. DOI: 10.1016/j.corsci.2013.01.017

[47] ZView software. Version 3.4d (Scribner Associates Inc.), Southern Pines, NC, USA.

[48] Johnston S, Shi ZM, Atrens A. The influence of pH on the corrosion rate of high-purity Mg, AZ91 and ZE41 in bicarbonate buffered Hanks' solution. Corrosion Science. 2015; 101: 182–192. DOI: 10.1016/j.corsci.2015.09.018

[49] Feliu Jr. S, El-Hadad AA, Barranco V, Llorente I, García-Galván FR, Jiménez-Morales A, Galván JC. Native oxide films on AZ31 and AZ61 commercial magnesium alloys: Corrosion behaviour, effect on isothermal oxidation and sol–gel thin film formation.

In: Ahmad Z, editor. New Trends in Alloy Development, Characterization and Application. Croatia: Intech; 2015. p. 97–122. DOI: 10.5772/60721

[50] Zander D, Schnatterer C. The influence of manufacturing processes on the microstructure and corrosion of the AZ91D magnesium alloy evaluated using a computational image analysis. Corrosion Science. 2015; 98: 291–303. DOI: 10.1016/j.corsci.2015.05.032

[51] Lebouil S, Duboin A, Monti F, Tabeling P, Volovitch P, Ogle K. A novel approach to online measurement of gas evolution kinetics: Application to the negative difference effect of Mg in chloride solution. Electrochimica Acta. 2014; 124: 176–182. DOI: 10.1016/j.electacta.2013.07.131

[52] Fajardo S, Frankel GS. Effect of impurities on the enhanced catalytic activity for hydrogen evolution in high purity magnesium. Electrochimica Acta. 2015; 165: 255–267. DOI: 10.1016/j.electacta.2015.03.021

[53] Murray JN, Moran PJ, Gileadi E. Utilization of the specific pseudocapacitance for determination of the area of corroding steel surfaces. Corrosion. 1988; 44: 533–538.

[54] Brooks EK, Der S, Ehrensberger MT. Corrosion and mechanical performance of AZ91 exposed to simulated inflammatory conditions. Materials Science and Engineering: C. 2016; 60: 427–436. DOI: 10.1016/j.msec.2015.11.059

[55] Wang J, Jang Y, Wan GJ, Giridharan V, Song GL, Xu ZG, Koo Y, Qi PK, Sankar J, Huang N, Yun YH. Flow-induced corrosion of absorbable magnesium alloy: *In-situ* and real-time electrochemical study. Corrosion Science. 2016; 104: 277–289. DOI: 10.1016/j.corsci.2015.12.020s

[56] Yeganeh M, Saremi M. Corrosion inhibition of magnesium using biocompatible Alkyd coatings incorporated by mesoporous silica nanocontainers. Progress in Organic Coatings. 2015; 79: 25–30. DOI: 10.1016/j.porgcoat.2014.10.015

[57] Hou RQ, Ye CQ, Chen CD, Dong SG, Lv MQ, Zhang S, Pan JS, Song GL, Lin CJ. Localized corrosion of binary Mg–Ca alloy in 0.9 wt% sodium chloride solution. Acta Metallurgica Sinica (English Letters). 2016; 29: 46–57. DOI: 10.1007/s40195-015-0361-2

[58] Wu CS, Zhang Z, Cao FH, Zhang LJ, Zhang JQ, Cao CN. Study on the anodizing of AZ31 magnesium alloys in alkaline borate solutions. Applied Surface Science. 2007; 253: 3893–3898. DOI: 10.1016/j.apsusc.2006.08.020

[59] Metikos-Hukovic M, Babic R, Grubac Z, Brinic S. Impedance spectroscopic study of aluminum and Al-alloys in acid-solution – inhibitory-action of nitrogen-containing compounds. Journal of Applied Electrochemistry. 1994; 24: 772–778. DOI: 10.1007/BF00578093

[60] Song YW, Shan DY, Chen RS, Han EH. Corrosion characterization of Mg–8Li alloy in NaCl solution. Corrosion Science. 2009; 51: 1087–1094. DOI: 10.1016/j.corsci.2009.03.011

[61] Liang J, Srinivasan PB, Blawert C, Dietzel W. Influence of pH on the deterioration of plasma electrolytic oxidation coated AM50 magnesium alloy in NaCl solutions. Corrosion Science 2010; 52: 540–547. DOI: 10.1016/j.corsci.2009.10.011

[62] Zhang YJ, Yan CW, Wang FH, Li WF. Electrochemical behavior of anodized Mg alloy AZ91D in chloride containing aqueous solution. Corrosion Science. 2005; 47: 2816–2831. DOI: 10.1016/j.corsci.2005.01.010

[63] Kannan MB. Influence of microstructure on the *in vitro* degradation behaviour of magnesium alloys. Materials Letters. 2010; 64: 739–742. DOI: 10.1016/j.matlet.2010.01.022

[64] Jin S, Amira S, Ghali E. Electrochemical impedance spectroscopy evaluation of the corrosion behavior of die cast and thixocast AXJ530 magnesium alloy in chloride solution. Advanced Engineering Materials. 2007; 9: 75–83. DOI: 10.1002/adem.200600199

[65] Alvarez-Lopez M, Pereda MD, del Valle JA, Fernandez-Lorenzo M, Garcia-Alonso MC, Ruano OA, Escudero ML. Corrosion behaviour of AZ31 magnesium alloy with different grain sizes in simulated biological fluids. Acta Biomaterialia. 2010; 6: 1763–1771. DOI: 10.1016/j.actbio.2009.04.041

[66] Srinivasan PB, Liang J, Balajeee RG, Blawert C, Stormer M, Dietzel W. Effect of pulse frequency on the microstructure, phase composition and corrosion performance of a phosphate-based plasma electrolytic oxidation coated AM50 magnesium alloy. Applied Surface Science. 2010; 256: 3928–3935. DOI: 10.1016/j.apsusc.2010.01.052

[67] Lim TS, Ryu HS, Hong SH. Electrochemical corrosion properties of CeO_2-containing coatings on AZ31 magnesium alloys prepared by plasma electrolytic oxidation. Corrosion Science. 2012; 62: 104–111. DOI: 10.1016/j.corsci.2012.04.043

[68] Baril G, Blanc C, Pebere N. AC impedance spectroscopy in characterizing time-dependent corrosion of AZ91 and AM50 magnesium alloys: Characterization with respect to their microstructures. Journal of the Electrochemical Society. 2001; 148: b469–b496. DOI: 10.1149/1.1415722

[69] Baril G, Galicia G, Deslouis C, Pebere N, Tribollet B, Vivier V. An impedance investigation of the mechanism of pure magnesium corrosion in sodium sulfate solutions. Journal of the Electrochemical Society. 2007; 154: C108–C113. DOI: 10.1149/1.2401056

[70] Chang JW, Guo XW, Fu PH, Peng LM, Ding WJ. Effect of heat treatment on corrosion and electrochemical behaviour of Mg–3Nd–0.2Zn–0.4Zr (wt.%) alloy. Electrochimica Acta. 2007; 52: 3160–3167. DOI: 10.1016/j.electacta.2006.09.069

[71] Brett CMA, Dias L, Trindade B, Fischer R, Mies S. Characterisation by EIS of ternary Mg alloys synthesised by mechanical alloying. Electrochimica Acta. 2006; 51: 1752–1760. DOI: 10.1016/j.electacta.2005.02.124

[72] Chang JW, Guo XW, He SM, Fu PH, Peng LM, Ding WJ. Investigation of the corrosion for Mg–xGd–3Y–0.4Zr (x = 6, 8, 10, 12 wt%) alloys in a peak-aged condition. Corrosion Science. 2008; 50: 166–177. DOI: 10.1016/j.corsci.2007.06.003

[73] Turhan MC, Lynch R, Killian MS, Virtanen S. Effect of acidic etching and fluoride treatment on corrosion performance in Mg alloy AZ91D (MgAlZn). Electrochimica Acta. 2009; 55: 250–257. DOI: 10.1016/j.electacta.2009.08.046

[74] Breslin CB, Rudd AL. Activation of pure Al in an indium-containing electrolyte — an electrochemical noise and impedance study. Corrosion Science. 2000; 42:1023–1039. DOI: 10.1016/S0010-938X(99)00128-6

[75] Barbucci A, Bruzzone G, Delucchi M, Panizza M, Cerisola G. Breakdown of passivity of aluminium alloys by intermetallic phases in neutral chloride solution. Intermetallics. 2000; 8: 305–312. DOI: 10.1016/S0966-9795(99)00114-4

[76] He WW, Zhang EL, Yang K. Effect of Y on the bio-corrosion behavior of extruded Mg–Zn–Mn alloy in Hank's solution. Materials Science and Engineering: C. 2010; 30: 167–174. DOI: 10.1016/j.msec.2009.09.014

[77] Asmussen RM, Binns WJ, Jakupi P, Shoesmith D. Microstructural effects on corrosion of AM50 magnesium alloys. Journal of the Electrochemical Society. 2014; 161: C501–C508. DOI: 10.1149/2.0781410jes

[78] Song GL, Bowles AL, St John DH. Corrosion resistance of aged die cast magnesium alloy AZ91D. Materials Science and Engineering: A. 2004; 366: 74–86. DOI: 10.1016/j.msea.2003.08.060

[79] Song GL, Atrens A, Wu XL, Zhang B. Corrosion behaviour of AZ21, AZ501 and AZ91 in sodium chloride. Corrosion Science. 1998; 40: 1769–1791. DOI: 10.1016/S0010-938X(98)00078-X

[80] Tekin KC, Malayoglu U, Shrestha S. Electrochemical behavior of plasma electrolytic oxide coatings on rare earth element containing Mg alloys. Surface and Coatings Technology. 2013; 236: 540–549. DOI: 10.1016/j.surfcoat.2013.10.051

[81] Lim TS, Ryu HS, Hong SH. Electrochemical corrosion properties of CeO2-containing coatings on AZ31 magnesium alloys prepared by plasma electrolytic oxidation. Corrosion Science 2012; 62: 104–111. DOI: 10.1016/j.corsci.2012.04.043

[82] Shen DJ, Ma HJ, Guo CH, Cai JR, Li GL, He DL, Yang QX. Effect of cerium and lanthanum additives on plasma electrolytic oxidation of AZ31 magnesium alloy. Journal of Rare Earths. 2013; 31: 1208–1213. DOI: 10.1016/S1002-0721(12)60428-1

[83] Einkhah F, Lee KM, Sani MAF, Yoo B, Shin DH. Structure and corrosion behavior of oxide layer with Zr compounds on AZ31 Mg alloy processed by two-step plasma electrolytic oxidation. Surface and Coatings Technology. 2014; 238: 75–79. DOI: 10.1016/j.surfcoat.2013.10.042

[84] Singh IB, Singh M, Das S. A comparative corrosion behavior of Mg, AZ31 and AZ91 alloys in 3.5% NaCl solution. Journal of Magnesium and Alloys. 2015; 3: 142–148. DOI: 10.1016/j.jma.2015.02.004

[85] Zhang JF, Yan CW, Wang FH. Electrodeposition of Al–Mn alloy on AZ31B magnesium alloy in molten salts. Applied Surface Science. 2009; 255: 4926–4932. DOI: 10.1016/j.apsusc.2008.12.039

[86] Shi ZM, Cao FY, Song GL, Liu M, Atrens A. Corrosion behaviour in salt spray and in 3.5% NaCl solution saturated with $Mg(OH)_2$ of as-cast and solution heat-treated binary Mg–RE alloys: RE = Ce, La, Nd, Y, Gd. Corrosion Science. 2013; 76: 98–118. DOI: 10.1016/j.corsci.2013.06.032

[87] Cao FY, Shi ZM, Song GL, Liu M, Dargusch MS, Atrens A. Influence of hot rolling on the corrosion behavior of several Mg–X alloys. Corrosion Science. 2015; 90: 176–191. DOI: 10.1016/j.corsci.2014.10.012

[88] Su YC, Li GY, Lian JS. A chemical conversion hydroxyapatite coating on AZ60 magnesium alloy and its electrochemical corrosion behaviour. International Journal of Electrochemical Science. 2012; 7: 11497–11511.

[89] Rajabalizadeh Z, Seifzadeh D. The effect of copper ion on microstructure, plating rate and anticorrosive performance of electroless Ni-P coating on AZ61 magnesium alloy. Protection of Metals and Physical Chemistry of Surfaces. 2014; 50: 516–523. DOI: 10.1134/S2070205114040157

[90] Rajabalizadeh Z, Seifzadeh D. Environmentally-friendly method for electroless Ni–P plating on magnesium alloy. Surface and Coatings Technology. 2013; 218: 119–126. DOI: 10.1016/j.surfcoat.2012.12.039

[91] Cao FY, Shi ZM, Song GL, Liu M, Atrens A. Corrosion behaviour in salt spray and in 3.5% NaCl solution saturated with Mg(OH)2 of as-cast and solution heat-treated binary Mg–X alloys: X = Mn, Sn, Ca, Zn, Al, Zr, Si, Sr. Corrosion Science. 2013; 76: 60–97. DOI: 10.1016/j.corsci.2013.06.030

Chapter 6

Corrosion Protection of Magnesium Alloys: From Chromium VI Process to Alternative Coatings Technologies

Jérôme Frayret, Jean Charles Dupin and
Sébastien Pommiers

Additional information is available at the end of the chapter

http://dx.doi.org/10.5772/66604

Abstract

Magnesium and its alloys present several advantages such as a high strength/weight ratio and a low density. These properties allow them to be used for many aeronautical applications but they are very sensitive to corrosion. In order to solve this problem, chromium VI conversion coatings (CCC) are deposited on the surface before a protective top coat application. This process is now limited by several environmental laws due to the high toxicity of hexavalent chromium. However the chemical mechanisms of CCC deposition will be detailed in this chapter in order to understand the chemical properties of this coating. Pre-treatment steps allow cleaning and preparing the surface for improving the coating deposition. A final layer of chromium (III) oxide and magnesium hydroxide composes the coating allowing the protective properties. Orthorhombic potassium chromate clusters trapped on the coating surface give self-healing property to the coating. Alternative conversion coatings are based onto solutions containing chromium (III), permanganate, phosphates, Rare Earth Elements (REEs) or vanadium. The second part of this chapter will detail the deposition and the protection mechanisms of these promising processes of CrVI substitution. Among them, permanganate/phosphate-based coating presents a better corrosion resistance than CCC and REEs have very efficient self-healing properties.

Keywords: corrosion protection, magnesium, alternative coatings, chromium VI, electrochemistry, surface analysis

1. Introduction

Magnesium alloys possess the lowest density of all metallic constructional materials and have good mechanical properties [1, 2]. They are suitable for the partial replacement of aluminium alloys for motorsport and aerospace applications. This could imply non-negligible weight and fuel savings in the aeronautical sector. Unfortunately, these materials are also very vulnerable to corrosion, do not resist wear and are highly chemically reactive [3]. Other metals, such as aluminium and lithium, have been added to pure magnesium to decrease its shortcomings [4], which limit the use of magnesium and its alloys for computer parts and in the aerospace and the automotive industries [5]. The poor corrosion resistance of magnesium is the consequence of its standard potential ($E° = -2.363$ V/SHE (standard hydrogen electrode)), which makes it extremely susceptible to galvanic corrosion [5].

In order to decrease their corrosion sensitivity, several processes can be used to deposit a protective film on the surface of the alloy: anodizing, sol-gel process, gas-phase deposition process and chemical conversion [6]. Among these methods, chemical conversion is the least expensive and the easiest to perform [6, 7]. Protection is achieved by the immersion of the concerned substrate in pre-treatment and treatment baths. The classical pre-treatment process comprises a degreasing bath to clean the surface up [8] and several acidic baths to make the alloy rough and reactive enough before the coating deposition. Afterwards, the piece is immersed in the treatment bath for several minutes in order to deposit a protected coating on its surface. Until now, chromate conversion coatings (CCCs) have been widely used, despite the health concerns raised by the use of hexavalent chromium, because they offer the best protection against corrosion [9, 10]. The implementation, in June 2017, of the Regulation on Registration, Evaluation, Authorization and Restriction of Chemicals (REACH) will prohibit the use of such toxic compounds [11].

In the first part of this chapter, the chemical mechanism of chromate conversion is explored on magnesium alloy EV31A as it appears to be a key point in the protection process. Some assumptions suggested that polar oxo-Cr(VI) anions, present inside of the CCC coating, annihilate the adsorption of depassivating anions such as chloride ions [12]. The presence of trapped hexavalent chromium is responsible of the 'self-healing' ability of the coating under corrosion which remains a tremendous advantage driven by this species [9]. The determination and the control of the chemical form of hexavalent chromium in the coating are then the key to understand the properties of the chromate films and to find out some equivalent coatings. The mechanism is monitored over the whole pre-treatment and treatment steps. To describe the coating composition during its deposition and its chemical properties, both the etching solutions (pre-treatment) and the coatings (CCC treatment) were monitored by electrochemical and spectrometric techniques. The chemical composition and the microstructure features of the protective coated layers were next examined by scanning electron microscopy (SEM) and X-ray photoelectron spectroscopy (XPS).

The second part of this chapter aims to compare the performance of some potential CCC alternative coatings built up from different salts. The properties of these coatings are presented in **Table 1**.

Coating	Properties	References
Cr(III)	• Cracked coating • Small improvement of the corrosion resistance for bare alloy	[13]
Phosphate	• Longer stability under salt-spray conditions than zinc phosphate coatingall • Better corrosion resistance after painting than zinc phosphate coating	[14–16]
Zinc phosphate	• Thicker than magnesium phosphate coating • Fine zinc particles surrounded phosphate crystals and filled in the interstice of the insoluble phosphate • Microporous structure of the coating • Better adhesion of paint on the zinc phosphate coating than on chromate conversion coating • Very susceptible under salt-spray conditions • Better corrosion resistance than phosphate coating	[16–18]
PCC (permanganate conversion coating)	• Nearly crack-free coating • Average thickness of 230 nm after 90 s of immersion and increasing with continued immersion • Sufficient electrical conductivity • Poor crystallinity, comparable to chromate conversion coatings • Corrosion resistance inferior to chromate conversion coatings	[19, 20]
Permanganate/ phosphate	• Typical coating thickness, 4–6 μm • Equivalent or slightly better passive capability than the conventional Cr (VI)-based conversion treatment	[20–23]
Vanadium	• Self-healing ability similar to CCCs for magnesium alloys • Poor protection against corrosion	[24–28]
REEs (rare earth elements–based coating)	• Self-healing effect • Two-layer structure for cerium conversion coating • Adhesive weakness of the coating • Homogeneous and uniform conversion coating • Good adhesion to substrate with thickness of about 10 μm for lanthanum-based conversion coatings	[7, 28–30]

Table 1. Existing CCC (chromate conversion coating) alternative coatings.

On the basis of the identified chemical protection events, some existing and new alternative solutions have been classified in this study into two categories, A and B. This classification is done according to the chemical properties of the constitutive compounds of the treatment solutions used for the coating deposition. Ten own-made solutions and three industrial solutions commonly used for protection of light metal (Al-based) alloys have then been considered in order to test their efficiency on specific EV31A magnesium alloy. The corrosion protection properties in relation with the coating chemical composition were compared to CCC ones on the basis of stationary voltammetry and X-ray photoelectron spectroscopy collected data. They were evaluated in terms of their formation mechanism.

2. Chemical mechanism of chromate conversion coating

The well-known and currently used chemical process for magnesium protection consists in three different steps: degreasing, pickling and treatment. The process synoptic is detailed in **Figure 1**.

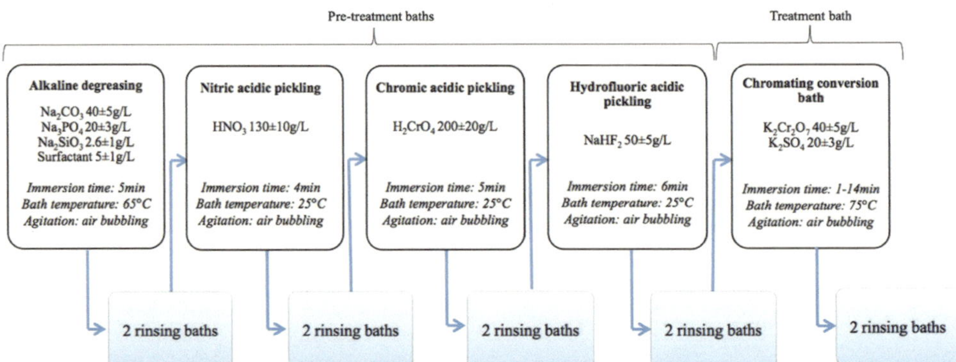

Figure 1. Etching process synoptic for the EV31A protection.

Each pre-treatment step participates in a chemical modification of the surface of the alloy (**Table 2**). These chemical compositions, determined by XPS after a rinsing step, are particularly important for the whole understanding of the deposit and the application of the future coating and on the protection properties against corrosion.

For bare EV31A, only carbon (C 1s), oxygen (O 1s) and magnesium (Mg 2p) signals were detected. The magnesium at the surface exits into two oxidation states: (0) and (+II), with Mg 2p components at 49.2 eV (Mg metal [31]) and around 50.5 eV corresponding to both $Mg(OH)_2(s)$ and MgO(s) [32]. The examination of O 1s peak confirms the co-existence of the oxide and hydroxide forms with two corresponding components at 530.7 eV (MgO(s)) and 532.6 eV ($Mg(OH)_2(s)$). A small roughness (0.62 μm) is recorded in the tri-dimensional representation of the bare EV31A alloy and a very low corrosion potential (-1.61 $V_{/SCE}$) is measured which confirms the fact that EV31A alloy has to be protected against corrosion.

%At.		Bare EV31A	Alkaline degreasing	Nitric acid pickling	Chromic acid pickling	Hydrofluoric acid pickling
C		34.1	25.3	14.7	14.0	15.6
Cr	*$Cr_2O_3(s)$*	*	*	*	6.5	2.7
	$Cr(OH)_3(s)$	*	*	*	5.1	1.3
	$K_2CrO_4(s)$	*	*	*	4.7	0.4
F		*	*	*	*	14.0
Gd		*	*	0.1	*	0.1
Mg	*MgO(s)*	20.7	4.7	6.8	7.2	10.8
	$Mg(OH)_2(s)$					
	$MgF_2(s)$	*	*	*	*	7.1
	Mg (in PO_4 environment)	*	15.1	*	*	*
	Mg^0	9.9	0.7	21	*	1.1
N		*	*	2.4	*	*
Na		*	2.3		*	4.4
O		35.3	46.3	53.2	62.3	42.2
P		*	5.5	*	*	*
Zn		*	0.1	0.8	*	0.2
Zr		*	*	1	0.2	0.1

*Not present.

Table 2. XPS atomic composition (%At) of the coating during the pre-treatment process.

2.1. Cleaning of the surface

During the degreasing step, a redox reaction between the magnesium (0) of the EV31A alloy surface and water causes the apparition of Mg^{2+} (reaction (1))

$$2H_2O + Mg \rightleftarrows Mg^{2+} + H_2(g) + 2OH^- \quad (1)$$

XPS results clearly evidenced this trend as atomic percentage of Mg^0 (BE ~49.5 eV) drastically went down (~10% for bare alloy and 0.7% after degreasing). Under alkaline conditions (pH = 11), Mg^{2+} ions then react with the hydroxide ions OH^- to form $Mg(OH)_2(s)$.

Once withdrawn from the bath and before being rinsed, the surface of the alloy is kept in contact with the atmosphere and $Mg(OH)_2(s)$ dehydrates into magnesium oxide MgO(s).

Na_3PO_4 of the degreasing bath seems to interact directly with the alloy as Mg in a PO_4 environment was identified (BE ~51.4 eV). This detection should indicate the formation of $Mg_3(PO_4)_2(s)$, a compound responsible for the protection of the magnesium surface, in the case of phosphate coatings [14, 15]. But no significant improvement or diminution of the corrosion resistance was observed: E_{corr} = -1.65 ± 0.04 V after the piece is dipped in the degreasing alkaline bath and the two rinsing baths, whereas E_{corr} = -1.61 ± 0.08 V for the bare alloy. It means that the layer of magnesium oxide /hydroxide formed at the surface would not act as a direct protective layer of the alloy against the corrosion. The observation of the microstructure confirms the absence of a surface modification (**Figure 2b**).

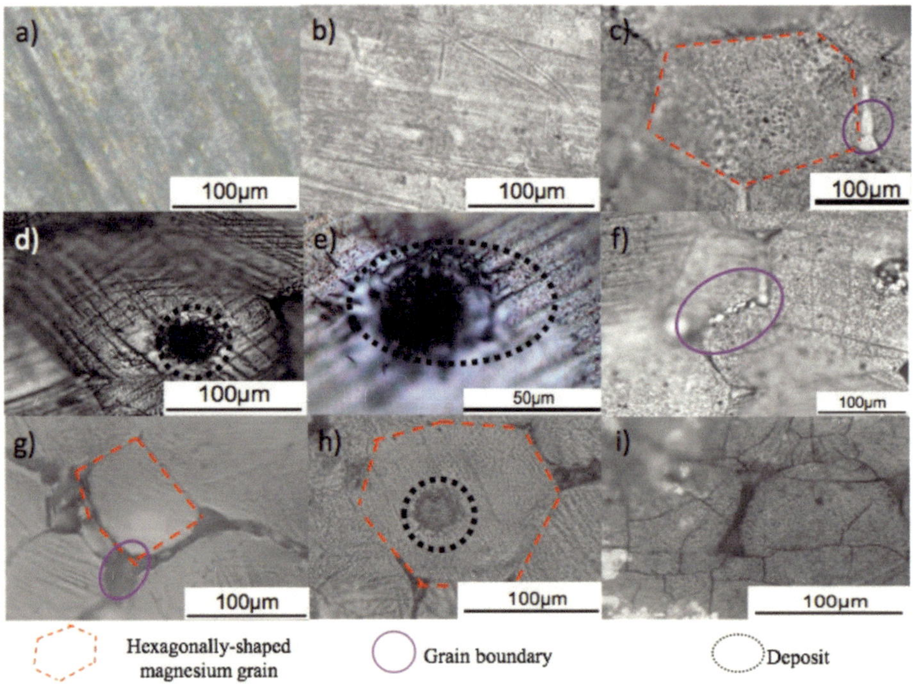

Figure 2. Surface of the EV31A alloy after being dipped in (a) bare alloy, (b) the alkaline-degreasing bath, (c) nitric acidic-pickling bath, (d)–(f) chromic acidic-pickling bath, (g) hydrofluoric acidic-pickling bath, (h) treatment bath (after 5 s) and (i) treatment bath (after 14 min).

The role of the alkaline degreasing bath is to remove the contamination on the surface of the alloy. The surfactant present in the bath is responsible for the removal of carbonated impurities which is confirmed by the XPS data (see **Table 2**, at.%C clearly decreases under degreasing process).

2.2. Increase of the surface roughness

With the nitric acidic-pickling bath, the magnesium in phosphate environment disappears (absence of Mg 2p component at 51.4 eV) and the oxide/hydroxide layer is highly damaged allowing the observation of the substrate (re-appearance of metal magnesium with Mg2p component at 49.2eV). In these acidic conditions, a main redox reaction occurs between the nitrate ions (E^0 = 0.96 V) and the magnesium (0) (E^0 = -2.36 V) at the surface of the alloy. This oxidative process dissolves Mg and some minor elements of EV31A are then detectable (e.g.,

Gd, Zn and Zr) (**Table 2**). Reaction (2), describing the oxidation of magnesium (0) by nitrate ions, leads to the solubilization of magnesium (II) in the bath and the formation of gaseous nitric oxide:

$$2NO_3^- + 8H_{aq}^+ + 3Mg \rightleftarrows 3Mg^{2+} + 4H_2O + 2NO(g) \quad (2)$$

The important difference of potential between these two species explains high reactivity in this bath and consequently the high surface roughness of the alloy surface (3.20 ± 0.40 μm instead of 0.62 ± 0.07 μm for the bare alloy) and the important weight loss of 130 g/m^2 measured after immersion in this bath.

Afterwards, with the immersion in the consecutive rinsing baths at neutral pH, the formation of $Mg(OH)_2(s)$ is renewed at the surface of the EV31A alloy.

Only a small improvement of the corrosion resistance is observed: ΔE_{corr} = +0.06 V in comparison with the bare alloy. The annual corrosion thickness increases to 227.3 μm/year and indicates a protection loss. This value could be correlated with the Mg^0 content, which went up to 21%. These data are in accordance with the previous conclusion discussing the minimized role of Mg oxide/hydroxide layer in the alloy protection.

Surface modification can be confirmed by **Figure 2c** with the appearance of hexagonally shaped grains of magnesium (red-dashed lines), the grain boundaries, rich in the previously cited minor elements (purple circle), appearing clearer than the grain.

The role of the nitric acidic-pickling bath is to dissolve the grains of magnesium, to make the grain boundaries denser by increasing the roughness; this would facilitate the deposition of a future layer. Indeed, when the two phases of the alloy are bare, a micro-current forms at the substrate surface due to a potential [33]. This difference of potential allows the preferential dissolution of one of the phases of the coating.

2.3. Activation of the surface in order to make it adhesive

In a solution, hexavalent chromium could exist as hydrochromate ($HCrO_4^-$), chromate (CrO_4^{2-}) and dichromate ($Cr_2O_7^{2-}$) ionic species. The proportion of each ion in the solution would depend on the pH and the concentration. During the chromic treatment step, the pH of the bath is 1.0 and the main species is $Cr_2O_7^{2-}$ [34]. This species reacts with magnesium (0) of the alloy (reaction (3))

$$Cr_2O_7^{2-} + 14H_{aq}^+ + 3Mg \rightleftarrows 2Cr^{3+} + 7H_2O + 3Mg^{2+} \quad (3)$$

The formed ions Cr^{3+} then react with H_2O (due to the acidic conditions) to form $Cr(OH)_3(s)$ which precipitates at the surface of the alloy due to its low K_s ($10^{-29.8}$) [35]. This precipitation begins at the centre of the magnesium grains (black-dotted circle) as seen in **Figure 2d** and **e** confirming the preferential reduction of chromium (VI) by magnesium.

The precipitation of $Cr(OH)_3(s)$ is minimal due to the low pH (below 2.0). For this reason, the chromium-based layer would not cover the entire surface of the alloy but it is thick enough to hide metal magnesium of the alloy (no XPS Mg (0) 2p component detected at 49.5 eV). This coating reduces the weight loss to 13 g/m^2 only and decreases the surface roughness (R_a): 1.80 ± 0.20 μm strongly, whereas it was 3.20 ± 0.40 μm after the nitric bath. Once rinsed twice, $Cr(OH)_3(s)$ dehydrates into $Cr_2O_3(s)$. An oxide/hydroxide magnesium layer is also formed in the same rinsing bath conditions. The atomic concentration in magnesium oxide/hydroxide is about 7.2%.

The role of the chromic-pickling bath is to deposit a first layer of chromic species ($Cr(OH)_3(s)$, $Cr_2O_3(s)$ and $K_2CrO_4(s)$) at the centre of the grains of magnesium. These species are responsible for the improvement of the corrosion resistance: $\Delta E_{corr} = +0.17\ V_{/SCE}$ in comparison with the bare alloy and the corrosion thickness decreases to 57.4 μm/year.

This first deposition acts as a nucleation layer that improves further the deposition of the protective coating.

Finally, in the last step of the pre-treatment process (hydrofluoric acidic-pickling bath), the sodium bifluorure $NaHF_2(s)$ is dissociated into sodium Na^+ and fluoride ions F^- (XPS BE Na 1s = 1071.7 eV [31]) and hydrofluoric acid HF. HF is classified as a weak acid (pKa = 3.20 [36]) and releases F^- and H^+ ions. The fluoride ions react with the dissolved magnesium (reaction (1)) to form $MgF_2(s)$ (B.E.= 50.9 eV) at the surface of the coating.

Due to weak value of their respective K_s ($10^{-11.15}$ for $Mg(OH)_2(s)$ and $10^{-8.15}$ for $MgF_2(s)$) [35], both compounds are formed on the surface of the alloy (7.1 %At for $MgF_2(s)$ in the Mg 2p spectrum beside 10.8 %At for $Mg(OH)_2(s)$ and $MgO(s)$ components). The apparition of $MgO(s)$ occurs later as explained previously.

The decrease of the corrosion potential of the immersed specimen in the hydrofluoric acidic-pickling bath ($E_{corr} = +1.54 \pm 0.06\ V_{/SCE}$) corresponds to the apparition of $MgF_2(s)$ into the coating and the diminution of $Cr(OH)_3(s)$, $Cr_2O_3(s)$ and $K_2CrO_4(s)$ contents (species responsible for the improvement of the corrosion resistance).

The detection in the XPS analysis of magnesium (0) and of minor elements (Gd, Zr and Zn) constitutive of the EV31A and the rise of the alloy surface roughness (2.30 ± 0.50 μm) indicate that the hydrofluoric-pickling bath has a similar role than the nitric acidic-pickling bath exposing the grain boundary (purple circle in **Figure 2g**). However, it is also responsible of a partial replacement of the layer of magnesium oxide/hydroxide with MgF_2 without a complete elimination of the chromic layer, which explains the very low weight loss (4.6 g/m^2). MgF_2 deposited on the surface is completely dissolved in the treatment bath, facilitating the deposition of the protective coating.

The different steps of the pre-treatment process assessed the preparation of the surface for the treatment step in order to make the coating adhesive on the alloy surface. They participated in cleaning the surface and making it rough and reactive for the coating deposition.

2.4. Coating deposition

The formation of the chromium coating needs several minutes to cover the entire magnesium surface. The coating begins to appear after an immersion time of 5 s at the centre of the hexagonally shaped magnesium grain, confirming the attack of dichromate on magnesium (**Figure 2h**). This first deposit is the result of a redox reaction that occurs very quickly (reaction (3)). This reaction induces a local pH raise at the alloy surface, favouring the precipitation of magnesium hydroxide and oxide with the deposition of chromium hydroxide and oxide. This coating expands to the grain boundaries as seen during the immersion in the chromic bath (**Figure 2i**)). The chemical composition of the coating is a function of the immersion time (**Table 3**).

%At.		2 min	6 min	8 min	12 min	14 min
Fe			0.4			
C		31.0	34.1	35.0	36.7	37.1
Cr	*$Cr_2O_3(s)$*	3.4	4.6	4.6	4.7	4.0
	$Cr(OH)_3(s)$	2.6	3.3	3.1	2.9	3.0
	$K_2CrO_4(s)$	1.1	1.7	2.2	1.1	1.4
F		7.8	0.3	0.1	0.4	0.4
K		0.2	0.6	0.4	0.1	0.4
Mg	*$MgO(s)$*	5.9	3.3	3.3	3.1	3.3
	$Mg(OH)_2(s)$					
	$MgF_2(s)$					
	Mg^0	3.2				
N		0.6	0.5	0.5	0.9	0.6
O		40.9	47.1	47.2	46.6	46.8
S		3.2	4.0	3.5	3.4	2.9
Zn		0.1	0.1	0.1	0.1	0.1

*Not present

Table 3. Atomic composition (%At) of the coating during the chromium conversion process on EV31A pieces after the pre-treatment step.

The quick growth of the chromium coating leads to the disappearance of magnesium (0) and the magnesium content tends to decrease from 17.9% (*t* = 0 min) to 5.9% after 2 min of immersion and is down to 3.3% after 6 min of immersion in the treatment bath (**Table 3**) and its replacement by chromium. Different chromium species exist ($Cr(OH)_3(s)$, $Cr_2O_3(s)$ and

CrO_3(s)/K_2CrO_4(s) (hard to differ with the XPS as all these environments get the same XPS Cr 2p BE)) and become predominant after 6 min of immersion in the bath. Peak assignments are made with respect to the reference compounds analysed in the same conditions, namely Cr_2O_3(s), $Cr(OH)_3$(s), CrO_3(s), K_2CrO_4(s) and $K_2Cr_2O_7$(s) [37, 38]. The concentrations of the chromium species increase during the first 6 min of immersion before stabilizing, indicating a complex surface chemical process and a homogeneous repartition into the coating.

Cr_2O_3(s) is issued from the dehydration of $Cr(OH)_3$(s) after the rinsing and drying steps. This dehydration gives the coating a surface morphology with plane domains separated with large cracked frontiers (**Figure 3**, zone 1).

Figure 3. (a) High-resolution SEM image of large cracked frontiers (zone 1) and clusters (zone 2) on the surface after 6 min of treatment. (b) High-resolution SEM image of zoom in a cluster (zone 2).

The presence of K_2CrO_4(s) or CrO_3(s) in the coating is attributed to re-crystallization at the surface (**Figure 3**, zone 2). Actually, the Cr(VI) species trapped in the protective coating over alloys are generally known to be responsible for the 'self-healing' effect of the considered coating [12]. The presence of zinc in the coating remains minimal (around 0.1%).

At final, the deposition of the coating seems to occur during the first 6 min of immersion and induces the decrease of the surface roughness from 2.3 ± 0.5 μm before treatment to 1.4 ± 0.2 μm after 6 min of immersion. During the two first minutes of immersion, the coating spreads across the surface and covers it entirely. During the next 4 min, the layer thickens and keeps growing to reach 11 μm after 14 min of immersion in the bath.

The corrosion resistance is improved to +0.2 V after 2 min of immersion in the chromium bath. The final corrosion resistance presents a ΔE_{corr} = +0.3 V after 6 min of immersion and stabilizes around this value with increasing immersion times. This corrosion potential and the annual corrosion thickness stay low (E_{corr} = -1.4 ± 0.1 $V_{/SCE}$, e = 50 ± 10 μm/year) and the application of a painting should be necessary to obtain a total protection against corrosion. Another way could be the replacement of chromium by an alternative coating.

3. Alternative conversion coatings

New alternative conversion coatings have been proposed since few years to replace CCCs due to the recent limitations and the future interdiction of the use of chromium (VI) (**Figure 4**). These coatings present proper action modes and their own mechanisms and properties for the protection of Mg.

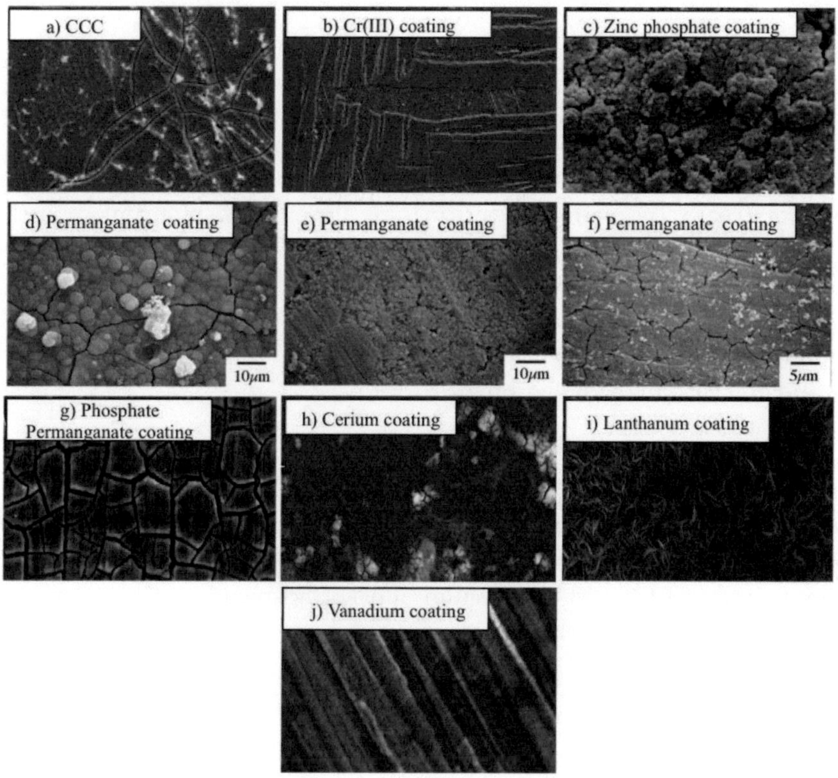

Figure 4. SEM (scanning electron microscopy) images of coatings obtained with treatment baths containing (a) Cr (VI) [12], (b) Cr (III) [12], (c) zinc phosphate [10], (d) permanganate with HNO_3 [19], (e) permanganate with HF [19], (f) permanganate with HCl [19], (g) phosphate permanganate [23], (h) cerium [39], (i) lanthanum [30] and (j) vanadium [24].

3.1. Cr(III)-based coatings

Cr(III)-based coatings involve a redox reaction: the metal is oxidized by an oxidant added to the bath. This oxidant (H_2O, NO_3^-, etc.) is simultaneously reduced as shown in reactions (1) and (4) [40, 41]

$$Mg + NO_3^- + 2H^+_{aq} \leftrightharpoons Mg^{2+} + NO_2^- + H_2O \tag{4}$$

A local pH increase is caused by the hydronium ion consumption (reaction 4) or the OH^- formation (reaction 1). This pH variation causes the precipitation of trivalent chromium as an insoluble hydroxide and then oxide.

An example of Cr(III) treatment gives a 90-nm coating made of 60% $Cr(OH)_3(s)$ and 40% $Cr_2O_3(s)$ (**Figure 4b**) [12]. The film presents a smooth and continuous structure with no cracks [13]. Generally, the corrosion resistance is less effective with Cr(III) coatings than with CCCs [6, 12]. This can be attributed to the difference in thickness of the two films and to the presence of mobile Cr(VI) species in CCCs that allow the repassivation of flaws and corrosion pits [6, 12].

Actual commercial Cr(III) solutions are made with CrF_3 and $(NH_4)_2ZrF_6$. These solutions present a better protection against corrosion: $\Delta E_{corr} = +0.60 \pm 0.1$ V. The annual corrosion depth is about 11 ± 1 μm/year, whereas it is 50 ± 10 μm/year for chromium VI coatings.

Cr content is lower than Cr(VI) solution: 6–7 At% for commercial Cr(III) solution and 9–10 At % for chromium VI bath. The main difference in the coating composition remains in the fluoride content: 13–14 At% for commercial Cr(III) solution and 0.1–0.2 At% for chromium VI bath. Could fluoride compounds with chromium or magnesium increase the protection against corrosion? An MgF_2 layer does not have this property as seen in the previous section, but a mix of chromium and magnesium fluorides could increase this protection in comparison to chromium and magnesium oxides and hydroxides.

3.2. Phosphate-based conversion coatings

Phosphate coatings are more environmentally friendly than CCCs, and many scientists have tested films deposited from phosphate solutions. Currently, this process is one of the most studied alternatives to CCCs on magnesium alloys. Phosphate coating on zinc, steel and aluminium is already a well-known process [42].

In contrast to Cr(VI), phosphate cannot oxidize the surface of an alloy. Phosphate can only passivate alloy surfaces. Without the presence of an oxidant, the oxidation of magnesium only occurs due to the reduction of water according to reaction (1).

To create an efficient film, it is necessary to add an oxidant to the bath to increase the rate of magnesium alloy oxidation [43]. The main oxidants used are NO_3^- or H_2O_2 [43]. Nitrites can be added with nitrates to further accelerate oxidation.

In the phosphate bath, an equilibrium exists among all forms of the orthophosphoric acid that can dissociate by successively liberating protons. The dominant form of orthophosphoric acid depends on the solution pH. Phosphate baths are mainly around pH 3.0 [14, 17, 18]. In these baths, H_3PO_4 and $H_2PO_4^-$ are the primary species according to their pKa values: $pKa_1 = 2.15$, $pKa_2 = 7.20$ and $pKa_3 = 12.35$ [35].

Reactions (1) and (4) lead to a local pH increase at the alloy surface and, consequently, to the modification of the phosphate species to HPO_4^{2-} and PO_4^{3-}. The phosphate ions PO_4^{3-} react with Mg^{2+} to form $Mg_3(PO_4)_2(s)$ ($K_s = 10^{-25.2}$ [35]) as shown in reaction (5) [14]

http://dx.doi.org/10.5772/66604

$$3Mg^{2+} + 2PO_4^{3-} \leftrightarrows Mg_3(PO_4)_2(s) \tag{5}$$

Another possibility for the formation of $Mg_3PO_4(s)$ has been considered [42]. The chemical reactions between the oxidized base metal Mg and the phosphate can be described by reactions (6)–(8):

$$Mg + 2H_2PO_4^- + 2H^+_{aq} \leftrightarrows Mg(H_2PO_4)_2(s) + H_2(g) \tag{6}$$

$$Mg(H_2PO_4)_2(s) \leftrightarrows MgHPO_4(s) + H_3PO_4 \tag{7}$$

$$3MgHPO_4(s) \leftrightarrows Mg_3(PO_4)_2(s) + H_3PO_4 \tag{8}$$

The value of the thermodynamic function H^0 for $Mg_3PO_4(s)$ and $Mg(OH)_2(s)$ is 3780.6 and 924.2 kJ/mol, respectively. Consequently, Mg^{2+} preferentially bonds with PO_4^{3-}, which explains the absence of $Mg(OH)_2(s)$ in the coating [14]. $Mg_3PO_4(s)$ is so the main compound of the coating and responsible for the corrosion protection in these kinds of coatings. Its density (2.74 g/cm^3) is two times lower than the Cr_2O_3 (5.22 g/cm^3) [36]. This property could be a hypothesis to explain the lower corrosion resistance of these coatings in comparison to CCCs (**Table 4**).

In the case of Mg-Li alloys, the deposition occurs mainly on the β-phase where the redox reactions (1) and (4) occur. A conversion film forms as shown in reactions (5)–(8) [14, 15]. The film grows until an equilibrium is established between film dissolution and formation [15].

If other divalent cations, such as Ca^{2+}, are added to the bath and hydroxide ions are present at the alloy surface, then HPO_4^{2-} and PO_4^{3-} preferentially bond to form insoluble $CaHPO_4{,}2H_2O(s)$ [14, 44].

To strengthen the corrosion resistance of the coating, Zn(II) salts are added to create an insoluble layer of $Zn_3(PO_4)_2(s)$. Sometimes, ZnO is added to the bath because the addition of a metallic oxide can influence the microstructure of a phosphate coating and make the coating denser and thinner [18]. Generally, the compositions of all the zinc phosphate baths are similar: a buffer made with orthophosphoric acid H_3PO_4, dihydrogen phosphate $H_2PO_4^-$ and NO_3^- with NO_2^- as an accelerating agent. The differences stay in the nature and the concentration of the zinc salt.

The deposition mechanism of zinc phosphate coatings begins with the phosphate precipitation on anodic areas [42, 47]. In these areas, metal alloy is oxidized while water is reduced from the surface, leading to the increase of pH [42]. The local increase of pH allows the precipitation of $Mg_3(PO_4)_2(s)$ ($K_s = 10^{-25.2}$ [35]) and more especially of insoluble $Zn_3(PO_4)_2{,}4H_2O(s)$ ($K_s = 10^{-35.3}$ [35]) as shown in Eq. (9) [42]

Alloy	Bath characteristics			Coating characteristics			Ref
	Compounds	Concentration	pH	Composition	j_{corr} (μA/cm²)	E_{corr} ($V_{/SCE}$)	
Mg-8.8Li	$NH_4H_2PO_4$	25 g/L	3	$CaHPO_4.2H_2O$(s)	0.2	-1.58	[14]
	$Ca(NO_3)_2$	25 g/L		$Ca_3(PO_4)_2$(s) $Mg_3(PO_4)_2$(s)			
AM60	H_3PO_4	7.4 mL	3	*n.d*	10,000	-0.90	[42]
	Na_2HPO_4	20 g/L					
	$NaNO_3$	3 g/L					
	$NaNO_2$	1.84 g/L					
	$Zn(NO_3)_2$	1.84 g/L					
	NaF	1 g/L					
AZ91D	H_3PO_4	17.5 g/L	3	$Zn_3(PO_4)_{2\cdot4}H_2O$(s)			[17]
	ZnO	3.2 g/L		Zn $AlPO_4$(s)			
	NaF	1.7 g/L		$MgZn_2(PO_4)_2$(s)			
	$NaNO_3$	0.17 g/L		$Mg_3(PO_4)_2$(s)			
	$NaNO_2$	0.83 g/L					
	$C_4H_4O_6$	2.2 g/L					
	Amine	0.18 g/L					
AZ91D	H_3PO_4	0.065 mol/L	2.4	$Zn_3(PO_4)_2.4H_2O$(s)	*n.d*	-1.17	[18]
	ZnO	0.0029 mol/L		Zn(s)			
	$Zn(NO_3)_2$	0.102 mol/L					
	NaF	0.040 mol/L					
	$NaClO_3$	0.0028 mol/L					
	NH_3	0.0034 mol/L					
	Amine	0.007 mol/L					
AZ31	H_3PO_4	7.4 mL/L	3	$Zn_3(PO_4)_2.xH_2O$(s)	*n.d*	-1.37	[45]
	$NH_4H_2PO_4$	20 g/L					
	$NaNO_3$	1.84 g/L					
	$NaNO_2$	3 g/L					
	$Zn(NO_3)_2$	5 g/L					
	NaF	1 g/L					
AZ31	H_3PO_4	0.10 mol/L	3.07	$Zn_3(PO_4)_2\cdot4H_2O$(s)	0.54×10^{-5}	*n.d*	[46]
	$NH_4H_2PO_4$	0.034 mol/L		$MgHPO_4\cdot3H_2O$(s)			
	$NaNO_3$	0.021 mol/L		$Mg_3(PO_4)_2$(s)			
	$NaNO_2$	0.042 mol/L		$AlPO_4$(s)			
	$Zn(NO_3)_2$	0.068 mol/L		Al_2O_3(s)			
	NaF	0.024 mol/L		$Al(OH)_3$(s)			
				MgO(s) and/or $Mg(OH)_2$ (s)			

Table 4. Composition of phosphate baths and their respective coating compositions and properties.

$$3Zn^{2+} + 3Mg + 4H_2PO_4^- + 4H_2O \leftrightarrows Zn_3(PO_4)_2(s),4H_2O(s) + Mg_3(PO_4)_2(s) + 2H^+_{aq} + 3H_2(g) \quad (9)$$

More precisely, the resulting amorphous layer consists of mixed phosphates of zinc and alloy magnesium (reaction 10). This first layer is the base for the development of crystal nuclei of zinc phosphate $Zn_3(PO_4)_2,4H_2O(s)$. This theory is based on the existence of $ZnPO_4^-$ resulting from the formation (reaction (11)) and then the dissociation (reaction (12)) of $Zn(H_2PO_4)_2,2H_2O$ in the bath [17]

$$Mg + 2H^+_{aq} + 2ZnPO_4^- \leftrightarrows MgZn_2(PO_4)_2(s) + H_2(g) \quad (10)$$

$$Zn^{2+} + 2H_2PO_4^- + H_2O \leftrightarrows Zn(H_2PO_4)_2,2H_2O \quad (11)$$

$$Zn(H_2PO_4)_2,2H_2O \leftrightarrows ZnPO_4^- + H_2PO_4^- + 2H^+_{aq} + 2H_2O \quad (12)$$

The coating (**Figure 4c**) has a double-layer structure, an inner amorphous layer and a crystal outer layer. The inner layer consists of an amorphous dense inner layer made of $MgZn_2(PO_4)_2(s)$, $Mg(OH)_2(s)$ and/or $MgO(s)$ and some small amounts of $AlPO_4(s)$, $Al_2O_3(s)$ and/or $Al(OH)_3(s)$. The outer porous crystal layer is composed of crystal hopeite, $Zn_3(PO_4)_2(s)$ [46]. This layer is generally porous due to the presence of cracks and flaws [48].

The detection of aluminium in the coating comes from its incorporation in several alloys. Its presence in the coating slows down the formation of the phosphate films on magnesium surfaces. Fluoride ions can be added to release Al complexes AlF_6^{3-} from the cathodic sites. This complexion influences the zinc phosphate film formation by increasing the number of nuclei and allowing the formation of a more compact layer [42]. Sodium fluoride NaF precipitates aluminium ions in the solution to form $Na_3AlF_6(s)$ [42].

Zinc phosphate improves the corrosion resistance relative to a simple phosphate coating [17, 18, 42, 45, 47]. Higher corrosion potentials are obtained with zinc phosphate baths (**Table 4**). The density of this compound is 3.998 g/cm^3 [36]. The presence of this compound in the coating of magnesium phosphate increases the density of the layer and could be responsible for better corrosion resistance.

The protection provided by phosphate alone is not equivalent to a CCC due to the formation of a low dense layer of $Mg_3(PO_4)_3$. On the other hand, the zinc phosphate coating seems to provide better corrosion resistance than the CCC. Indeed, the corrosion potential is from -1.37 to -0.9 $V_{/SCE}$ for this coating, and it is of -1.4 ± 0.1 $V_{/SCE}$ for the Cr(VI) coating.

3.3. Permanganate-based conversion coatings

The coatings based on permanganate solutions are named PCCs (permanganate conversion coatings). These solutions contain Mn(VII) species acting as oxidant agents on the alloy. The reduced Mn(IV) species coat and passivate the substrate as shown in Eqs. (13) and (14) [49].

$$2MnO_4^- + 3Mg + 8H^+_{aq} \leftrightarrows 2MnO_2(s) + 3Mg^{2+} + 4H_2O \quad (13)$$

$$2MnO_4^- + 4Mg + 10H^+_{aq} \leftrightarrows Mn_2O_3(s) + 4Mg^{2+} + 5H_2O \quad (14)$$

The protective species are MnO_2 or Mn_2O_3 and have similar or superior densities (respectively 5.0 and 9.6 g/cm^3) than Cr_2O_3 formed in the case of CCC [36]. The corrosion protection properties are close but not better than CCC (E_{corr} = -1.5 ± 0.1 $V_{/SCE}$ in comparison to E_{corr} = -1.4 ± 0.1 $V_{/SCE}$ for Cr(VI) coating). The annual corrosion depth is worse in the case of PCC (e = 100 ± 10 μm/year in comparison to e = 50 ± 10 μm/year for Cr(VI) coating).

PCC baths still possess the advantage of not requiring the need to be heated [49]. Moreover, PCC process does not form protective inorganic polymer coatings like CCC, which makes it less affected by heat than CCCs and are more convenient for painting [49].

Table 5 groups the PCC baths that are usually tested on magnesium alloys. These baths contain $KMnO_4$ associated with different strong acids [19, 49]. The nature and the concentration of the acid affect the structure and the composition of the coating [19]. **Figure 4d–f** shows the different morphologies of PCCs. The addition of HNO_3 affects the surface of the coating, forming clusters of particles (**Figure 4d**) [19]. With the addition of HF (**Figure 4e**), the coating has an amorphous structure. This film is thinner than the coating deposited in a bath containing HNO_3, and the deposition rate is slower [19]. The thinner film can be explained by the reaction of fluorine ions with magnesium to produce insoluble magnesium fluoride MgF_2(s), creating a passivating layer that prevents any further dissolution of magnesium [19]. Coatings dipped in the HF and the $KMnO_4$ solutions are composed mainly of MgF_2(s) and manganese oxides (MnO_2(s), Mn_2O_3(s) and Mn_3O_4(s)), while the coatings formed in the HNO_3/$KMnO_4$ bath contain Mg and Mn oxi-hydroxides. More cracks in the coating can be observed when HCl is present in the solution (**Figure 4f**). The role of an addition of an acid to a permanganate bath on the protective properties of the final coating has not been quantified but the coatings characteristics (thinner film with HF, cracks with HCl and presence of clusters with HNO_3) may decrease the corrosion resistance of the coating.

Permanganate coatings provide an alternative to CCC but they do not provide good corrosion resistance. Another problem that prevents PCCs from replacing CCCs is the stability of the bath pH. Important pH variations occur when PCC baths are used and the alloys are dipped in the solution. The attack of MnO_4^- on Mg alloy surfaces consumes H^+_{aq} ions (reactions 13 and 14), increasing the pH of the solution. The addition of $Na_2B_4O_7$ to an HCl bath has a buffering effect and stabilize the pH of the bath [19, 49]. However, the Registration, Evaluation, Authorization and Restriction of Chemicals added tetraborate products to the list of substances of very high concern (SVHC), and their use is now limited and will be completely forbidden in a few years. Other compounds must be found to stabilize the pH of PCC baths.

Alloy	Bath characteristics		Coating characteristics	Ref
	Compounds	Concentration mol/L	Composition	
AZ91D	$KMnO_4$	0.02	$MnO_2(s)$ $Mn_2O_3(s)$	[19]
	HNO_3	0.02–0.2	$Mn_3O_4(s)$	
			Mg oxide/hydroxide	
AZ91D	$KMnO_4$	0.02	Mn oxide/hydroxide	[19]
	HCl	0.02–0.2	B oxide/hydroxide Mg oxide/hydroxide	
	$Na_2B_4O_7$	unknown		
AZ91D	$KMnO_4$	0.02	$MnO_2(s)$ $Mn_2O_3(s)$	[19]
	HF	0.02–0.2	$Mn_3O_4(s)$ $MgF_2(s)$	
			Mg oxide/hydroxide	

Table 5. Composition of permanganate baths and their respective coating compositions and properties.

3.4. Permanganate/phosphate-based conversion coatings

The combination of phosphate and permanganate has been considered as a solution to avoid the pH increase in baths. In a potassium permanganate bath, monopotassium dihydrogenophosphate (KH_2PO_4) or manganese hydrogenophosphate ($MnHPO_4$) is added as a buffer [3]. The reactions of the deposit are the same as discussed previously: the oxidation of magnesium and the reduction of Mn(VII) consume the H^+_{aq} responsible for the pH increase, allowing the phosphate species to precipitate at the metal surface (reactions (5), (13) and (14)) [23, 50].

The grain boundaries act as cathodes, and grains function as anodes, forming local cell effects. Meanwhile, hydrogen and phosphate ions are consumed at the substrate/solution interface, causing a pH increase. This phenomenon favours the formation of $Mg_3(PO_4)_2(s)$ deposits (**Figure 4g**).

$Mg(OH)_2(s)$, $MgO(s)$, $MnO_2(s)$ and $Mn_2O_3(s)$ can be found in the coatings as detailed on reactions (15) and (16) (**Table 6**). The presence of Al in the alloys leads to the formation of $Al(OH)_3(s)$, $Al_2O_3(s)$ and $MgAl_2O_4(s)$ as shown in reactions (17) (**Table 6**) [22]

$$2MnO_4^- + 5Mg + 12H_{aq}^+ \leftrightarrows 2MnO(s) + 5Mg^{2+} + 6H_2O \tag{15}$$

$$2MnO_4^- + 4Mg + 10H_{aq}^+ \leftrightarrows Mn_2O_3(s) + 4Mg^{2+} + 5H_2O \tag{16}$$

$$MgO(s) + Al_2O_3(s) \leftrightarrows MgAl_2O_4(s) \tag{17}$$

MnO_4^- species is also detected in the coating (**Table 6**). It has been trapped during the formation of the coating. Its presence could be responsible for a 'self-healing' effect as in the case of chromium VI coatings. MnO_4^- ions have not been detected in the 'permanganate-only' coatings.

Bath characteristics					Coating characteristics					
Alloy	Compounds	Concentration (g/L)	Immersion time (s)	pH	Composition	j_{corr} (μA/cm²)	E_{corr} (V_{SCE})	*e* (mm/year)	Thickness (μm)	Ref
Mg alloy (10%Li 1%Zn)	**$KMnO_4$** **KH_2PO_4**	40 50	1200	*n.d*	$Mg(OH)_2$(s) MgO(s) Mn_2O_3(s) MnO_2(s) K and P detected	*n.d*	-1.57	*n.d*	*n.d*	[33]
AZ91D	**$KMnO_4$** **KH_2PO_4** **H_3PO_4**	40 150 *n.d*	600	3–6	MgO(s) $Mg(OH)_2$(s) MnO_2(s) Mn_2O_3(s) $KMnO_4$	585.8	–1.40	13.2033	7–10	[5]
EV31A	**$KMnO_4$** **$MnHPO_4$**	40 75	300	3.5	MgO(s) $Mg(OH)_2$(s) MnO_2(s) Mn_2O_3(s) $KMnO_4$	0.3	–0.38	7.0×10^{-3}	*n.d*	
EV31A	**$KMnO_4$** **$MnHPO_4$**	40 150	300	3.5	MgO(s) $Mg(OH)_2$(s) MnO_2(s) Mn_2O_3(s) $KMnO_4$	0.06	-0.24	1.3×10^{-3}	*n.d*	
AZ91D	**$KMnO_4$** **$MnHPO_4$**	20 60	600	*n.d*	MgO(s) $Mg(OH)_2$(s) MnO_2(s) Mn_2O_3(s) $MgAl_2O_4$(s) Al_2O_3(s) $Al(OH)_3$(s)	18	-1.50	*n.d*	*n.d*	[22]
AZ91D	**$MnHPO_4.2H_2O$**	*n.d*	1800	*n.d*	Mn, O, P, Mg, Al	*n.d*	*n.d*	*n.d*	10	[3]
AM60B	**$NH_4H_2PO_4$** **$KMnO_4$** **H_3PO_4**	100 20 *n.d*	1200	3.5	*n.d*	*n.d*	*n.d*	*n.d*	*n.d*	[21]

Table 6. Composition of permanganate/phosphate baths and their respective coating compositions and properties.

The coating thicknesses are between 7 and 10 μm [5]. The conversion coating thickness decreased gradually with the increase of the pH value and the concentration of KH_2PO_4 in the bath. Below pH 3, a non-compact surface layer is formed because manganese is under the soluble Mn^{2+} form and not the protective MnO_2. Around pH 5, the formation of the coating is too slow due to the quick main precipitation of $Mg_3(PO_4)_2$ [5, 51]. The conditions are optimum when the concentration of KH_2PO_4 is maximum (150 g/L) and the pH fixed between 3.5 and 4.0 to form a coating made essentially with MnO_2 and $Mg_3(PO_4)_2$.

The good corrosion resistance of the permanganate-phosphate coatings allows considering this solution as a serious alternative to chromium coatings. Indeed, the immersion for 10 h in conventional corrosive electrolyte-artificial seawater (3.5 wt.% NaCl solution) does not present any trace of corrosion whereas only 300 s of immersion of the bare alloy in the same solution is sufficient to observe corrosion pits on the surface of the alloy [33]. Chromium VI coating presents several corrosion pits after less than 4 h in the same conditions.

The corrosion protection properties are clearly better than CCC (E_{corr} = -0.3 ± 0.1 $V_{/SCE}$ in comparison to E_{corr} = -1.4 ± 0.1 $V_{/SCE}$ for Cr(VI) coating) (**Table 6**). The annual corrosion is also better in the case of permanganate/phosphate coatings (e = 1–7 ± 10 μm/year in comparison to e = 50 ± 10 μm/year for Cr(VI) coating).

The presence of aluminium in the final coating (when this element participates in the composition of the Mg alloys studied) does not disturb its corrosion resistance properties.

These results clearly indicate that permanganate/phosphate coatings present better corrosion resistance properties than CCCs for the protection of magnesium alloys due to the dense MnO_2/Mn_2O_3 layer mixed with $Mg_3(PO_4)_2(s)$

3.5. Vanadium-based coatings

Self-healing properties similar to the CCCs can also be obtained by adding vanadium-based oxyanions to the coating. Vanadium solutions are generally used as corrosion inhibitors in many paints or pigments [27].

The coatings obtained with a vanadium solution of 50 g/L are uniform and compact (**Figure 4j**). Their thickness is about 2.0 ± 0.5 μm [27]. The vanadium oxides that composed these coatings are mostly unidentified.

Vanadium coatings present good corrosion protection properties. Indeed, the vanadium coatings immersed in a 3.5% NaCl solution present, respectively, 15, 7 and 2 pits/cm^2 for the samples treated in 10, 30 and 50 g/L solution, whereas 50 pits/cm^2 can be observed on the surface of the polished alloy [26]. The size of the pits decreases also on vanadium coatings. The vanadium species responsible for the corrosion resistance are vanadium oxides [40].

The optimal conditions are obtained with a 50 g/L vanadium solution and pH 7. An increase of the pH from 7 to 9 had a negative effect on the corrosion protection performance of the coating [27]. The vanadium layer loosely adhered to the substrate alloy and the surface was severely corroded and covered with pits.

The increased corrosion resistance of the coated samples at 50 g/L and pH 7 is explained by the self-healing action that blocked the pitted areas from corrosive attack and other surface defects [27]. It seems that the formation of a vanadium oxide layer plays a distinct role in healing cracks in coating surfaces and repairing pits, and hence improving the overall localized corrosion resistance. Coatings formed in a 50 g/L vanadium solution are more effective than the other treatments with lower vanadium concentrations in reducing the number of pits due to the self-healing ability of the films and the 'buffer effect' of the vanadium-rich oxide layers that reject the corrosive chloride ions from the surface of the magnesium substrate [27].

Vanadium coating seems to be one serious option for the replacement of CCC. The self-healing ability of vanadium coatings due to their rich oxide layer is an important advantage. However, vanadium coatings have been responsible for worsening the corrosion of the magnesium alloy

EV31A [52]. This unexpected behaviour has been explained by the formation of multi-oxide layers of vanadium in addition to the alloying elements Zr, Nd and Zn at the surface, resulting in heterogeneous coatings [52]. The effectiveness of the coating directly depends on the composition of the alloy. This phenomenon has been only observed for vanadium coatings and not for the other alternatives to CCCs

3.6. Rare earth elements–based coatings

Among the 17 REEs, cerium and lanthanum are the most commonly used for conversion coating. The corrosion resistance of magnesium is improved by adding a small amount of REEs, although an excessive addition of REEs detract from the corrosion resistance. The optimum REEs content is between approximately 0.3 and 0.5wt% of the alloy [29].

When the alloy is dipped in one of the conversion solutions (listed in **Table 7**), the preformed hydroxide film on the substrate surface immediately dissolves. After that, the primary anodic dissolution reaction of magnesium occurs simultaneously with the reduction of hydronium ions [53]. The addition of oxidants, such as NO_3^- or H_2O_2, can favour oxidation [40, 54]. At the same time, the formation of OH^- increases the pH at the interface between the substrate and the solution.

Alloy	Bath characteristics				Coating characteristics			
	Compounds	Concentration	Immersion time (s)	pH	Composition	j_{corr} (μA/cm²)	E_{corr} (V_{SCE})	Ref
Mg-8.5Li	$Ce(NO_3)_3$ $La(NO_3)_3$ $KMnO_4$	2 g/L 2 g/L 2 g/L	300	4.0	La_2O_3(s) CeO_2(s) Mn_2O_3(s) MnO_2(s)	*n.d*	-1.5	[53]
Mg-8Li	$La(NO_3)_3$	5 g/L	1200	5.0	$La(OH)_3$(s)		-1.3	[30]
AZ63	$CeCl_3$ H_2O_2	10 mg/L 50 mL/L	6 × 30	*n.d*	*n.d*	1.05 × 10³	-1.49	[39]
AZ31	$Ce(NO_3)_3$	0.05 mol/L	*n.d*	3.6	CeO_2(s)	*n.d*	*n.d*	[29]
WE43	$Ce(NO_3)_3$	0.05 mol/L	300	3.6	*n.d*	*n.d*	*n.d*	[56]
WE43	$La(NO_3)_3$	0.05 mol/L	300	3.6	*n.d*	*n.d*	*n.d*	[56]
WE43	$Pr(NO_3)_3$	0.05 mol/L	300	3.6	*n.d*	*n.d*	*n.d*	[56]
AZ31	$Ce(NO_3)_3$ H_2O_2	10 g/L 20 mL/L	300	*n.d*	CeO_2(s) CeO(s) Ce_2O_3(s) MgO(s) $Mg(OH)_2$(s) Al^2(s)	*n.d*	*n.d*	[40]

Table 7. Composition of REEs (rare earth elements–based coating) baths and their respective coating compositions and properties.

For cerium baths, if the interfacial pH value is high enough, then Ce^{3+} precipitates on cathodic sites to form $Ce(OH)_3$(s) [40, 54]. Coating formation is fast, and it is mainly controlled by the production of OH^- at the surface of the alloy. As coating formation proceeds, the surface is covered gradually with cerium hydroxide and then oxide. The rate of coating formation gradually slows down. A dense Ce-based conversion coating is obtained on the surface of magnesium alloys. The conversion coating consists of a mixture of trivalent and tetravalent cerium oxides. Exposure to the atmosphere causes the oxidation of $Ce(OH)_3$(s) to $Ce(OH)_4$(s) and the dehydration of the hydroxides to oxides. Consequently, the Ce(IV) content is higher at the coating surface than at the inside of the coating [29]. The conversion coating (**Figure 4h**) consists of a mixture of CeO_2(s) and Ce_2O_3(s) [30, 53]. The inside and surface layers have different morphologies. It has been calculated that approximately 61% of the cerium species in the coating surface and 45% of the cerium species in the inside area exist in a tetravalent state [29].

When lanthanum (La) solutions are used, they contain $La(NO_3)_3$ in the range of 2–16.3 g/L [30, 55, 56]. The deposition mechanisms of these coatings are similar to those of the cerium coatings. The conversion films (**Figure 4i**) consist of mixtures of $La(OH)_3$(s), La_2O_3(s), $Mg(OH)_2$(s), MgO(s) and Al_2O_3(s) [30, 53]. Despite numerous studies on Ce coatings, the greatest corrosion resistance was obtained with La, although the reproducibility of those results was poor [55].

One of the important effects of the REEs on corrosion resistance is the 'scavenger effect'. Indeed, REEs create intermetallic compounds with impurities, cancelling the influence of some minor elements, such as Cl and Fe, on the corrosion resistance [55]. Indeed, the corrosion potential of these coatings varies between -1.50 and -1.30 V as shown in **Table 7**, which is comparable to the values found for CCCs (between -1.30 and -1.50 V) [29]. To improve the corrosion resistance, the presence of the same REE in the alloy and in the conversion bath is advised [55]. The corrosion resistance of the coated alloy is improved if the REEs content in the alloy equals 0.3 wt% Ce or is less than 0.1 wt% La [55]. REE coatings are the only baths that provide this 'scavenger effect' with properties as interesting as the self-healing ability of CCCs. However, an adhesive weakness of a cerium conversion coating on AZ31 alloy was noted where the surface layer was easily peeled off with an adhesive tape [29]. The adhesion between the surface layer and the inside layer of the coating was much weaker than between the inside layer and the metallic substrate.

4. Conclusion

The actual process for the protection of magnesium alloys against corrosion uses chromium VI baths. This process needs an initial four-step pre-treatment to obtain a high-performance coating. This step is essential for preparing the surface to promote the anchorage of the protective coating in the treatment bath. Then, the nitric acidic bath is responsible for a strong attack of the alloy surface to increase the roughness of the surface in order to improve adhesion. The chromic acid–pickling bath initiates the chromate coating deposition (thin layer) at the centre of the magnesium grains. The hydrofluoric acid–pickling bath allows the deposition of an MgF_2 layer, making the surface more reactive for further coating deposition. The first deposition of chromium (III) oxide is not completely removed by the hydrofluoric acidic pickling and enhances further deposition of the coating during the immersion in the chromate conversion bath.

The description of the coating deposition mechanism shows that the species responsible for the protection of the alloy are trivalent chromium compounds: $Cr(OH)_3(s)$ and $Cr_2O_3(s)$. The presence of $K_2CrO_4(s)$ or $CrO_3(s)$ trapped in the coating allows a unique 'self-healing' property of the coating.

The chromate conversion coatings are actually the reference for the protection of magnesium alloys, and their only disadvantage is the toxicity of its main compound. Alternative coatings exist and present some difference with the chromium VI process. Naturally, considering the data from the Cr(VI) coating deposition mechanism, Cr(III) could be considered as the replacement for CCC. However, less corrosion resistance is obtained with Cr(III) coatings. This is due to the deposition of a thinner layer and the absence of hexavalent species in the coating that are responsible for the 'self- healing' effect. The phosphatization of metals is a well-known process, and zinc can be added to the bath to increase the protection of magnesium to form mixed $Mg_3(PO_4)_2$ and $Zn_3(PO_4)_2$ layers. Zinc phosphate has superior corrosion resistance, but it presents no evidence of having the ability to 'self-heal'.

REE coatings could be considered as a solution, but only alloys that contain REEs could benefit from protection similar to CCCs. REEs and vanadium-based coatings possess a 'scavenger' and a 'self-healing' effect, respectively, that makes them comparable to the CCC. However, REE salts are expensive, and the efficiency of vanadium coating is highly dependent on the substrate composition.

Permanganate can be used alone or with phosphate to create an efficient protective coating made essentially by a MnO_2 layer with $Mg_3(PO_4)_2$ when phosphate has been added in the solution. The properties of the coating could also be linked to the manganese speciation in the coating: Mn (IV) and Mn (VII). Mn (VII) gives the coating a self-healing property, like the chromate conversion coatings. The protection against corrosion is also better than chromium VI coatings when optimum conditions on phosphates concentration and pH are respected. **Table 8** presents a summary of the advantages and disadvantages of each alternative solution.

Coating	Advantages	Disadvantages
CCC	Good corrosion resistance	High toxicity of its main compound
Cr(III)	Resistance to heating superior to CCC	Less corrosion resistance than CCC Thinner layer
Phosphate	Less affected by heat than CCC	Small improvement of the corrosion resistance
Zinc phosphate	Corrosion resistance equal to CCC	Problem of adhesion of the coating to the substrate
PCC (permanganate conversion coating)	No bath heating necessary, less affected by heat than CCC	Need to stabilize the pH
Permanganate/phosphate	Better corrosion resistance than CCC	Need to stabilize the pH
Vanadium	Scavenger effect	Expensive
REEs (rare earth elements–based coating)	Self-healing	Corrosion protection sensitive to alloy composition

Table 8. Summary of the advantages/disadvantages for each coating in comparison to the CCC (chromate conversion coating).

http://dx.doi.org/10.5772/66604

Such coatings mixing an oxidizing agent and a precipitating agent should be developed and tested with different combinations to optimize the alternatives to chromate conversion coatings for the protection of magnesium against corrosion. Moreover, in order to develop more efficient coatings without any chromate, it will be necessary to take into account the entire process with the pre-treatment steps and the possible interactions with the magnesium substrate. They should then be validated by tests at industrial scale.

Author details

Jérôme Frayret*, Jean Charles Dupin and Sébastien Pommiers

*Address all correspondence to: jerome.frayret@univ-pau.fr

University of Pau and Adour Countries, IRPEM, UMR, Pau, France

References

[1] B.L. Mordike, T. Ebert, Magnesium: properties — applications — potential, Materials Science and Engineering: A, 302 (2001) 37–45.

[2] R. Ferragut, F. Moia, F. Fiori, D. Lussana, G. Riontino, Small-angle X-ray scattering study of the early stages of precipitation in a Mg–Nd–Gd (EV31) alloy, Journal of Alloys and Compounds, 495 (2010) 408–411.

[3] W. Zhou, D. Shan, E.-H. Han, W. Ke, Structure and formation mechanism of phosphate conversion coating on die-cast AZ91D magnesium alloy, Corrosion Science, 50 (2008) 329–337.

[4] H. Ardelean, I. Frateur, P. Marcus, Corrosion protection of magnesium alloys by cerium, zirconium and niobium-based conversion coatings, Corrosion Science, 50 (2008) 1907–1918.

[5] M. Zhao, S. Wu, J. Luo, Y. Fukuda, H. Nakae, A chromium-free conversion coating of magnesium alloy by a phosphate–permanganate solution, Surface and Coatings Technology, 200 (2006) 5407–5412.

[6] J.E. Gray, B. Luan, Protective coatings on magnesium and its alloys — a critical review, Journal of Alloys and Compounds, 336 (2002) 88–113.

[7] C. Wang, S. Zhu, F. Jiang, F. Wang, Cerium conversion coatings for AZ91D magnesium alloy in ethanol solution and its corrosion resistance, Corrosion Science, 51 (2009) 2916–2923.

[8] M. Menta, J. Frayret, C. Gleyzes, A. Castetbon, M. Potin-Gautier, Development of an analytical method to monitor industrial degreasing and rinsing baths, Journal of Cleaner Production, 20 (2012) 161–169.

[9] S.A. Kulinich, A.S. Akhtar, D. Susac, P.C. Wong, K.C. Wong, K.A.R. Mitchell, On the growth of conversion chromate coatings on 2024-Al alloy, Applied Surface Science, 253 (2007) 3144–3153.

[10] R. Amini, A.A. Sarabi, The corrosion properties of phosphate coating on AZ31 magnesium alloy: The effect of sodium dodecyl sulfate (SDS) as an eco-friendly accelerating agent, Applied Surface Science, 257 (2011) 7134–7139.

[11] C. Askham, A.L. Gade, O.J. Hanssen, Combining REACH, environmental and economic performance indicators for strategic sustainable product development, Journal of Cleaner Production, 35 (2012) 71–78.

[12] X. Zhang, C. van den Bos, W.G. Sloof, A. Hovestad, H. Terryn, J.H.W. de Wit, Comparison of the morphology and corrosion performance of Cr(VI)- and Cr(III)-based conversion coatings on zinc, Surface and Coatings Technology, 199 (2005) 92–104.

[13] W.-K. Chen, C.-Y. Bai, C.-M. Liu, C.-S. Lin, M.-D. Ger, The effect of chromic sulfate concentration and immersion time on the structures and anticorrosive performance of the Cr(III) conversion coatings on aluminum alloys, Applied Surface Science, 256 (2010) 4924–4929.

[14] Y. Song, D. Shan, R. Chen, F. Zhang, E.-H. Han, A novel phosphate conversion film on Mg–8.8Li alloy, Surface and Coatings Technology, 203 (2009) 1107–1113.

[15] Y. Song, D. Shan, R. Chen, F. Zhang, E.-H. Han, Formation mechanism of phosphate conversion film on Mg–8.8Li alloy, Corrosion Science, 51 (2009) 62–69.

[16] N. Van Phuong, S. Moon, Comparative corrosion study of zinc phosphate and magnesium phosphate conversion coatings on AZ31 Mg alloy, Materials Letters, 122 (2014) 341–344.

[17] L.Y. Niu, Z.H. Jiang, G.Y. Li, C.D. Gu, J.S. Lian, A study and application of zinc phosphate coating on AZ91D magnesium alloy, Surface and Coatings Technology, 200 (2006) 3021–3026.

[18] G.Y. Li, J.S. Lian, L.Y. Niu, Z.H. Jiang, Q. Jiang, Growth of zinc phosphate coatings on AZ91D magnesium alloy, Surface and Coatings Technology, 201 (2006) 1814–1820.

[19] H. Umehara, M. Takaya, S. Terauchi, Chrome-free surface treatments for magnesium alloy, Surface and Coatings Technology, 169–170 (2003) 666–669.

[20] S.-Y. Jian, Y.-R. Chu, C.-S. Lin, Permanganate conversion coating on AZ31 magnesium alloys with enhanced corrosion resistance, Corrosion Science, 93 (2015) 301–309.

[21] D. Hawke, D.L. Albright, A phosphate-permanganate conversion coating for magnesium, Metal Finishing, 93 (1995) 34–38.

[22] K.Z. Chong, T.S. Shih, Conversion-coating treatment for magnesium alloys by a permanganate–phosphate solution, Materials Chemistry and Physics, 80 (2003) 191–200.

[23] Y.L. Lee, Y.R. Chu, W.C. Li, C.S. Lin, Effect of permanganate concentration on the formation and properties of phosphate/permanganate conversion coating on AZ31 magnesium alloy, Corrosion Science, 70 (2013) 74–81.

[24] K.H. Yang, M.D. Ger, W.H. Hwu, Y. Sung, Y.C. Liu, Study of vanadium-based chemical conversion coating on the corrosion resistance of magnesium alloy, Materials Chemistry and Physics, 101 (2007) 480–485.

[25] A.S. Hamdy, I. Doench, H. Möhwald, Vanadia-based coatings of self-repairing functionality for advanced magnesium Elektron ZE41 Mg–Zn–rare earth alloy, Surface and Coatings Technology, 206 (2012) 3686–3692.

[26] A.S. Hamdy, I. Doench, H. Möhwald, Smart self-healing anti-corrosion vanadia coating for magnesium alloys, Progress in Organic Coatings, 72 (2011) 387–393.

[27] A.S. Hamdy, I. Doench, H. Möhwald, Assessment of a one-step intelligent self-healing vanadia protective coatings for magnesium alloys in corrosive media, Electrochimica Acta, 56 (2011) 2493–2502.

[28] X. Jiang, R. Guo, S. Jiang, Evaluation of self-healing ability of Ce–V conversion coating on AZ31 magnesium alloy, Journal of Magnesium and Alloys, 4 (2016) 230–241.

[29] L. Li, J. Lei, S. Yu, Y. Tian, Q. Jiang, F. Pan, Formation and characterization of cerium conversion coatings on magnesium alloy, Journal of Rare Earths, 26 (2008) 383–387.

[30] L. Yang, J. Li, X. Yu, M. Zhang, X. Huang, Lanthanum-based conversion coating on Mg–8Li alloy, Applied Surface Science, 255 (2008) 2338–2341.

[31] J.F. Moulder, W.F. Stickle, P.E. Sobol, K. Bomben, Handbook of X-ray Photoelectron Spectroscopy, Perkin-Elmer Corporation (Physical Electronics), 2nd ed., 1992.

[32] T.L. Barr, The nature of the relative bonding chemistry in zeolites: an XPS study, Zeolite, 10 (1990) 760–765.

[33] H. Zhang, G. Yao, S. Wang, Y. Liu, H. Luo, A chrome-free conversion coating for magnesium–lithium alloy by a phosphate–permanganate solution, Surface and Coatings Technology, 202 (2008) 1825–1830.

[34] M. Pourbaix, Electrochemical Equilibrium Atlas at 25°C, Gauthier-Villar, 1963.

[35] R. Smith, A.E. Martell, Critical Stability Constants, Plenum Press, New York and London, 1976.

[36] W.M. Haynes, CRC Handbook of Chemistry and Physics, 83re Edn ed., CRC Press, Floride, 2013.

[37] H. Ma, Y. Berthier, P. Marcus, AES, XPS, and TDS study of the adsorption and desorption of NH_3 on ultra-thin chromium oxide films formed on chromium single crystal surfaces, Applied Surface Science, 153 (1999) 40–46.

[38] M. Oku, S. Suzuki, N. Ohtsu, T. Shishido, K. Wagatsuma, Comparison of intrinsic zero-energy loss and Shirley-type background corrected profiles of XPS spectra for quantitative surface analysis: Study of Cr, Mn and Fe oxides, Applied Surface Science, 254 (2008) 5141–5148.

[39] M. Dabalà, K. Brunelli, E. Napolitani, M. Magrini, Cerium-based chemical conversion coating on AZ63 magnesium alloy, Surface and Coatings Technology, 172 (2003) 227–232.

[40] X. Cui, Y. Yang, E. Liu, G. Jin, J. Zhong, Q. Li, Corrosion behaviors in physiological solution of cerium conversion coatings on AZ31 magnesium alloy, Applied Surface Science, 257 (2011) 9703–9709.

[41] X. Wang, L. Zhu, X. He, F. Sun, Effect of cerium additive on aluminum-based chemical conversion coating on AZ91D magnesium alloy, Applied Surface Science, 280 (2013) 467–473.

[42] L. Kouisni, M. Azzi, M. Zertoubi, F. Dalard, S. Maximovitch, Phosphate coatings on magnesium alloy AM60 part 1: study of the formation and the growth of zinc phosphate films, Surface and Coatings Technology, 185 (2004) 58–67.

[43] T.S.N.S. Narayanan, Surface pretreatment by phosphate conversion coatings—A review, Reviews on Advanced Materials Science, 9 (2005) 130–177.

[44] X.B. Chen, N. Birbilis, T.B. Abbott, Effect of [Ca^{2+}] and [] levels on the formation of calcium phosphate conversion coatings on die-cast magnesium alloy AZ91D, Corrosion Science, 55 (2012) 226–232.

[45] Y. Cheng, H. Wu, Z. Chen, H. Wang, L. Li, Phosphating process of AZ31 magnesium alloy and corrosion resistance of coatings, Transactions of Nonferrous Metals Society of China, 16 (2006) 1086–1091.

[46] N.V. Phuong, K.H. Lee, D. Chang, S. Moon, Effects of Zn^{2+} concentration and pH on the zinc phosphate conversion coatings on AZ31 magnesium alloy, Corrosion Science, 74 (2013) 314–322.

[47] L. Kouisni, M. Azzi, F. Dalard, S. Maximovitch, Phosphate coatings on magnesium alloy AM60: Part 2: electrochemical behaviour in borate buffer solution, Surface and Coatings Technology, 192 (2005) 239–246.

[48] Q. Li, S. Xu, J. Hu, S. Zhang, X. Zhong, X. Yang, The effects to the structure and electrochemical behavior of zinc phosphate conversion coatings with ethanolamine on magnesium alloy AZ91D, Electrochimica Acta, 55 (2010) 887–894.

[49] S.A. Kulinich, A.S. Akhtar, P.C. Wong, K.C. Wong, K.A.R. Mitchell, Growth of permanganate conversion coating on 2024-Al alloy, Thin Solid Films, 515 (2007) 8386–8392.

[50] C.S. Lin, C. Lee, W. Li, Y. Chen, G. Fang, Formation of phosphate/permanganate conversion coating on AZ31 magnesium alloy, Journal of The Electrochemical Society, 153 (2006) B90–B96.

[51] X.-B. Chen, X. Zhou, T.B. Abbott, M.A. Easton, N. Birbilis, Double-layered manganese phosphate conversion coating on magnesium alloy AZ91D: insights into coating formation, growth and corrosion resistance, Surface and Coatings Technology, 217 (2013) 147–155.

[52] A.S. Hamdy, I. Doench, H. Mö ;hwald, The effect of vanadia surface treatment on the corrosion inhibition characteristics of an advanced magnesium Elektron 21 alloy in chloride media, International Journal of Electrochemical Science, 7 (2012) 7751–7761.

[53] X. Yang, G. Wang, G. Dong, F. Gong, M. Zhang, Rare earth conversion coating on Mg–8.5Li alloys, Journal of Alloys and Compounds, 487 (2009) 64–68.

[54] S.-X. Yu, J.-Y. Cao, L. Chen, J. Han, R.-J. Zhang, Corrosion resistance, composition and structure of RE chemical conversion coating on magnesium alloy, Transactions of Nonferrous Metals Society of China, 18, Supplement 1 (2008) s349–s353.

[55] T. Takenaka, T. Ono, Y. Narazaki, Y. Naka, M. Kawakami, Improvement of corrosion resistance of magnesium metal by rare earth elements, Electrochimica Acta, 53 (2007) 117–121.

[56] A.L. Rudd, C.B. Breslin, F. Mansfeld, The corrosion protection afforded by rare earth conversion coatings applied to magnesium, Corrosion Science, 42 (2000) 275–288.

Section 4

Application

Chapter 7

Cardiovascular Applications of Magnesium Alloys

Tobias Schilling, Michael Bauer, Leslie LaLonde,
Hans Jürgen Maier, Axel Haverich and
Thomas Hassel

Additional information is available at the end of the chapter

http://dx.doi.org/10.5772/66182

Abstract

Therapy in cardiovascular medicine often relies on implantation of prosthetic materials or application of stents. The diseases of many cardiovascular structures require their complete and immediate repair by utilising prosthetic materials. The ideal cardiovascular prosthesis involves good functional properties, capability of regeneration and does not activate the host's immune system. Ideally, the graft can be applied for a temporary use and degrades after a predefined period according to controlled degradation kinetics. Only biological grafts would provide this spectrum of properties by today's level of knowledge. However, biological prostheses exhibit some relevant drawbacks as well, such as insufficient mechanical stability or restricted availability. Implants or supporting structures of magnesium alloys would bridge this gap and would either provide a substrate for innovative and temporary grafts or would—as supporting structures—transiently add some missing properties to regenerative biological prostheses. This chapter reviews the different fields of cardiovascular therapeutic applications of magnesium alloys. The required properties of magnesium alloys and their preparation, fabrication and testing will be discussed under the specific cardiovascular perspective.

Keywords: cardiovascular prostheses, cardiac surgery, aortic surgery, aortic aneurysm, congestive heart failure, coronary heart disease, sternum cerclage wires, cardiology, stents, degradation kinetics, magnesium alloys, coating, in vitro test, in vivo test

1. Introduction

The treatment of cardiovascular disease often requires the implantation of prosthetic material [1–7] or insertion of stents [8–14]. The ideal cardiovascular prosthesis has good functional properties and the ability to regenerate, without activating the immune system of the host organism [15]. In some cases, the implant should only remain in place for a limited period of time in the patient, followed by controlled degradation. Up to date, only biological grafts meet most of the spectrum of these relevant requirements (see **Figure 1** and **Table 1**). However, biological prostheses have specific disadvantages, such as insufficient mechanical stability or limited availability (see **Figure 1** and **Table 1**).

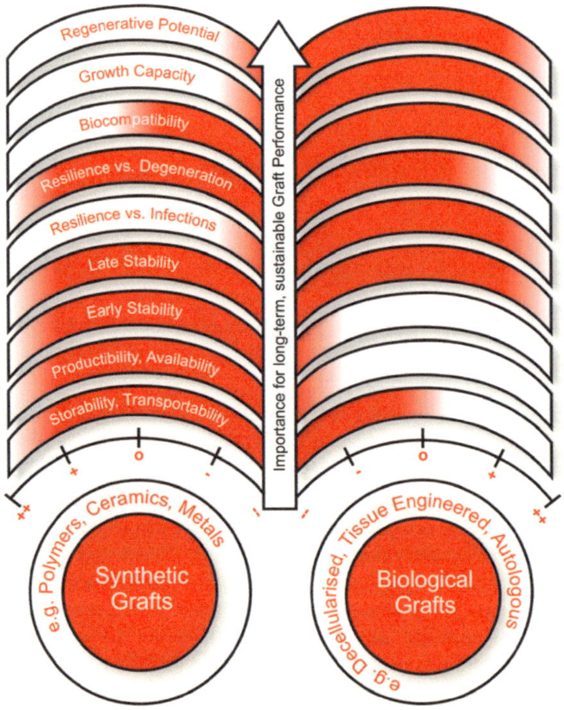

Figure 1. Rough estimation of the average characteristic and extent of relevant properties of grafts made of synthetic or biological materials that influence sustainable and long-term graft performance in cardiovascular applications (Properties of individual grafts may of course deviate). + +, excellent/very high; +, good/high; o, medium/average; –, bad/low; – –, insufficient/very low.

Tissue	**Application, success (+) and limitation (–)**	**Year**	**Authors**
Cardiac muscle	Reconstruction RVOT (rat model) + Good functional and histomorphological outomes – Not applied in left ventricular high pressure system	2012	Wainwright et al. [16]
Skeletal muscle	Epicardial fixation of skeletal muscle (human) + Better vascularisation of myocardium – No transmural reconstruction	1935	Beck et al. [17]
Pericardium	Transmural reconstruction of right ventricle (rat model) + Vascularisation of patch and presence of cardiomyocytes	2007	Chang et al. [18]

Tissue	Application, success (+) and limitation (–)	Year	Authors
	– Not applied in left ventricular high pressure system		
Diaphragm	Epicardial fixation of autologous diaphgram (dog model) + Optimised hemodynamics – No transmural reconstruction	1973	Kusaba et al. [19]
Urinary bladder	Transmural right ventricular reconstruction (pig model) + Improved contractility, repopulation of patch with host cells – Not applied in left ventricular high pressure system	2008	Ota et al. [20]
Peritoneum	Epicardial fixation of stem cells and peritoneum (rat model) + Improved left ventricular function, immigration of cells – No transmural reconstruction	2006	Huang et al. [21]
Myometrium	Epicardial fixation of uterine myometrium (rabbit model) + Good biological integration and angiogenesis of patch – No transmural reconstruction	2008	Taheri et al. [22]
Small intestine	Transmural reconstruction of right ventricle (pig model) + Improved contractility, cardiomyocytes found in patch – Not applied in left ventricular high pressure system	2009	Tudorache et al. [5]
Stomach	Epicardial fixation of stomach on left ventricle (pig model) + Improved angiogenesis of infarcted area – No transmural reconstruction	2003	Ruel et al. [23]
Aorta	Decellularisation of xenogeneic aorta segments (mouse model) + Adhesion of cells – Small diameter, small animal model, no implantation into circulation	2016	Song et al. [24]
Natural polymers	Engineering of small diameter bypass grafts with degradable polymers (rabbit model) + Cell infiltration, good graft patency – Implantation in carotid position (lower stress than in aorta)	2016	Antonova et al. [25]
Fibrin	Engineering of vascular grafts with compacted fibrin (human) + On-the-fly generation of regenerative grafts – Mechanical stability up to 230 mm of mercury	2016	Aper et al. [26]
Induced pluripotent stem cells	Abdominal aorta interposition of engineered blood vessels from induced pluripotent stem cells (rat model) + No rupture of grafts, cellular proliferation – 2 weeks observation period, small animal model	2016	Gui et al. [27]

Table 1. Overview of exemplary studies to test diverse biological materials as grafts for selected cardiovascular applications.

Structures of degradable magnesium alloys could temporarily compensate for such disadvantages, through means of complementing some of those missing features of biological prostheses (see **Figure 2**).

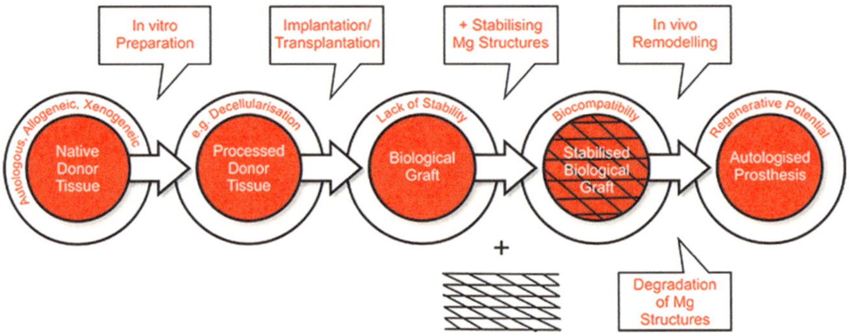

Figure 2. Schematic principle of temporarily stabilising biological grafts with degradable magnesium structures, resulting in highly biocompatible, autologised, and therefore regenerative prostheses.

However, implants made of magnesium alloys remain in an outsider position with regard to cardiovascular medicine so far. While the experimental application of coronary stents made of magnesium alloys was performed for some decades [11], degradable magnesium has hardly ever played a role in the development of other cardiovascular implants.

This chapter introduces scenarios for a beneficial application of degradable magnesium implants or supporting structures in clinical settings. Desirable and essential requirements of the materials are presented here as well as research emanating from experimentally well studied alloys in other disciplines. This chapter does not claim to be exhaustive, but rather to convey the essential principles that need to be considered in the development of cardiovascular implants made of magnesium alloys for various application scenarios.

2. Fields of potential applications for magnesium alloys in cardiovascular medicine

2.1. Coronary heart disease

Coronary heart disease (CHD) is associated with the pathological reduction in the lumen of the coronary arteries, which can eventually lead to a total closure and thus the clinical picture of the life-threatening myocardial infarction. Cardiovascular diseases are among the most common causes of death. According to WHO, in 2012, about 17.5 million people died from cardiovascular disease, of which approximately 7.4 million can be attributed to a coronary heart disease [28].

The treatment of coronary heart disease is dependent on the extent of the disease. The common guidelines of the European Society of Cardiology and the European Society for Cardiothoracic

Surgery recommend interventional balloon dilation of stenotic vessels or the application of coronary stents for patients with one- or two-vessel-CHD [29]. For this purpose, a tubular metal structure under X-ray fluoroscopy is positioned into the affected coronary artery, usually through a femoral arterial puncture. Subsequently, the initially crimped stent is dilated with an inflatable balloon via the same catheter. The specific framework design and a partial plastic deformation of the stent ensure that this does not collapse.

According to current studies, patients with left main coronary artery disease or two- or three-vessel-disease benefit more in the long term from surgical coronary artery bypass grafting [12–14, 30, 31]. Segments of autologous saphenous vein, radial artery or internal mammary artery are herewith connected via a proximal anastomosis to the ascending aorta and via a distal anastomosis to the affected coronary vessel behind the stenosis. This intervention takes place, in most cases, using a heart-lung machine. In heart-lung machine interventions, systemic anticoagulation with heparin is necessary to prevent clotting of blood along the long synthetic tube surfaces of the machine. Patients who have to undergo a surgical revascularisation due to a secondary closure of the inserted stent are subject to a higher perioperative risk of bleeding because of chronic, medical, dual anticoagulant therapy with platelet inhibitors, and vitamin K antagonists [32]. Furthermore, the surgeon is faced with a sealed coronary in such cases, where—because of the inserted metal stents—only a few sites to connect the bypass graft remain. But even in cases where no secondary surgical intervention is required, permanent stents may have a chronically negative impact. This may lead to a long-term endothelial dysfunction, a permanent physical irritation, the release of toxic metal ions or a local chronic inflammation [11]. Removal of these stents poses a significant risk to the patient and is associated with high additional costs.

In the last two decades, therefore, innovative stents from polymers [33, 34] and degradable metals such as iron and magnesium were developed, which initially ensure the openness of the coronary vessel and are degraded without residue after a sufficient regeneration phase [35–40]. The coating of degradable metal stents should slow the corrosion [35]. The hydrolytic degradation in vivo of certain polymer coatings as PLA or PGA, however, leads to an acceleration of the metal corrosion because of the acidic degradation products, such as carboxylic acid acidifying the milieu, which promotes the oxidation of the magnesium alloys [41, 42]. Bioabsorbable stents made of magnesium alloys [43] with an absorbable polymer coating, from which antiproliferative drugs are released represent a promising future treatment alternative. With this combination, the in-stent restenosis, which occurs in 20–30% of patients should be prevented whilst exploiting the advantages of a transient implant.

2.2. Congestive heart failure

Heart failure is associated with an insufficient function of the heart whereby the heart is no longer able to provide the required output. Between 800,000 and 1,600,000 people in Germany suffer from heart failure [44]. According to the Federal Statistics Office, heart failure is currently the third leading cause of death in Germany [45], resulting in costs of €3.2 billion in 2008 [46], showing that heart failure is also of economic relevance. As a result of demographic change and improved chances of survival from triggering diseases—for example, after a heart attack

—this number is expected to increase in the future, leading to an increase in related health expenditures [47].

Surgical treatment of severe heart failure focuses on regaining a sufficient pumping function primarily of the left ventricle, which results from a physiological volume and the natural geometry of the left ventricle [48]. Since the 1980s, therefore, myocardial reconstruction according to Dor has been applied [6, 7] whereby, the damaged, non-functioning heart tissue is resected and reinforced by synthetic grafts such as Dacron. The use of grafts currently available allows for reconstruction of the physiological volume of the left ventricle, but not the ellipsoidal shape, which is required for an optimal pumping function [4]. In addition, the currently available synthetic materials are neither capable of growth nor regeneration and do not contribute actively to the left ventricular function. The dyskinetic cardiac tissue is therefore replaced by equally dyskinetic grafts. Innovative surgical procedures that address the use of regenerative materials, the physiological volume, and the physiological ellipsoidal shape are therefore necessary in order to offer the patients a sufficient and sustainable therapy.

In this sense, the replacement of damaged myocardium has been tested with decellularised bladder [49], skeletal muscle [50], intestinal mucosa [5], stomach [23, 51] (see **Figure 3** and **Table 1**) amongst others [52].

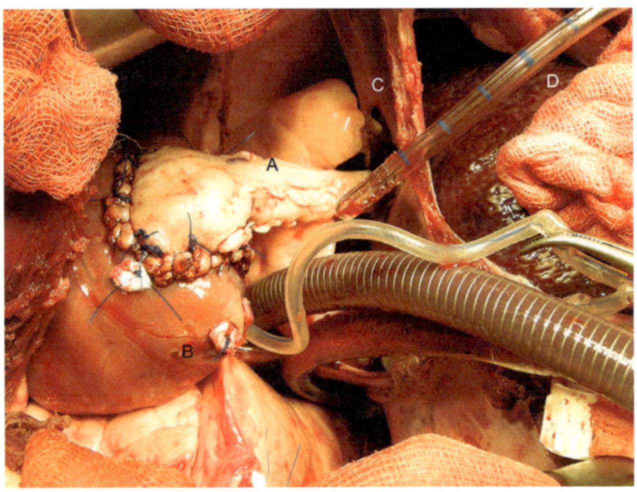

Figure 3. Transdiaphragmatic autologous transplantation of a vascularised segment of the stomach as a myocardial patch for the anterolateral facies of the left ventricle. (A) Gastric patch; (B) apex cordis; (C) diaphgram; (D) spleen.

The feasibility of a clinical application of these potentially regenerative grafts has already been shown for the replacement of right ventricular or atrial cardiac muscle tissue. However, the mechanical strength of most of these biological grafts is too low for use in the left ventricular high-pressure area with up to 240 mmHg of blood pressure. Especially in the early phase after implantation, there is a great risk for an aneurysm or even a fatal rupture of the delicate, prosthetic tissue. Nevertheless, a physiological transformation process of heterotopically applied tissue has been ascertained, which raises hope for an acquisition of specific cardiac functions, but also an increase in the mechanical stability [5]. Stabilising structures from degradable magnesium structures could support biological grafts by attaining sufficient mechanical stability through the physiological transformation, thus making regenerative therapy options available to all needy patients [53].

Implants such as the Paracor HeartNet or the Acorn CorCap can be used for left ventricular support and reduction in the wall load. The aim is to reduce the mechanical stress acting on the (damaged) heart wall with reticular structures that span the epicardial layer of the myocardium, so that the high intraventricular blood pressure cannot further dilate the myocardial wall. The Acorn CorCap is a polyester net that is placed circumferentially from the apex to the atrioventricular fossa around the heart. The network supports end-diastolic resistance and reduces left ventricular wall stress, thus preventing further dilatation of the ventricle [54, 55]. The nitinol elastic Paracor HeartNet device acts in a similar manner [56, 57].

2.3. Aneurysm and dissection of the aorta

Congenital (Marfan syndrome, Ehlers-Danlos Syndrome) and acquired (atherosclerosis) disease of the aorta lead to about 7000 surgical procedures annually in Germany and thus to a high demand for prosthetic material [58]. In acute aortic dissection, the true number of cases is often underestimated because many patients die from unknown causes of death before reaching the hospital [59]. The incidence of diseases of the aorta in 2011 consisted of 34 hospital-based main diagnoses per 100,000 inhabitants. Of this amount, 17.4% belonged to the group of aortic dissection, which corresponds to an incidence of 5 per 100,000 inhabitants in Germany.

In particular, in many cases, aortic aneurysms (ballooning of the aortic wall) or aortic dissection (tearing of the aorta) require surgical therapy. However, the surgical treatment of aortic disease does not always lead to complete cure of patients. In a study over a period of 14 years, it was shown that reoperation is required in approximately 13% of all performed surgical procedures of the thoracic aorta. Reoperation may be required for progressive aneurysms, persistent or recurrent aortic dissection or false aneurysms at the suture lines, on the one hand, while on the other hand, stenoses and infections of implanted grafts may occur and proximal or recurrent aneurysms distal to the grafts may be observed, requiring a reoperation [60].

2.3.1. Aortic prostheses and their limitations

In the last 50 years, aortic prostheses were produced essentially of polyethylene terephthalate (PET, Dacron), polytetrafluoroethylene (PTFE, Gore-Tex®) or polyurethanes (PU) [61]. The disadvantages of synthetic vascular grafts include the risk of thromboembolism, a subsequent stenosis and occlusion of the prosthesis by intimal hyperplasia. Infections resulting from synthetic implants are difficult to handle. In addition, prostheses consisting of artificial materials have no regenerative potential and are not able to grow [62, 63]. Finally, synthetic grafts have a low elasticity, so that the function of Windkessel of the aorta, which is responsible for the uniform and muted forwarding of the pulse wave, is restricted. Especially in cases of the application of very extensive vascular grafts, there may occur refractory hypertension, which can lead to further organ damage in the long term (see **Figure 4**).

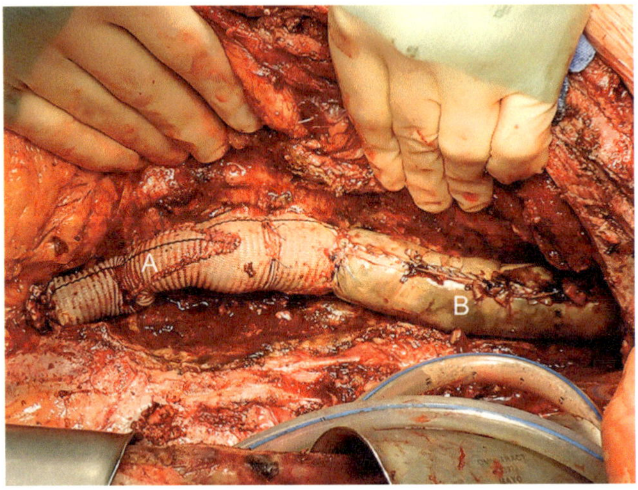

Figure 4. Operative situs following replacement of thoracic (with Dacron prostheses (A), left side) and abdominal (Pericardium graft (B), right side) aorta. Now re-operation because of graft infection.

Bio prostheses or hybrid grafts represent an alternative to synthetic implants [3, 64, 65]. The gold standard for biological grafts is allogeneic native vessels made from fresh and cryopreserved human donor tissue. Fresh homografts, which tend to have a later expiration than the cryopreserved homograft, bear the risk of forming aneurysms and are only of limited availability. The cryopreserved homograft is prone to calcific degeneration approximately 10–15 years following implantation. Usage in younger patients therefore entails at least one reoperation [66, 67]. The long-term degeneration of biological implants is thus far, a significant, unresolved limitation of these prostheses [68].

2.3.2. The ideal vascular prosthesis

Artificial blood vessels, which consist of viable tissue, constitute the ideal vascular prosthesis. An ideal prosthesis must have physiological mechanical resistance and compliance. The graft should be non-toxic and non-immunogenic. The graft must be biocompatible and always available in various sizes. In addition, the graft should be surgically easy to handle. It should not, if possible, trigger any thrombogenic effects. The graft must integrate into the host tissue and have the ability to grow, when used in children [69–71]. Furthermore, an easy manufacturing process, the shelf life of the vascular prostheses and readily available supply for an acute emergency, all play very important roles. According to the principles of tissue engineering, innovative biological aortic prostheses can be produced, which can overcome the previous limitations of available grafts. One option is the implantation of a decellularised allogeneic or xenogeneic donor aorta. After the original cells have been removed from the donor tissue chemically or enzymatically, these can colonise after implantation in vivo with cells of the host organism. In this way, a regenerative, biological implant, which will not be attacked immunologically by the host, is converted successively to truly endogenous tissue in the course of

http://dx.doi.org/10.5772/66182

the physiological remodelling. Tubular structures of fibrin and collagen can be constructed in vitro as an alternative to decellularisation of preformed donor tissue [26, 72].

However, both grafts have a limited mechanical stability at the time of implantation. To avoid premature rupture or aneurysm formation, a stabilising biodegradable clip of magnesium could be placed around the biological graft until the latter reaches sufficient stability. Subsequently, this support structure of magnesium should be degraded, preferentially in a biologically inert manner.

2.4. Closure of sternotomy following cardiac surgery

In cardiac surgery, many procedures are already performed using minimally invasive approaches that do not require traditional median sternotomy. But the indication for minimally invasive access paths is still conservative due to the difficulty involved in the operation [73, 74]. In addition, most cardiac surgical interventions are not isolated, but combined procedures. The average age of cardiac surgical patients is steadily increasing, and in many cases, the requirement is not just for a single heart valve or bypass surgery, but rather a combination of both procedures or the replacement of several valves all taking place in a single operative session [75]. In such cases, the safest and most effective access path is the median sternotomy. Here, the sternum is opened with an oscillating bonesaw starting from the xiphoid process—the lowest point of the sternum—to the manubrium sterni. After the surgical treatment of the heart, the separate halves of the sternum are brought together with non-degradable metal wires and fixed under tension. For this purpose, the wires are guided laterally from the sternum with a needle through the intercostal spaces. Thereafter, the free ends of the wires are twisted with pliers. The twisted ends are placed as flat as possible along the bony structures (see **Figures 5** and **6**).

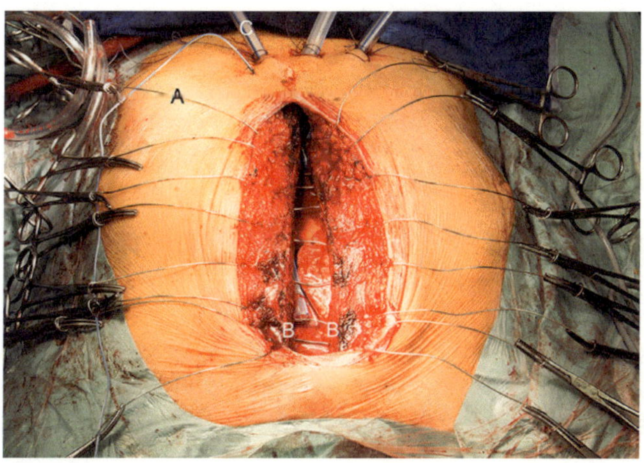

Figure 5. Intraoperative situs. Closure of sternotomy (B) with conventional sternum cerclage wires (A). Cerclages already inserted and just prior to fixing and drilling. Chest tubes in position (C).

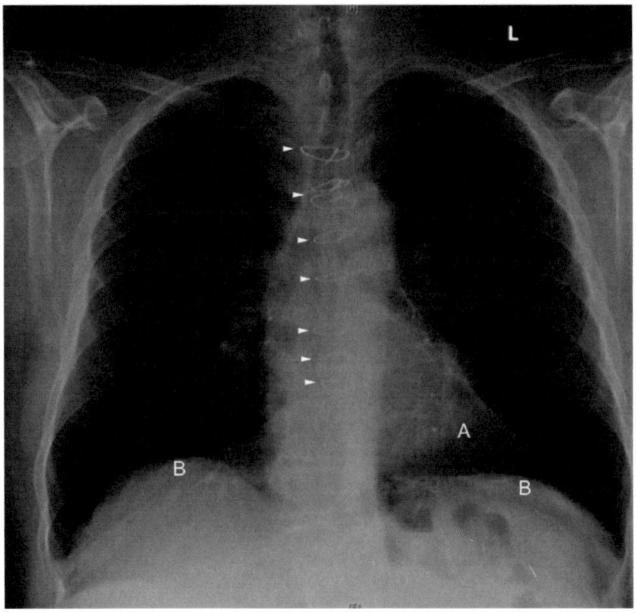

Figure 6. X-ray of thorax in posterior-anterior beam path with clearly visible sternum cerclage wires (arrow heads). (A) Apex cordis; (B) diaphragm.

2.4.1. Limitations of current available sternal cerclage wires

In some cases, these wires lead to postoperative complications. Some patients perceive the twisted ends of sternal cerclage as truly painful [76]. Other patients can get the twisted ends with its tip to curve in the direction of body skin, and even penetrate it by mechanical stress or remodelling of the bony skeleton. This leads to impairment of wound healing, wound pain and wound infections, which, under unfavourable configuration can lead to secondary bone or mediastinal infections. The latter require an extremely burdensome, tedious and sometimes highly complex wound therapy often using reserve antibiotics. Multiple re-operations for rehabilitating serious findings in such cases constitute a very high-risk stress factor for patients. Even the development of sternal cerclage systems made of polyester did not lead to an improvement of the postoperative complication rate or the subjective well-being of the patients [77].

2.4.2. Degradable sternal cerclage and release of active agent

Against this background, the ideal cerclage wire material consists of controllable degradable metal. The healing of the bone to physiological stability following sternotomy requires about 3–6 months. A degradation process of resorbable sternal cerclage should lie roughly within this time frame.

It is conceivable that the coating of degradable magnesium wires would not only provide an additional protective layer for the magnesium but would allow the possibility for controlled local delivery of drugs. First, cardiac surgical patients immediately receive painkillers after surgery to alleviate the pain [78]. It is conceivable that for a limited period of time, local anaesthetics could be released from the coatings of magnesium wires. This could lead to considerable local relief of postoperative pain and perhaps reduce the administration of systemically acting analgesics. Second, the infection of the sternotomy wound is a common complication after cardiac surgery, which is often associated with mediastinitis and the most protracted wound healing problems with instability of the sternum [79]. A reoperation for revision of the infected and unstable sternum is essential in such cases and not only provides an extremely stressful engagement with additional risks and pain for the patient himself but is usually directly associated with increased postoperative mortality [80]. Although sternal infections are largely reduced to a minimum following utmost care and hygiene, this burdensome and costly complication cannot be avoided in all cases [81]. Particularly, elderly patients or patients with an impaired immune system and diabetics are at an increased risk of infection where wound healing processes are complicated due to microcirculatory disorders [82]. The immediate postoperative period, commencing with the release of antibiotic or at least bacteriostatic substances from the wire cerclages used in the closure of the sternum, could further reduce the likelihood of wound infection [83]. Robinson et al. could show that—at least in in vitro experiments—the degradation of magnesium and the degradation products exert an inhibitory effect on the growth of bacteria. The local increase of pH in the medium as a result of degradation corresponded approximately to the effect of an antibiotic [84].

3. Magnesium alloys for cardiovascular implants

Magnesium alloys are currently being investigated for use as biodegradable implants. The implants should take over the desired function for the needed duration and then biologically degrade as biocompatibly as possible. Magnesium, although a very promising material is, in principle, far from being perfect material for implants because it is electrochemically highly active and easily corrodes in biological milieus [85]. The corrosion of magnesium is therefore to be controlled by alloying with other components. The most-studied alloy components for magnesium are calcium, aluminium, lithium and rare earth (RE) elements [39, 86]. Cardiovascular implants will also benefit from the modification of magnesium by these alloys, because a uniform and temporally controllable corrosion as well as a long-term mechanical stability or even elasticity is a common requirement for most implants.

Calcium is the most abundant mineral in the human body. It is mainly found in the bones and teeth [39]. In the alloy with magnesium, there is a reduction in the tendency to dissolve. It also leads to a refinement of the grain structure [39, 87]. Wan et al. found an increase in the compressive and elastic moduli. Wu et al. were able to show that the addition of calcium to a magnesium-aluminium-zinc alloy improved the corrosion properties of this alloy [88]. With an alloy of 9 ma.-% Al, 1 ma.-% Zn, 1 ma.-% REE and 1 ma.-% Ca, they were able to achieve the best results and to reduce the corrosion rate compared to the AZ91 alloy without calcium by 19% [88]. Due to their good biocompatibility, good mechanical properties and desirable influence on the corrosion kinetics, magnesium-calcium alloys constitute a particularly attractive material for implant development.

Aluminium is used for alloying of magnesium, because it increases the corrosion resistance of magnesium [89]. A corrosion product of aluminium in physiological solution is Al_2O_3, which forms an insoluble outer protective layer on the alloy [87]. Moreover, aluminium tends to precipitate and reduces the tendency to dissolve [39].

Alloys with rare earth elements also often show positive results, although these are not always easy to reproduce. This is also because rare earth elements are often used as so-called master alloys for the production of a magnesium alloy, which already contains a mixture of many rare earths [39]. While this system is used for industrial applications without major limitations, the exact composition of the material must be known in the development of medical implants, as even minor changes can dramatically impair the function and longevity of the product that is used in a biological environment [90]. The effects of rare earth on solubility and precipitation behaviour are mediated by the formation of intermetallic phases. This results in an improved corrosion resistance, creep resistance, as well as an optimisation of the mechanical properties [39, 88, 91]. Following the implantation of magnesium alloys, a foreign body reaction involving macrophages and giant cells (giant cells) takes place. The pH in phagolysosomal vacuoles of macrophages and giant cells can be as low as 3, therefore, their activity results in an acidification of the granulation tissue surrounding the magnesium implant after a short time. A low pH of the environment in turn accelerates the oxidation of the magnesium. Therefore, lithium is used in magnesium alloys in order to stabilise the corrosion layer on its alkalising effect [86]. This effect during the degradation of magnesium is further enhanced by the solution and accumulation of magnesium hydroxide as the main degradation product [92].

4. Degradation of cardiovascular implants

The degradation of magnesium implants influences the host's organism through the (corroded) surface, the release of degradation products such as magnesium ions, hydroxide ions and hydrogen gas [87, 93]. Both the gas development and the alkalisation of the medium by the released hydroxide ions lead to an enhanced localised corrosion as well as to a reduced cell adhesion [85, 94].

There are four basic types of corrosion possible for Mg alloy biomaterials [95]. **Galvanic corrosion**, also known as couple corrosion, occurs when two metals with differing electrochemical potentials are in electronic contact in an ionic conduction fluid such as serum or interstitial fluid [95]. This kind of corrosion plays only a minor role in cardiovascular applications, because cardiovascular implants are basically made of only one alloy. **Pitting corrosion** occurs when small, distinct areas of a material are rapidly corroded while most of the surface remains unaffected. Corrosion pits can also cause crack formation, leading to further breakdown. Magnesium is very susceptible to pitting corrosion. **Corrosion fatigue** takes place in materials undergoing cyclic loading, such as the pressures that would occur under normal blood flow past a stent. This form of corrosion is heavily dependent on a material's microstructure. Whether testing is carried out in a dry or aqueous environment strongly influences a material's performance under cyclic loading conditions. **Erosion corrosion** occurs due to the mechanical action of surrounding liquids or particles. Although not as influential as the other corrosion types listed here, erosion corrosion must also be considered when discussing biodegradable implants [95].

Corrosion rate and type are very important when evaluating an implant's performance. It is of paramount importance that localised corrosion attack in stents is avoided, as localised corrosion in the thin portions of the device can lead to premature loss of mechanical integrity [96].

Moreover, the rapid corrosion of magnesium implants leads to undesirable formation of gas bubbles in the tissue [37, 86, 97–99]. Song et al. determined a rate of 0.01 $ml/cm^2/day$ as a conservative, acceptable level for the development of hydrogen gas in the human body [100]. The effects of gas generation vary with the implantation site. In a hardly flexible bony environment, the rapid development of high volumes of hydrogen gas undoubtedly has serious effects on bone remodelling due to an increase in pressure. In an epicardial application of magnesium structures, negative effects due to the development of higher gas volumes are hardly to be expected because of the relatively larger space available in the pericardium and mediastinum. In intraluminal stents, there is an immediate and effective removal of the released gas through the blood. Adventitially fixed clips for stabilising biological, large lumen aortic prostheses find sufficient thoracic as well as abdominal space into which the released gas can diffuse without damaging the surrounding tissue [101].

5. Biocompatibility

Many of the early Mg-based implants researched were pure magnesium. In truth, "pure" Mg will always have some level of impurity. The characteristic impurities in magnesium alloys are copper, nickel, iron and beryllium. The composition and production of the alloys will influence the levels of these impurities [39]. In general, it is best to have as low an impurity level as possible, not only because some of these impurities are considered carcinogenic or can become toxic at higher levels, but also because having higher levels of uncontrolled impurities can skew results. With impurities as well as alloying or coating elements, it is important to keep in mind the tolerable upper intake levels of these elements in humans so that improved strength or corrosion resistance does not come at the expense of biocompatibility. Pure Mg alloys for biomedical applications have mostly been replaced by other alloys, though some studies still contain pure magnesium, often as a control [102–105].

Pure magnesium is biocompatible in the highest degree. It is ubiquitous in the human body and plays an important role as a cofactor in many enzymatic reactions [106]. Calcium is also a cofactor for biochemical reactions and an essential component of bone metabolism. Therefore, it shows a comparably higher biocompatibility as an alloying component.

Magnesium aluminium alloys were initially developed for non-biomedical applications. As such, their biocompatibility was initially not important. Naturally occurring aluminium in the human body is practically non-existent. The inclusion of high concentrations of aluminium, for example, in the metal working industry or with drugs, is associated with the aetiology of neurodegenerative diseases [39, 107–109]. The required concentrations for those effects are not to be expected during the degradation of relatively small cardiovascular implants made of alloys that usually contain only small amounts of aluminium. Above all, the degradation process ideally extends over a period of months, and only small amounts of aluminium are released in given intervals [101]. Following the implantation of epicardial degradable support structures of the magnesium alloy LA63, which contains lithium and aluminium, no adverse effects on the surrounding myocardium were determined in a pig model [101]. Feyerabend et al. found that aluminium and lithium exert toxic effects on perivascular cells only at concentrations above 1000 micromole [90]. Except for gadolinium, rare earth elements are only toxic at high concentrations, although calcium antagonistic effects could be shown, which in principle could affect many biochemical reactions, particularly muscle and heart muscle contractions as well as cardiac conduction [110]. Many RE elements have even been said to show some anti-carcinogenic properties [39].

In general, even cardiovascular implants made of magnesium alloys, and their degradation products cause the usual foreign body reactions of the host. The foreign body reaction, which is characterised as an inflammatory healing process through the activity of macrophages and giant cells, represents the natural reaction to the implantation of magnesium alloys [111].

6. Developing innovative implants made of magnesium alloys for cardiovascular application

6.1. Design and shapes

Magnesium is a versatile material that can be used for different cardiovascular implants of almost any shape. Using finite element simulation, various geometric shapes and designs of the stent or support structures can be analysed for possible weak spots [112]. The most promising approaches can be identified with these simulations in many cases prior to costly and tediously manufacturing them. Even the degradation of magnesium alloys can be simulated to a certain extent, at least for in vitro experiments [113].

In support of biological myocardial prosthetic devices, degradable magnesium lattices can be individually adapted to the patient's own geometry and anatomy of the damaged heart region. The preforming of implants reduces the plastic deformation of the implants in vivo, which induces high stress and consequently premature breakage of the implants [114]. Cardiac MRI or CT data could thus be employed to adjust the geometry of magnesium structures to the individual anatomy of each patient [114].

Magnesium alloys can be cut with a very high resolution into almost any shape by means of water abrasive jet cutting, laser or plasma beam. Nevertheless, it is mandatory to take into account the specific properties of each cutting process and its impact on the material. It is certainly possible to achieve finest structures down to 0.1 mm kerf width with a laser beam

[115]. However, the laser and plasma cutting bring about a significant heating of the surrounding material. The resulting melting and oxidation processes can significantly affect the integrity of the structures, so that the a-thermic water abrasive jet cutting, although considerably more complex and expensive, certainly is the gentlest method. Hence, the most precise cuts can be performed with water abrasive jet cutting due to the lack of heat.

6.2. Increasing the corrosion resistance

Three experimental routes are essentially taken in order to control the magnesium corrosion and to optimise the mechanical properties of the implants: The **alloying of magnesium**, the **manufacturing process** such as the reduction of grain size and phase transitions and the **surface modifications and coatings** [85]. Wang et al. provide a compact overview of the basic methods available for surface treatment [35]. They hint to microarc oxidation [116], anodising [117], evaporation [118], alkaline heat treatment [119], fluoride coating [120], electro deposition [121], phosphate coating [122], shock peening [123], ion implantation [124], physical vapour enrichment [125] and polymer coating [126]. The aim of the process is always to reduce the oxidation tendency of the surface of the magnesium alloy or to provide the surface with a protective layer. This protective layer may occur as the result of a reaction product of the alloy, such as magnesium fluoride via immersion of the implant in hydrofluoric acid, or a dedicated polymer layer, which is applied to the metal. To optimise cardiovascular implants made of magnesium alloys, procedures should be used, which are satisfactory for the field of use and are biocompatible. The adhesion of cells should also be made more difficult by an appropriate surface modification, which can be unequivocally desirable for coronary stents as well as for other applications.

6.3. In vitro and in vivo tests

As different alloys and coatings are developed to improve the mechanical and corrosion behaviour of these implants, in vivo and in vitro testing techniques must be developed. Unfortunately, since these techniques have not been standardised, it is difficult to compare results from different studies. There is difficulty in establishing a set corrosion rate for magnesium implants, because the experimental degradation varies based on so many variables, including composition of alloy, manufacturing method, coating, implant shape, in vivo location of implant, type of surrounding tissue, species of testing subject, temperature, local pH, ions or proteins present in an in vitro solution and more [37, 85, 126–128]. The same alloy tested in different in vitro scenarios can yield different corrosion rates, and often, in vitro corrosion tests do not accurately predict in vivo behaviour [93, 129].

Although in vitro experiments are helpful to get a first indication of the desired properties of the alloy, large animal experiments are essential to characterise the degradation of magnesium alloys in vivo before it can be used clinically for humans.

Scientific discoveries in medicine have to be gradually shifted into practical application. This path extends from basic research in the laboratory to preclinical models and ultimately to clinical trials. Preclinical models allow a forecast of possible future clinical reality without the risks and costs associated with clinical trials [130]. Careful review of the biological responses to the implant, and the surgical procedure is necessary especially with complex, in vitro-generated products. It is important to be aware of the overall reaction of the organism to the implant, which may differ significantly from the tissue reactions that had taken place previously, or that have been detected in in vitro tests. This applies in particular to possible immunological reactions.

When choosing an animal model, many factors must be considered, including ease of operation, cost of acquisition and upkeep, susceptibility to disease or infection and similarity to human physiological mechanisms. Researchers may also need to consider the ethics of their animal model choice; some animals, such as dogs, cats, non-human primates, and perhaps horses might fulfil the other requirements of an animal model, but are considered inappropriate for medical research. No one animal model will be appropriate for all purposes, but no model should be dismissed entirely [131, 132]. Animal model size is an important consideration. Larger animals may have more in common with humans, but they will also be more difficult and expensive to care for.

To study cardiovascular implants, the pig can well be used as an experimental animal. The vascular systems of humans and pigs show similar behaviour under acute stress, and the fine structural elements of the porcine heart are particularly useful for morphological studies, given the analogy to the human morphology. Another suitable animal model is provided by the sheep, which serves as a model for rapid calcific degeneration. Degenerative changes can thus be observed particularly well in the sheep model.

In the field of implant biodegradation, there is a further concern: the total amount of fluid in the body seems to influence degradation rate, with less fluid leading to less corrosion. Yamamoto et al. used this line of thinking to discourage the use of small animals as models for implant corrosion studies [133]. Thus, they recommend not using small animal models to perform corrosion studies.

Once a more accurate in vitro testing protocol is designed and implemented by as many research groups as possible, magnesium alloy implant technology will finally be able to move forward as a united front, rather than a plethora of loosely connected individual lines.

7. Conclusion and outlook

This chapter gives a brief overview of potential cardiovascular applications of different magnesium alloys. Stents, grafts, support structures and sternal cerclage of degradable magnesium provide innovative and promising approaches for combating cardiovascular diseases, which are not only among the leading causes of death worldwide, but significantly reduce quality of life of affected patients and represent a high financial burden on healthcare systems worldwide. Cardiovascular implants made of magnesium alloys are currently still in an experimental or preclinical stage, but have a high potential in the foreseeable future for clinical applications, since magnesium can easily be processed and offers a good biocompatibility.

http://dx.doi.org/10.5772/66182

There is a significant need for temporary, degradable implants in cardiovascular medicine. The drug-eluting degradable magnesium stent is just the next logical development of the currently available magnesium stents. Support structures using magnesium catalyse on the clinical use of renewable biological myocardial prostheses enabling the survival of patients with terminal heart failure who cannot receive a donor organ due to donor organ shortage. The same applies to patients who have to undergo an aortic intervention. Biological aortic prostheses that might exhibit low mechanical stability in the early phase after implantation could be made available to a large number of patients through stabilising, degradable magnesium clips. Finally, some months after successful bony healing of the sternum following sternotomy, it is in principle no longer necessary that the sternal cerclages remain in the patient and bear risks such as pain, infection and dermal puncturing. The development of degradable wires made of magnesium alloys is therefore a predictable consequence of the ongoing highly active magnesium research.

The high variety of available alloys, coatings, manufacturing processes and the high variability of the myriad of potential in vitro and in vivo test methods—including a plethora of different result parameters to be investigated—currently complicate the comparability of studies for the analysis of magnesium implants. The development of analytical standards would not only unify magnesium research, making studies comparable, but would give an immense boost to the field. With the expected advances in magnesium research, there is hope to offer reliable, renewable and sustainable therapeutic options to the millions of patients with cardiovascular disease in the near future.

Author details

Tobias Schilling[1*], Michael Bauer[2], Leslie LaLonde[2], Hans Jürgen Maier[2], Axel Haverich[1] and Thomas Hassel[2]

*Address all correspondence to: schilling.tobias@mh-hannover.de

1 Department of Cardiothoracic, Transplantation and Vascular Surgery, Hannover Medical School, Hannover, Germany

2 Institut für Werkstoffkunde (Materials Science), Leibniz Universität Hannover, Garbsen, Germany

References

[1] Martens A, Beckmann E, Kaufeld T, Umminger J, Fleissner F, Koigeldiyev N, et al. Total aortic arch repair: risk factor analysis and follow-up in 199 patients. Eur J Cardiothorac Surg. 2016.

[2] Hagl C, Khaladj N, Peterss S, Bonz A, Pichlmaier M, Haverich A, et al. Treatment of acute aortic dissection type A (AADA): technical considerations. Vasa. 2010;39(3): 212–8.

[3] Kirsch ME, Ooka T, Zannis K, Deux JF, Loisance DY. Bioprosthetic replacement of the ascending thoracic aorta: what are the options? Eur J Cardiothorac Surg. 2009;35(1):77–82.

[4] Calafiore AM, Iaco AL, Abukoudair W, Penco M, Di MM. Left ventricular surgical remodeling after the STICH trial. Thorac Cardiovasc Surg. 2011;59(4):195–200.

[5] Tudorache I, Kostin S, Meyer T, Teebken O, Bara C, Hilfiker A, et al. Viable vascularized autologous patch for transmural myocardial reconstruction. Eur J Cardiothorac Surg. 2009;36(2):306–11.

[6] Dor V. Surgery for left ventricular aneurysm. Curr Opin Cardiol. 1990;5(6):773–80.

[7] Dor V, Saab M, Coste P, Kornaszewska M, Montiglio F. Left ventricular aneurysm: a new surgical approach. Thorac Cardiovasc Surg. 1989;37(1):11–9.

[8] Tenekecioglu E, Bourantas C, Abdelghani M, Zeng Y, Silva RC, Tateishi H, et al. From drug eluting stents to bioresorbable scaffolds; to new horizons in PCI. Expert Rev Med Devices. 2016;13(3):271–86.

[9] Lindholm D, James S. Bioresorbable stents in PCI. Curr Cardiol Rep. 2016;18(8):74.

[10] Testa L, Latib A, Montone RA, Colombo A, Bedogni F. Coronary bioresorbable vascular scaffold use in the treatment of coronary artery disease. Circ Cardiovasc Interv. 2016;9(7). DOI: 10.1161/CIRCINTERVENTIONS.116.003978.

[11] Bowen PK, Shearier ER, Zhao S, Guillory RJ, 2nd, Zhao F, Goldman J, et al. Biodegradable metals for cardiovascular stents: from clinical concerns to recent Zn-Alloys. Adv Healthcare Mater. 2016;5(10):1121–40.

[12] Serruys PW, Morice MC, Kappetein AP, Colombo A, Holmes DR, Mack MJ, et al. Percutaneous coronary intervention versus coronary-artery bypass grafting for severe coronary artery disease. N Engl J Med. 2009;360(10):961–72.

[13] Kappetein AP, Feldman TE, Mack MJ, Morice MC, Holmes DR, Stahle E, et al. Comparison of coronary bypass surgery with drug-eluting stenting for the treatment of left main and/or three-vessel disease: 3-year follow-up of the SYNTAX trial. Eur Heart J. 2011;32(17):2125–34.

[14] Weintraub WS, Grau-Sepulveda MV, Weiss JM, O'Brien SM, Peterson ED, Kolm P, et al. Comparative effectiveness of revascularization strategies. N Engl J Med. 2012;366(16): 1467–76.

[15] Vacanti JP, Langer R. Tissue engineering: the design and fabrication of living replacement devices for surgical reconstruction and transplantation. Lancet. 1999;354(Suppl. 1):SI32–4.

[16] Wainwright JM, Hashizume R, Fujimoto KL, Remlinger NT, Pesyna C, Wagner WR, et al. Right ventricular outflow tract repair with a cardiac biologic scaffold. Cells Tissues Organs. 2012;195(1–2):159–70.

[17] Beck CS. The development of a new blood supply to the heart by operation. Ann Surg. 1935;102(5):801–13.

[18] Chang Y, Lai PH, Wei HJ, Lin WW, Chen CH, Hwang SM, et al. Tissue regeneration observed in a basic fibroblast growth factor-loaded porous a cellular bovine pericardium populated with mesenchymal stem cells. J Thorac Cardiovasc Surg. 2007;134(1):65–73.

[19] Kusaba E, Schraut W, Sawatani S, Jaron D, Freed P, Kantrowitz A. A diaphragmatic graft for augmenting left ventricular function: a feasibility study. Trans Am Soc Artif Intern Organs. 1973;19:251–7.

[20] Ota T, Gilbert TW, Schwartzman D, McTiernan CF, Kitajima T, Ito Y, et al. A fusion protein of hepatocyte growth factor enhances reconstruction of myocardium in a cardiac patch derived from porcine urinary bladder matrix. J Thorac Cardiovasc Surg. 2008;136(5):1309–17.

[21] Huang W, Zhang DS, Millard RW, Wang T, Zhao TM, Fan GC, et al. Gene manipulated peritoneal cell patch repairs infarcted myocardium. J Mol Cell Cardiol. 2010;48(4):702–12.

[22] Taheri SA, Yeh J, Batt RE, Fang Y, Ashraf H, Heffner R, et al. Uterine myometrium as a cell patch as an alternative graft for transplantation to infarcted cardiac myocardium: a preliminary study. Int J Artif Organs. 2008;31(1):62–7.

[23] Ruel MA, Sellke FW, Bianchi C, Khan TA, Faro R, Zhang JP, et al. Endogenous myocardial angiogenesis and revascularization using a gastric submucosal patch. Ann Thorac Surg. 2003;75(5):1443–9.

[24] Song L, Duan P, Zhou Q. Preparation and characterization of a de-cellularized rabbit aorta as a promising scaffold in vascular tissue engineering. Cell Mol Biol (Noisy-le-grand). 2016;62(3):31–8.

[25] Antonova LV, Seifalian AM, Kutikhin AG, Sevostyanova VV, Krivkina EO, Mironov AV, et al. Bioabsorbable bypass grafts biofunctionalised with RGD have enhanced biophysical properties and endothelialisation tested in vivo. Front Pharmacol. 2016;7:136.

[26] Aper T, Wilhelmi M, Gebhardt C, Hoeffler K, Benecke N, Hilfiker A, et al. Novel method for the generation of tissue-engineered vascular grafts based on a highly compacted fibrin matrix. Acta Biomater. 2016;29:21–32.

[27] Gui L, Dash BC, Luo J, Qin L, Zhao L, Yamamoto K, et al. Implantable tissue-engineered blood vessels from human induced pluripotent stem cells. Biomaterials. 2016;102:120–9.

[28] WHO. Cardiovascular Diseases (CVDs). WHO. 2016 [cited 2016 28.6.2016]. Available from: http://www.who.int/mediacentre/factsheets/fs317/en/.

[29] Authors/Task Force m, Windecker S, Kolh P, Alfonso F, Collet JP, Cremer J, et al. 2014. ESC/EACTS guidelines on myocardial revascularization: the Task Force on Myocardial Revascularization of the European Society of Cardiology (ESC) and the European Association for Cardio-Thoracic Surgery (EACTS) Developed with the special contribution of the European Association of Percutaneous Cardiovascular Interventions (EAPCI). Eur Heart J. 2014;35(37):2541–619.

[30] Mohr FW, Morice MC, Kappetein AP, Feldman TE, Stahle E, Colombo A, et al. Coronary artery bypass graft surgery versus percutaneous coronary intervention in patients with three-vessel disease and left main coronary disease: 5-year follow-up of the randomised, clinical SYNTAX trial. Lancet. 2013;381(9867):629–38.

[31] Benedetto U, Gaudino M, Ng C, Biondi-Zoccai G, D'Ascenzo F, Frati G, et al. Coronary surgery is superior to drug eluting stents in multivessel disease: systematic review and meta-analysis of contemporary randomized controlled trials. Int J Cardiol. 2016;210:19–24.

[32] Biancari F, Myllyl M, Lepojrvi S, Kuttila K, Porela P, Laitio T, et al. Preoperative warfarin treatment and outcome of coronary artery bypass graft surgery. Ann Thorac Surg. 2010;89(4):1139–45.

[33] Rizas KD, Mehilli J. Stent polymers: do they make a difference? Circ Cardiovasc Interv. 2016;9(6). DOI: 10.1161/CIRCINTERVENTIONS.115.002943.

[34] Tamai H, Igaki K, Kyo E, Kosuga K, Kawashima A, Matsui S, et al. Initial and 6-month results of biodegradable poly-l-lactic acid coronary stents in humans. Circulation. 2000;102(4):399–404.

[35] Wang J, He Y, Maitz MF, Collins B, Xiong K, Guo L, et al. A surface-eroding poly(1,3-trimethylene carbonate) coating for fully biodegradable magnesium-based stent applications: toward better biofunction, biodegradation and biocompatibility. Acta Biomater. 2013;9(10):8678–89.

[36] Moravej M, Mantovani D. Biodegradable metals for cardiovascular stent application: interests and new opportunities. Int J Mol Sci. 2011;12(7):4250–70.

[37] Witte F. The history of biodegradable magnesium implants: a review. Acta Biomater. 2010;6(5):1680–92.

[38] Levesque J, Hermawan H, Dube D, Mantovani D. Design of a pseudo-physiological test bench specific to the development of biodegradable metallic biomaterials. Acta Biomater. 2008;4(2):284–95.

[39] Witte F, Hort N, Vogt C, Cohen S, Kainer KU, Willumeit R, et al. Degradable biomaterials based on magnesium corrosion. Curr Opin Solid State Mat Sci. 2008; 12(5–6):63–72.

[40] Heublein B, Rohde R, Kaese V, Niemeyer M, Hartung W, Haverich A. Biocorrosion of magnesium alloys: a new principle in cardiovascular implant technology? Heart. 2003;89(6):651–6.

[41] Taylor MS, Daniels AU, Andriano KP, Heller J. Sixbioabsorbable polymers:in-vitro acute toxicity of accumulated degradation products. J Appl Biomater. 1994;5(2):151–7.

[42] Chen Y, Song Y, Zhang SX, Li JN, Zhao CL, Zhang XN. Interaction between a high purity magnesium surface and PCL and PLA coatings during dynamic degradation. Biomed Mater. 2011;6(2). DOI: 10.1088/1748-6041/6/2/025005.

[43] Ramcharitar S, Serruys PW. Fully biodegradable coronary stents: progress to date. Am J Cardiovasc Drugs. 2008;8(5):305–14.

[44] DEGAM. National guideline no. 9: Congestive heart failure- long version [German]. Düsseldorf: Omikron Publ; 2006. 96 S p.

[45] Statistisches Bundesamt. Health - Causes of death in Germany 2011 [German]. Fachserie 12 Reihe 42012. Wiesbaden. Available from: https://www.destatis.de/DE/Publikationen/Thematisch/Gesundheit/Todesursachen/Todesursachen.html.

[46] Statistisches Bundesamt. Calculation of sickness costs: Sickness costs: Germany, years, sickness costs (ICD10) 2009. [German]. Available from: https://www-genesis.destatis.de/genesis/online/logon?language=de&sequenz=tabellen&selectionname=23631*.

[47] Bundesärztekammer, Kassenärztliche Bundesvereinigung, Arbeitsgemeinschaft der Wissenschaftlichen Medizinischen Fachgesellschaften. National guideline for congestive heart failure - long version [German]. 2009. Available from: http://www.versorgungsleitlinien.de/themen/herzinsuffizienz. DOI: 10.6101/AZQ/000166

[48] Athanasuleas CL, Stanley AW, Buckberg GD, Dor V, Di DM, Siler W. Surgical anterior ventricular endocardial restoration (SAVER) for dilated ischemic cardiomyopathy. Semin Thorac Cardiovasc Surg. 2001;13(4):448–58.

[49] Badylak SF, Kochupura PV, Cohen IS, Doronin SV, Saltman AE, Gilbert TW, et al. The use of extracellular matrix as an inductive scaffold for the partial replacement of functional myocardium. Cell Transplant. 2006;15:S29–40.

[50] Taheri SA, Ashraf H, Merhige M, Miletich RS, Satchidanand S, Malik C, et al. Myoangiogenesis after cell patch cardiomyoplasty and omentopexy in a patient with ischemic cardiomyopathy. Texas Heart Inst J. 2005;32(4):598–601.

[51] Robinson KA, Li JS, Mathison M, Redkar A, Cui JH, Chronos NAF, et al. Extracellular matrix scaffold for cardiac repair. Circulation. 2005;112(9):I135–I43.

[52] Schilling T, Cebotari S, Tudorache I, Haverich A. Tissue engineering of vascularized myocardial prosthetic tissue: biological and solid matrices. Chirurg. 2011;82(4):319–24.

[53] Bach FW, Haverich A, Cebotari S, Biskup C, Schuster B, inventors. Supporting element for tissue implants patent WO 2011/101142 A1. 2011.

[54] Oliveira GH, Al-Kindi SG, Bezerra HG, Costa MA. Left ventricular restoration devices. J Cardiovasc Trans Res. 2014;7(3):282–91.

[55] Walsh RG. Design and features of the Acorn CorCap Cardiac Support Device: the concept of passive mechanical diastolic support. Heart Fail Rev. 2005;10(2):101–7.

[56] Atluri P, Acker MA. Diastolic ventricular support with cardiac support devices: an alternative approach to prevent adverse ventricular remodeling. Heart Fail Rev. 2013;18(1):55–63.

[57] Magovern JA. Experimental and clinical studies with the Paracor cardiac restraint device. Semin Thorac Cardiovasc Surg. 2005;17(4):364–8.

[58] Bruckenberger E. 27. Heart Report 2015 [German]. Frankfurt: Deutsche Herzstiftung. 2015.

[59] Eckstein HH, Knipfer E, Ockert S. Aortic dissection type B. Gefaesschirurgie. 2011;16(8): 552–6.

[60] Carrel T, Pasic M, Jenni R, Tkebuchava T, Turina MI. Reoperations after operation on the thoracic aorta: etiology, surgical techniques, and prevention. Ann Thorac Surg. 1993;56(2):259–68; discussion 69.

[61] Chlupac J, Filova E, Bacakova L. Blood vessel replacement: 50 years of development and tissue engineering paradigms in vascular surgery. Physiol Res. 2009;58(Suppl. 2):S119–39.

[62] Breuer CK. The development and translation of the tissue-engineered vascular graft. J Pediatr Surg. 2011;46(1):8–17.

[63] Torikai K, Ichikawa H, Hirakawa K, Matsumiya G, Kuratani T, Iwai S, et al. A self-renewing, tissue-engineered vascular graft for arterial reconstruction. J Thorac Cardiovasc Surg. 2008;136(1):37–45, e1.

[64] Mazzola A, Di Mauro M, Pellone F, Faragalli F, Villani C, Di Eusanio M, et al. Freestyle aortic root bioprosthesis is a suitable alternative for aortic root replacement in elderly patients: a propensity score study. Ann Thorac Surg. 2012;94(4):1185–90.

[65] Miskovic A, Monsefi N, Doss M, Ozaslan F, Karimian A, Moritz A. Comparison between homografts and Freestyle(R) bioprosthesis for right ventricular outflow tract replacement in Ross procedures. Eur J Cardiothorac Surg. 2012;42(6):927–33.

[66] Vogt PR. Arterial allografts in treating aortic graft infections: something old, something new. Semin Vasc Surg. 2011;24(4):227–33.

[67] Aper T, Haverich A, Teebken O. The dream of ideal vascular graft material in surgery [German]. Gefaesschirurgie. 2008;13(2):87–98.

[68] Graeter T, Langer F, Nikoloudakis N, Wendler O, Demertzis S, Schafers HJ. Treatment of the aortic valve and ascending aorta: the significance of reconstructive methods on the aortic root. Dtsch Med Wochenschr. 1998;123(41):1195–200.

[69] Sarikouch S, Horke A, Tudorache I, Beerbaum P, Westhoff-Bleck M, Boethig D, et al. Decellularized fresh homografts for pulmonary valve replacement: a decade of clinical experience. Eur J Cardiothorac Surg. 2016;50(2):281–290.

[70] Da Costa ML, Ghofaili FA, Oakley RM. Allograft tissue for use in valve replacement. Cell Tissue Bank. 2006;7(4):337–48.

[71] Kakisis JD, Liapis CD, Breuer C, Sumpio BE. Artificial blood vessel: the Holy Grail of peripheral vascular surgery. J Vasc Surg. 2005;41(2):349–54.

[72] Wolf F, Vogt F, Schmitz-Rode T, Jockenhoevel S, Mela P. Bioengineered vascular constructs as living models for in vitro cardiovascular research. Drug Discov Today. 2016;21(9):1446–1455.

[73] Bruckenberger E, editor. Multidisciplinary Health Report on Cardiology and Cardiac Surgery. Frankfurt a. Main: Deutsche Herzstiftung Deutsche Herzstiftung e.V. 2013.

[74] Furukawa N, Kuss O, Aboud A, Schonbrodt M, Renner A, Hakim MK, et al. Ministernotomy versus conventional sternotomy for aortic valve replacement: matched propensity score analysis of 808 patients. Eur J Cardiothorac Surg. 2014;46(2):221–226.

[75] Beckmann A, Funkat AK, Lewandowski J, Frie M, Schiller W, Hekmat K, et al. Cardiac surgery in Germany during 2012: a report on behalf of the German Society for Thoracic and Cardiovascular Surgery. Thorac Cardiovasc Surg. 2014;62(1):5–17.

[76] Rashidi S, Elenbaas TW, Hamad MA, van Suijlekom HJ, van Straten AH. Does removal of steel wires relieve post-sternotomy pain after cardiac surgery? Asian Cardiovasc Thorac Ann. 2013;21(4):409–13.

[77] Malhotra A, Garg P, Bishnoi AK, Pendro V, Sharma P, Upadhyay M, et al. Is steel wire closure of sternotomy better than polyester suture closure? Asian Cardiovasc Thorac Ann. 2014;22(4):409–15.

[78] Raksamani K, Wongkornrat W, Siriboon P, Pantisawat N. Pain management after cardiac surgery: are we underestimating post sternotomy pain? J Med Assoc Thai. 2013;96(7):824–8.

[79] Yavuz SS, Tarcin O, Ada S, Dincer F, Toraman S, Birbudak S, et al. Incidence, aetiology, and control of sternal surgical site infections. J Hosp Infect. 2013;85(3):206–12.

[80] de Moraes AA, Abboud CS, Chammas AZ, Aguiar YS, Mendes LC, Melo NJ, et al. Long term mortality of deep sternal wound infection after coronary artery bypass surgery. Rev Bras Cir Cardiovasc. 2012;27(3):377–82.

[81] Graf K, Sohr D, Haverich A, Kuhn C, Gastmeier P, Chaberny IF. Decrease of deep sternal surgical site infection rates after cardiac surgery by a comprehensive infection control program. Interact Cardiovasc Thorac Surg. 2009;9(2):282–6.

[82] Vranken NP, Weerwind PW, Barenbrug PJ, Teerenstra S, Ganushchak YM, Maessen JG. The role of patient's profile and allogeneic blood transfusion in development of post-

cardiac surgery infections: a retrospective study. Interact Cardiovasc Thorac Surg. 2014;19(2):232–238.

[83] Kuehn C, Graf K, Mashaqi B, Pichlmaier M, Heuer W, Hilfiker A, et al. Prevention of early vascular graft infection using regional antibiotic release. J Surg Res. 2010;164(1):e185–e91.

[84] Robinson DA, Griffith RW, Shechtman D, Evans RB, Conzemius MG. In vitro antibacterial properties of magnesium metal against *Escherichia coli, Pseudomonas aeruginosa* and *Staphylococcus aureus*. Acta Biomater. 2010;6(5):1869–77.

[85] Hornberger H, Virtanen S, Boccaccini AR. Biomedical coatings on magnesium alloys: a review. Acta Biomater. 2012;8(7):2442–55.

[86] Witte F, Kaese V, Haferkamp H, Switzer E, Meyer-Lindenberg A, Wirth CJ, et al. In vivo corrosion of four magnesium alloys and the associated bone response. Biomaterials. 2005;26(17):3557–63.

[87] Xin Y, Hu T, Chu PK. In vitro studies of biomedical magnesium alloys in a simulated physiological environment: a review. Acta Biomater. 2011;7(4):1452–9.

[88] Wu G, Fan Y, Gao H, Zhai C, Zhu YP. The effect of Ca and rare earth elements on the microstructure, mechanical properties and corrosion behavior of AZ91D. Mat Sci Eng A. 2005;408(1–2):255–63.

[89] Makar GL, Kruger J. Corrosion of magnesium. Int Mater Rev. 1993;38(3):138–53.

[90] Feyerabend F, Fischer J, Holtz J, Witte F, Willumeit R, Drucker H, et al. Evaluation of short-term effects of rare earth and other elements used in magnesium alloys on primary cells and cell lines. Acta Biomater. 2010;6(5):1834–42.

[91] Gu X, Zheng Y, Cheng Y, Zhong S, Xi T. In vitro corrosion and biocompatibility of binary magnesium alloys. Biomaterials. 2009;30(4):484–98.

[92] Janning C, Willbold E, Vogt C, Nellesen J, Meyer-Lindenberg A, Windhagen H, et al. Magnesium hydroxide temporarily enhancing osteoblast activity and decreasing the osteoclast number in peri-implant bone remodelling. Acta Biomater. 2010;6(5):1861–8.

[93] Kirkland NT, Birbilis N, Staiger MP. Assessing the corrosion of biodegradable magnesium implants: a critical review of current methodologies and their limitations. Acta Biomater. 2012;8(3):925–36.

[94] Lorenz C, Brunner JG, Kollmannsberger P, Jaafar L, Fabry B, Virtanen S. Effect of surface pre-treatments on biocompatibility of magnesium. Acta Biomater. 2009;5(7):2783–9.

[95] Zeng R, Dietzel W, Witte F, Hort C, Blawert C. Progress and challenge for magnesium alloys as biomaterials. Adv Eng Mat. 2008;10(8):B3–B14. DOI: 10.1002/adem.200800035.

[96] Kirkland NT, Lespagnol J, Birbilis N, Staiger MP. A survey of bio-corrosion rates of magnesium alloys. Corrosion Sci. 2010;52:287–91.

[97] Lalk M, Reifenrath J, Angrisani N, Bondarenko A, Seitz JM, Mueller PP, et al. Fluoride and calcium-phosphate coated sponges of the magnesium alloy AX30 as bone grafts: a comparative study in rabbits. J Mater Sci Mater Med. 2013;24(2):417–36.

[98] Imwinkelried T, Beck S, Iizuka T, Schaller B. Effect of a plasmaelectrolytic coating on the strength retention of in vivo and in vitro degraded magnesium implants. Acta Biomater. 2013;9(10):8643–9.

[99] Hanzi AC, Gerber I, Schinhammer M, Loffler JF, Uggowitzer PJ. On the in vitro and in vivo degradation performance and biological response of new biodegradable Mg-Y-Zn alloys. Acta Biomater. 2010;6(5):1824–33.

[100] Song GL. Control of biodegradation of biocompatable magnesium alloys. Corrosion Science. 2007;49(4):1696–701.

[101] Schilling T, Brandes G, Tudorache I, Cebotari S, Hilfiker A, Meyer T, et al. In vivo degradation of magnesium alloy LA63 scaffolds for temporary stabilization of biological myocardial grafts in a swine model. Biomed Tech (Berl). 2013;58(5):407–16.

[102] Kim SM, Jo JH, Lee SM, Kang MH, Kim HE, Estrin Y, et al. Hydroxyapatite-coated magnesium implants with improved in vitro and in vivo biocorrosion, biocompatibility, and bone response. J Biomed Mater Res A. 2014;102(2):429–41.

[103] Iskandar ME, Aslani A, Liu H. The effects of nanostructured hydroxyapatite coating on the biodegradation and cytocompatibility of magnesium implants. J Biomed Mater Res A. 2013;101(8):2340–54.

[104] Xu L, Yamamoto A. Characteristics and cytocompatibility of biodegradable polymer film on magnesium by spin coating. Colloids Surf B Biointerfaces. 2012;93:67–74.

[105] Keim S, Brunner JG, Fabry B, Virtanen S. Control of magnesium corrosion and biocompatibility with biomimetic coatings. J Biomed Mater Res B Appl Biomater. 2011;96(1):84–90.

[106] Persaud-Sharma D, McGoron A. Biodegradable magnesium alloys: a review of material development and applications. J Biomim Biomater Tissue Eng. 2012;12:25–39.

[107] Ferreira PC, de Almeida Piai K, Takayanagui AMM, Megura-Munoz SI. Aluminum as a risk factor for Alzheimer's disease. Revista Latino-Americana Enfermagem. 2008;16(1):151–7.

[108] El-Rahman SS. Neuropathology of aluminum toxicity in rats (glutamate and GABA impairment). Pharmacol Res. 2003;47(3):189–94.

[109] Soni MG, White SM, Flamm WG, Burdock GA. Safety evaluation of dietary aluminum. Regul Toxicol Pharmacol. 2001;33(1):66–79.

[110] Drynda A, Deinet N, Braun N, Peuster M. Rare earth metals used in biodegradable magnesium-based stents do not interfere with proliferation of smooth muscle cells but

do induce the upregulation of inflammatory genes. J Biomed Mater Res A. 2009;91(2): 360–9.

[111] Anderson JM, Rodriguez A, Chang DT. Foreign body reaction to biomaterials. Semin Immunol. 2008;20(2):86–100.

[112] Weidling M, Besdo S, Schilling T, Bauer M, Hassel T, Haverich A, et al. Finite element simulation of myocardial stabilising structures and development of new designs. Biomed Tech (Berl). 2013;58(Suppl.1). DOI: 10.1515/bmt-2013-4061.

[113] Wu W, Chen SS, Gastaldi D, Petrini L, Mantovani D, Yang K, et al. Experimental data confirm numerical modeling of the degradation process of magnesium alloys stents. Acta Biomaterialia. 2013;9(10):8730–9.

[114] Bauer M, Schilling T, Weidling M, Hartung D, Biskup C, Wriggers P, et al. Geometric adaption of biodegradable magnesium alloy scaffolds to stabilise biological myocardial grafts. Part I. J Mater Sci Mater Med. 2014;25(3):909–16.

[115] Krajarz D. Comparison metal water jet cutting with laser and plasma cutting. Procedia Eng. 2014;69:838–43.

[116] Gu XN, Li N, Zhou WR, Zheng YF, Zhao X, Cai QZ, et al. Corrosion resistance and surface biocompatibility of a microarc oxidation coating on a Mg-Ca alloy. Acta Biomater. 2011;7(4):1880–9.

[117] Blawert C, Dietzel W, Ghali E, Song GL. Anodizing treatments for magnesium alloys and their effect on corrosion resistance in various environments. Adv Eng Mater. 2006;8(6):511–33.

[118] Hiromoto S, Yamamoto A. Control of degradation rate of bioabsorbable magnesium by anodization and steam treatment. Mat Sci Eng C-Mater. 2010;30(8):1085–93.

[119] Gu XN, Zheng W, Cheng Y, Zheng YF. A study on alkaline heat treated Mg-Ca alloy for the control of the biocorrosion rate. Acta Biomaterialia. 2009;5(7):2790–9.

[120] Drynda A, Hassel T, Hoehn R, Perz A, Bach FW, Peuster M. Development and biocompatibility of a novel corrodible fluoride-coated magnesium-calcium alloy with improved degradation kinetics and adequate mechanical properties for cardiovascular applications. J Biomed Mater Res A. 2010;93(2):763–75.

[121] Wang HX, Guan SK, Wang X, Ren CX, Wang LG. In vitro degradation and mechanical integrity of Mg-Zn-Ca alloy coated with Ca-deficient hydroxyapatite by the pulse electrodeposition process. Acta Biomaterialia. 2010;6(5):1743–8.

[122] Xu LP, Zhang EL, Yang K. Phosphating treatment and corrosion properties of Mg-Mn-Zn alloy for biomedical application. J Mater Sci Mater M. 2009;20(4):859–67.

[123] Bauer M, Biskup C, Schilling T, Haverich A, Bach FW, Maier HJ, et al. Influence of shot peening on surface roughness and in vitro load cycles of magnesium alloys. Biomed Tech (Berl). 2013;58(Suppl.1). DOI: 10.1515/bmt-2013-4060.

[124] Wu GS, Xu RZ, Feng K, Wu SL, Wu ZW, Sun GY, et al. Retardation of surface corrosion of biodegradable magnesium-based materials by aluminum ion implantation. Appl Surf Sci. 2012;258(19):7651–7.

[125] Salunke P, Shanov V, Witte F. High purity biodegradable magnesium coating for implant application. Mater Sci Eng B Adv. 2011;176(20):1711–7.

[126] Wong HM, Yeung KWK, Lam KO, Tam V, Chu PK, Luk KDK, et al. A biodegradable polymer-based coating to control the performance of magnesium alloy orthopaedic implants. Biomaterials. 2010;31(8):2084–96.

[127] Chen J, Wang JQ, Han EH, Ke W. In situ observation of pit initiation of passivated AZ91 magnesium alloy. Corrosion Sci. 2009;51(3):477–84.

[128] Rettig R, Virtanen S. Composition of corrosion layers on a magnesium rare-earth alloy in simulated body fluids. J Biomed Mater Res A. 2009;88(2):359–69.

[129] Witte F, Fischer J, Nellesen J, Crostack HA, Kaese V, Pisch A, et al. In vitro and in vivo corrosion measurements of magnesium alloys. Biomaterials. 2006;27(7):1013–8.

[130] Donnelly KB. Cardiac valvular pathology: comparative pathology and animal models of acquired cardiac valvular diseases. Toxicol Pathol. 2008;36(2):204–17.

[131] Pearce AI, Richards RG, Milz S, Schneider E, Pearce SG. Animal models for implant biomaterial research in bone: a review. Eur Cell Mater. 2007;13:1–10.

[132] Hazzard DG, Bronson RT, McClearn GE, Strong R. Selection of an appropriate animal model to study aging processes with special emphasis on the use of rat strains. J Gerontol. 1992;47(3):B63–4.

[133] Yamamoto A, Hiromoto S. Effect of inorganic salts, amino acids and proteins on the degradation of pure magnesium in vitro. Mater Sci Eng C. 2009;29(5):1559–68.

Chapter 8

Surface Modification of Magnesium and its Alloys Using Anodization for Orthopedic Implant Application

Hamouda M. Mousa, Chan Hee Park and
Cheol Sang Kim

Additional information is available at the end of the chapter

http://dx.doi.org/10.5772/66341

Abstract

Magnesium (Mg) as a biodegradable implant brings a revolution in medical field application, especially in bone implant and stent application. Biodegradability of Mg has attracted attentions of researchers to avoid secondary surgery to remove the implant materials after healing process. Various advantages of Mg make it suitable for medical application such as density, good mechanical properties and biodegradation. However, Mg biodegradability must be controlled to meet tissue-healing period of time because of the high degradation in a physiological environment. Fast corrosion and high alkalinity due to hydrogen release induce tissue inflammation, which limits its clinical applications. Many techniques are applied to the Mg surface to improve surface biocompatibility and control its biodegradability. This chapter focuses on anodization of Mg and its alloys to improve corrosion resistance and biocompatibility for orthopedic application. Mg coating with thin film apatite could enhance the biocompatibility and increase osseointegration formation in the bone fracture side. Evaluation of the required anodized film discussed in the chapter such as chemical composition, biodegradability and biocompatibility.

Keywords: magnesium, anodization, SBF, b106048iocompatibility, biodegradable metals

1. Introduction

Biodegradable metallic implant material has received considerable attention in biomedical field such as blood vessels or orthopedic application as load-bearing implant [1, 2]. Mg is suitable for implant application in human body, for example, Mg stent, bone fixation screw, microclips in laryngeal microsurgery, bone fixation and wound-closing devices, as shown in **Figure 1**. Mg has many appealing properties such as light weight, high strength-to-weight ratio, good castability and osteoconductivity [3]. However, Mg has limitations mainly due to its high surface chemical reactivity resulting in high degradation rate [4]. The poor corrosion resistance of Mg limits its clinical applications, as hydrogen evaluation is one of the corrosion products that increase alkalinity of the surrounded media and causing inflammation of the surrounding tissues due to the formation of gas pockets [5, 6]. The high degradation rate may eventually hinder the bone formation and hamper the long-term success of the implants and decrease its bioactivity as well as loss its mechanical properties [7]. Mg-based implants exhibited rough surfaces as well as shallow pits and small cavities after one day of implantation, which formed during the on-going corrosion process to form cracks until the implant totally dissolves [8]. The high purity of Mg finds to corrode uniformly *in vivo* [9]. Biodegradable metals (BMs) are typically degraded through the corrosion process when exposed to a corrosive media. For example, when BMs especially Mg implanted in human body, the corrosion/degradation process generated electrochemically in different reactions of metals with an electrolyte and produced metal oxides and hydroxides [7]. Moreover, hydrogen gas evaluation is a combined corrosion product; these reactions could be represented in the following chemical equations at anodic dissolution of Mg and the cathodic reaction [10].

$$Mg \rightarrow Mg^{2+} + 2e^{-} \text{ (anodic reaction)} \quad (1)$$

$$2H_2O + 2ne^{-} \rightarrow H_2 + 2OH \text{ (cathodic reaction)} \quad (2)$$

$$2H_2O + O_2 + 4e^{-} \rightarrow 4OH^{-} \text{ (cathodic reaction)} \quad (3)$$

$$Mg + 2H_2O \rightarrow Mg(OH)_2 + H_2 \text{ (overall reaction)} \quad (4)$$

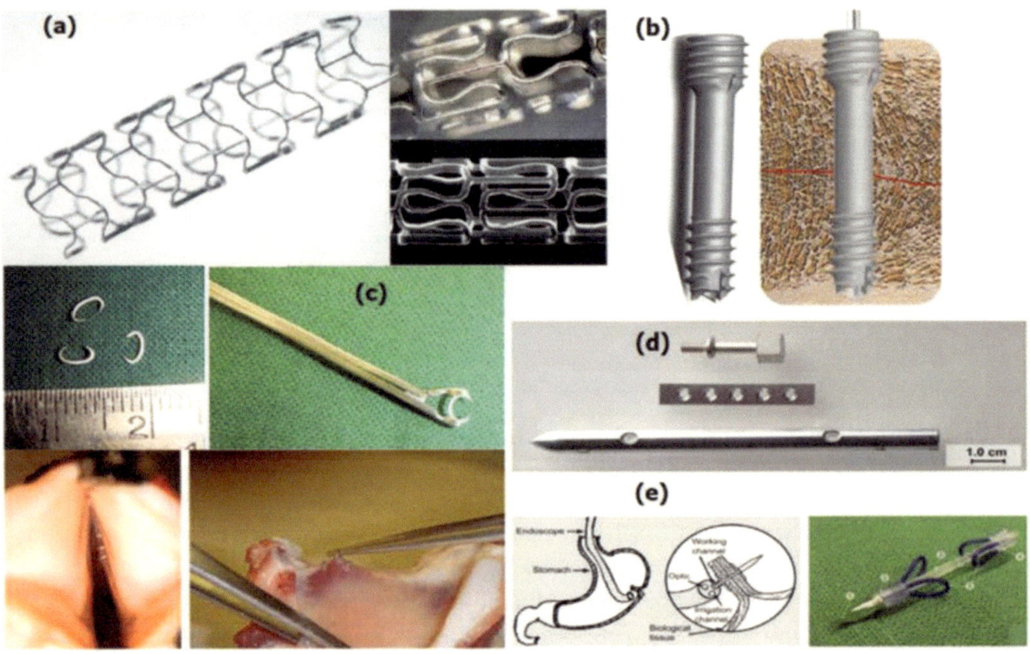

Figure 1. Different applications of Mg-based implant material: (a) cardiovascular Mg stents, (b) MAGNEZIX screw, (c) microclip for laryngeal microsurgery (pure magnesium), (d) biodegradable orthopedic implants and (e) wound-closing devices (WZ21) [11].

http://dx.doi.org/10.5772/66341

Surface modification is considered one of the most useful and effective methods to control the initial degradation of Mg and its alloys [12]. **Table 1** summarizes the previous research on Mg coating with different applied techniques and chemical composition. Among these techniques, anodization is a widely and traditional process for metal surface modification to improve the physiochemical properties of metals [13]. A suitable electrolyte of anodization for the specific application of Mg is one of the essential requirements when it employed. For example, hydroxyapatite (HA) is a bioactive ceramic material which widely used in bone application [14]. HA could be engineered to mimic the three-dimensional inorganic component of the bone which is composed of 65% of bone. The structure of HA could provide the space and area necessary for vascularization and tissue regeneration. In this chapter, HA coating with different nanostructures (nanoplates/nanospheres) by means of anodization is discussed with the associated mechanical integrity, biodegradability and biocompatibility. Formation of nanoplates could promote the osseointegration and eliminate the mismatching of the implant material. Accordingly, using stimulated body fluid (SBF) finds to form apatite film on the surface of Mg in a short duration.

Substrate	Experimental and coating type	Reference
Mg-Zn-Ca alloy	Fabrication of hydroxyapatite nanorod on MAO coating to increase bioactivity and improve the biodegradation behavior	[15]
Mg-1.0Ca alloys	Sodium phytate ($Na_{12}Phy$) used as an electrolyte with anodic coatings fabricated in an organic phosphate containing solution on the Mg-1.0Ca alloys. In order to achieve a proper degradation rate, acceptable biocompatibility and good antibacterial ability	[16]
AZ31B	Different electrolytes such as KOH, Na_2SiO_3 and $Na_2B_4O_7$ were used for pulsed DC micro-arc oxidation (MAO) process	[17]
Mg-Zn-Ca	A porous bioceramic containing tricalcium phosphate in (TCP) coating was prepared by (MAO) at different voltages	[18]
Pure Mg	Anodic oxide coatings were prepared using 0.3 M NaOH + 15 g/l ZrO_2 and 3 M NaOH + 15 g/l ZrO_2	[19]
AZ31B	A chemical conversion film on magnesium alloys is proposed based on the interaction of a deep eutectic solvent (DES) with the substrate	[20]
Mg-3Zn	A nanostructured hydroxyapatite (HA) coating was grown on through the electrophoretic deposition (EPD) technique	[21]
ZK61	MAO coating film with low crystallinity is composed of MgO, Mg_2SiO_4 and $Mg_2Si_2O_6$ employed	[22]
AZ31	A dopamine-induced hydroxyapatite coating was successfully developed on the AZ31 alloy	[23]
AZ31	Use of a microwave-assisted coating technology to improve the *in vitro* corrosion resistance and biocompatibility of AZ31Mg alloy	[24]
Pure Mg	A simple strontium phosphate (SrP) conversion coating process was developed to protect magnesium (Mg) from the initial degradation postimplantation	[25]
AZ31	A Si-doped calcium phosphate coating was achieved via pulse ED on the AZ31 alloy. A novel dual-layer structure was observed with a porous lamellar-like and outer block-like apatite layer	[26]

Table 1. Summary of Mg surface modification techniques.

2. Anodization process

Anodization is an electrochemical process that converts the metal surface into a decorative, durable, corrosion-resistant and anodic oxide finish [27]. The coating thicknesses can range from 5 to 200 μm. Typically, anodic oxide layers grow depending on the process time and applied voltages [28], leading to a direct dependence of the oxide thickness on the applied voltage as shown in **Figure 2**. For metals and alloys with barrier-type anodic oxide films, blocking electron conduction under anodic polarization an anodization can be carried out at high voltages in aqueous solution [29]. Therefore, thick oxides that can be grown on the conductive oxide layers on the metal surface by means of anodization are limited to the applied voltage. The applied voltage is lower than that at which water can dissociated with evaluation of oxygen, whereas, above that potential water tends to decompose rather than thickening of the oxide layer. For example, Mg has potential and conductivity; therefore, the resulting potential while anodization applied depends mainly on the electrolyte composition [29]. The incorporation of electrolyte materials with growing oxide/hydroxide layers can form an oxide layers that have higher blocking efficiency

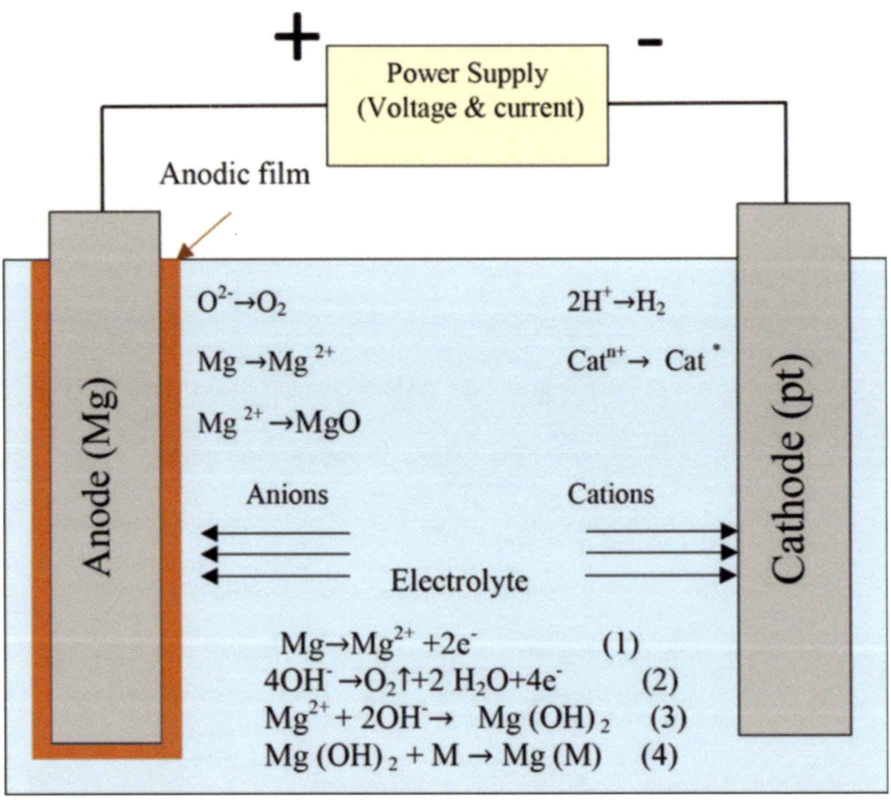

Figure 2. Illustrative diagram shows the mechanism of anodization technique. Mg acts as an anode where it converted to Mg^{2+} then reacts with O^{2-} to form MgO in the presence of OH ions, $Mg(OH)_2$ formation on the metal surface and hydrogen formed the surrounding cathode part.

toward the corrosive ions. Therefore, thick and compact film is a challenge for Mg interface anodization treatment, however, obtaining a low Pilling-Bedworth ratio for the formed anodized film [30]. This could cause an internal stresses on the generated anodic film and subsequently crack defects [31]. The degree of porosity and oxide layer quality could be enhanced by anodization parameter adjustment. These parameters include electrolyte composition, anodization voltage, current and time [32]. Anodization performed in different baths, for example an alkaline electrolyte is based on potassium hydroxide, phosphate, fluoride, or silicate-containing baths. Electrolyte composition plays a critical role not only to enable anodization at high voltage but also to reduce Mg dissolution during the process [33]. There are various methods and techniques with a wide range of patents to produce such anodic films [34]. In addition to anodization approaches which are mainly used to thicken the native oxide/hydroxide films on metal surfaces, dedicated anodization approaches have been explored to obtain nanoporous oxide layers. Therefore, the appropriate electrolyte composition leads to competition between Mg dissolution during anodization and anodic oxide film growth. Thus, optimized parameters based on the electrochemical process self-organized growth of nanoporous or nanotubular oxide layers could performed; however, it is still at early stage for Mg and its alloys [35, 36]. **Table 2** summarizes different Mg alloy anodization/PEO on different electrolytes with the resulted film thickness and chemical composition and the mainly electrochemical corrosion parameters (E_{corr} and i_{corr}).

Substrate	Electrolyte	Thickness	Layer composition	i_{corr}	E_{corr}	Refs.
AZ91D	PEO in NaOH + $(NaPO_3)_6$ + $Ca(H_2PO_2)_2$ solution	3–5 μm	Mg, Al, P and Ca and little crystallized MgO	X	X	[40]
AZ91D	PEO in Na_2SiO_3 + $(NaPO_3)_6$ + $Ca(H_2PO_2)_2$ solution	8–10 μm	Mg, Al, Si, P and Ca, crystallized Mg_2SiO_4 and MgO	X	X	[2]
AM50	PEO in $CaOH_2$ + Na_3PO_4 solution in different mass ratios	in the range of 20–70 μm	MgO, $Mg_3(PO4)_2$, amorphous Ca-phases, $CaH(PO_4)_2$, CaO_2	X	X	[41]
AZ91	NaOH	1–2 μm	MgO and Mg	X	X	[42]
ZK60	100 g/l NaOH + 20 g/l $Na_2B_4O_7{\cdot}10H_2O$ + 50 g/l $C_6H_5Na_3O_7{\cdot}2H_2O$ + 60 g/l $Na_2SiO_3{\cdot}9H_2O$	10–60 μm	MgO and Mg_2SiO_4	1.829 × 10^{-2} (mA/cm^2)	−1.46	[43]
AZ31	SBF solution	5–25 μm	MgO, Mg and amorphous apatite	103 to 0.9 μA/cm^2	−1.39 to −1.45	[44]
AZ31	(ZrO_2-NPs) dispersed in SBF	X	Mg, MgO, ZrO_2, and $Mg_2Zr_5O_{12}$	−1.46 to −1.38	2.796 to 1.9	[45]
AZ31	(SBF solution + HA) then hydrothermal in 5 M NaOH at 60°C for 2 h	X	Mg, MgO, CaO and HA	7.6 to 1025 nA/cm^2	1.52 to 1.31	[46]

Table 2. Anodization of Mg alloys in different electrolyte.

3. Surface morphology and composition

The design of surface morphology structure of biodegradable implant is an important factor since the interconnection of biomaterial interface with surrounding tissues is important for implant engagement and cell attachment [37]. In bone implant, nanoplate and nanosphere structure of HA coating as a biomimetic films are considered for the Mg coating interface, which is characterized by mimicking that of bone [38]. **Figure 3a** shows the nanoplates formation on the surface of AZ31 Mg alloy by the anodization method in SBF solution at 50 V and 30 mA with a process time of 10 min followed by the hydrothermal process in NaOH solution at 60°C for 2 h. However, adding HA powder to SBF solution resulted in nanosphere structure. Natural bone consists of HA nanocrystals in a plate-like shape with a length of 30–200 nm and a thickness of 2–7 nm [39]. As a result, designing HA films with the specific orientation and morphology is an important approach to improve Mg biological properties such as bioactivity and mimic that on natural bone. Furthermore, such nanoplates can promote the porosity of the implant interface, as a result avoid a mechanical mismatch between the hosts and implant interface, stress shield effect can be eliminated by altering surface porosity.

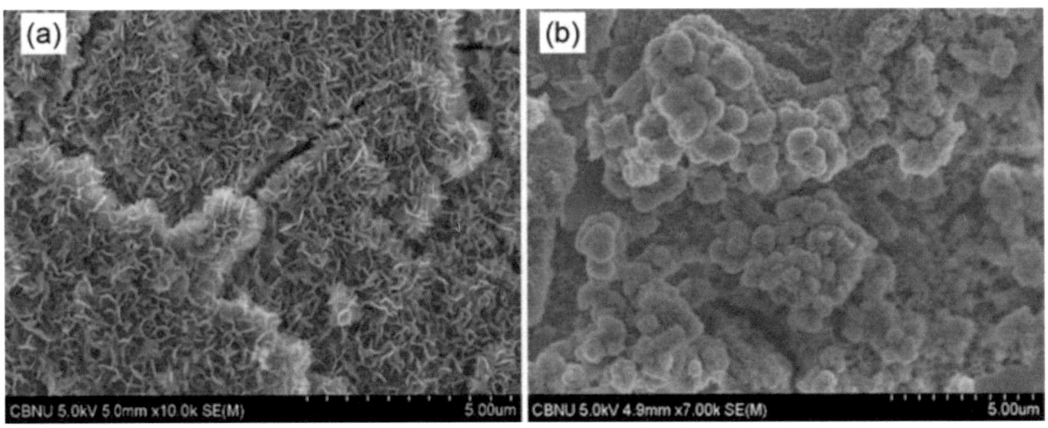

Figure 3. Surface morphology of anodized AZ31 Mg alloy in different electrolytes followed by the hydrothermal process in NaOH at 60°C for 2 h and SBF for 2 days at 37°C: (a) SBF as an electrolyte resulted in nanoplates morphology and (b) (SBF-10 g/l HA) resulted in nanospheres structure.

The chemical composition of HA coating finds is composed of Mg, MgO, HA and CaO peaks as shown in XRD peaks in **Figure 4a**. Furthermore, FT-IR spectra can indicate the outer HA film formation as shown in **Figure 4b**. The bands at a wave number of around 530 cm^{-1} is assigned to PO_4^{3-} in HA at the vibration of (ν_4) and at the vibration of (ν_3) around peaks of 1070 cm^{-1}. On the other hand, hydroxyl group absorption of HA is located at 3703 cm^{-1} [47], in addition to the stretching ν(OH) that is observed at 3550, 3489 and 3412 cm^{-1}. The bands at 1147, 1070, 986 and 877 cm^{-1} are assigned to the P-O stretching vibration in the HPO_4^{2-} groups, respectively. The CO_3^{2-} group is located at 1386 cm^{-1} [26].

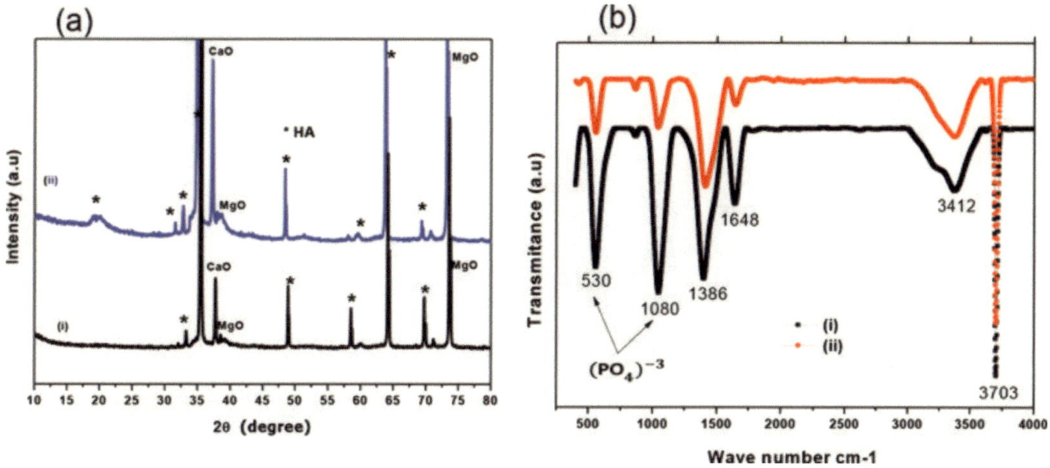

Figure 4. XRD patterns and FTIR spectra of the surface treatment samples, two anodized samples in (i) SBF solution and (ii) SBF/HA solution followed by the hydrothermal process according to reference [46].

4. Mechanical integrity

Mechanical tuning is one of the most effective factors for biodegradable Mg implant in load-bearing application and stent application [48]. Basically, implant materials act as a mechanical support during the healing process thereafter degrade and loss their mechanical properties. Because of that the chemical and mechanical stabilities of implant materials during the healing period are critically important. While implants are exposed to human body fluid, it often experiences considerable loadings and, thus, may undergo environmentally assisted cracking (stress corrosion cracking (SCC) and corrosion fatigue). **Figure 5** shows the mechanical behavior of Mg implant *in vivo* for 12 weeks of implantation and the resulted tensile strength. The coated samples with HA indicated a higher mechanical stability than uncoated samples. The degradation volumes of the bare and HA-coated Mg specimens after 6 and 12 weeks of implantation find that the coated samples have lower degradation with addition protection (**Figure 5b**). The concept of the mechanically tuned with degradation rate during tissue regeneration is illustrated in **Figure 5c**. In period of 1–7 days, inflammation process, hematoma formation with atypical inflammatory response, occurs. Next stage repairs, hematoma, granulation tissue, connective tissue, cartilage, mineralization, woven bone, continuous for 3–6 months depend on the fracture position and type. In the final stage, remodeling, woven bone is replaced by cortical bone and the medullary cavity is restored, which persists for several years.

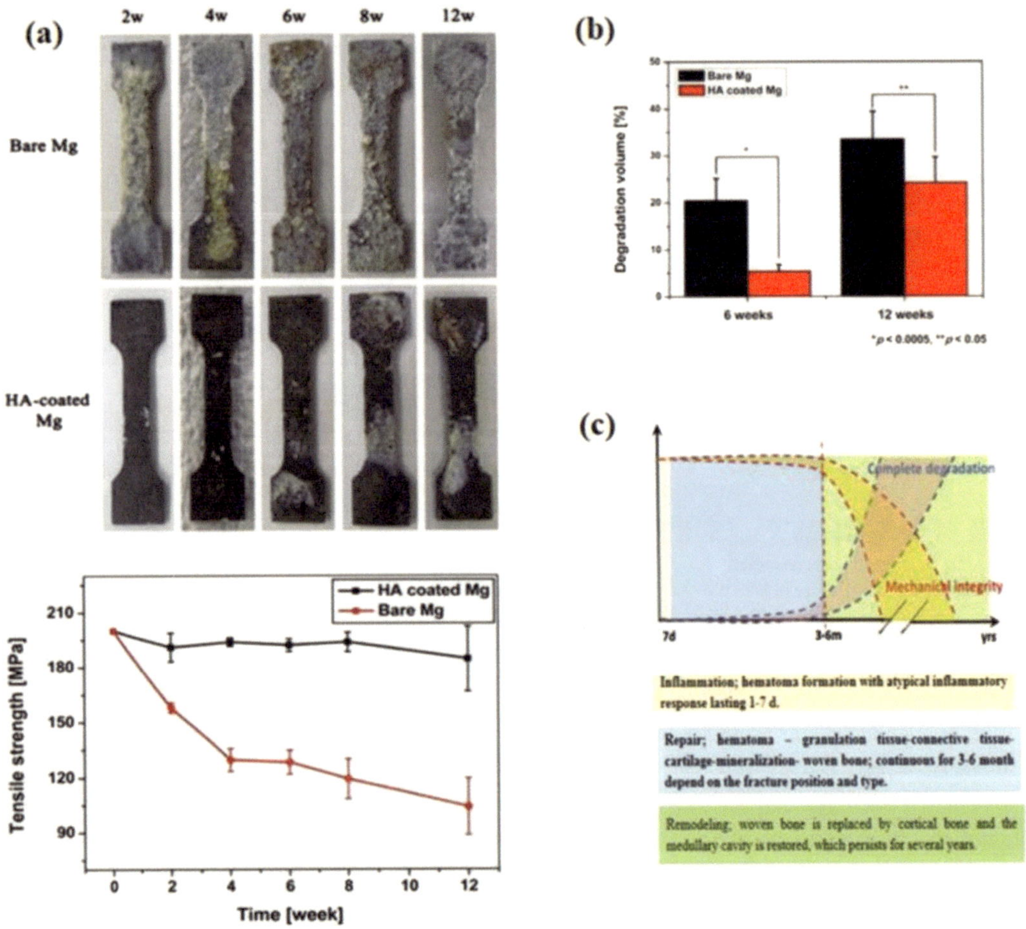

Figure 5. (a) *In vivo* evaluation of the mechanical and degradation properties of Mg coated with calcium orthophosphate coatings. (a) Optical images of the HA coated and bare samples at different implantation times at top and tensile strength of the HA-coated Mg samples comparing to the bare one after interval time. Reprinted from reference [49]. (b) Degradation volumes of the bare and HA-coated Mg specimens after 6 and 12 weeks of implantation [50]. (c) The schematic diagram of degradation behavior and the change of mechanical integrity of BM implants during the bone healing process [51].

5. Biodegradation evaluation

5.1. Electrochemical evaluation

Electrochemical polarization is an efficient technique used to evaluate metal corrosion potential in a short duration. Metals are commonly performed using electrochemical corrosion tests in SBF solution (pH = 7.4) at 37°C to mimic that of human blood plasma. The experimental setup consisted of three conventional electrodes within a cell, which named as working electrode, a saturated calomel electrode (SCE), or Ag/AgCl as a reference electrode, third is counter electrode such as a platinum wire. The experiment is conducted and monitored the current density as a function of the free open-circuit potential using the potentiostat of an electrochemical device. Initially, the samples are exposed to the solution for 10–20 min, a scan rate (mV/s) of the potentiodynamic polarization test is main parameter when test was performed. Corrosion current density (i_{corr}) could be estimated from the linear fit and Tafel extrapolation to the cathodic and anodic parts of the polarization curves. Thereafter, the corrosion rate can be calculated based on Faraday's laws.

$$CR = 3.2710^{-3}\frac{Ewi_{corr}}{\rho} \tag{5}$$

where Ew is the equivalent weight of the corroding species in grams and ρ is the density of the corroding material in g/cm^3.

Faraday's laws assume a uniform corrosion in terms of the penetration, here the corrosion current (i_{corr}) is an effective factor in the corrosion rate and therefore the resulted value does not typically indicate an absolute corrosion rate for Mg. However, it represents indication of the corrosion distortion, which occurs at a selected point in time, in terms of current density. It is seldom in Mg and its alloys to degrade uniformly. The corrosion rate expressed with a current density value is highly accurate and can be considered to have the highest resolution of all methods. Current could be originating from a different local site on the surface and the method is considered as short-term and destructive one. In addition to the potentiodynamic polarization technique, it is essential to understand the mechanism of the corrosion rate; it may not suitable as a good indication for long-term corrosion rates. Electrochemical impedance spectroscopy (EIS) is a powerful tool that is used to evaluate a different application such as biosensors and conductivity as well as corrosion resistance of different metals using the frequency response of AC polarization [52]. EIS conducted with a range of low magnitude polarizing voltages that cycle from a peak anodic to peak cathodic voltage spanning a spectra of voltage frequencies. However, the technique has different components to understand the corrosion resistance of a metal surface. The components such as capacitance and resistance are obtained for each frequency and can then be used to explain a number of phenomena and properties of the metal surface. A determination of the corrosion rate is possible when the EIS determined polarization resistance (R_{polar}) parameter at the zero frequency limit is used. The R_{polar} is inversely proportional to the current density (j_{corr}) as described by the Stern-Geary relationship [44].

$$j_{corr} = \frac{\beta a \beta c}{2.303R\text{polar}(\beta a + \beta c)} \tag{6}$$

where βa and βc are the anodic and cathodic Tafel slopes, respectively.

A corrosion rate can be determined by transferring the current using EIS, the primary function of performing EIS on Mg and its alloys in an electrolytic solution is the identification and quantification of the formation behavior of corrosion layers which produced by the corrosion process. However, EIS results have some limitations as it can be affected by the Mg dissolution at low frequencies and therefore the chosen equivalent circuit. As a result, to employ EIS properly, a deep understanding of the corrosion processes takes place through the process and the best model. **Figure 6a** shows the potentiodynamic polarization curve of bare samples and anodized ones in SBF and SBF/ZrO_2 NPs as an electrolyte with the resulted potential and current density. Moreover, EIS results in terms of Nyquist plot and bode diagrams are shown in **Figure 6b** and **c**. Both techniques find corrosion resistance in anodized samples comparing to the bare samples.

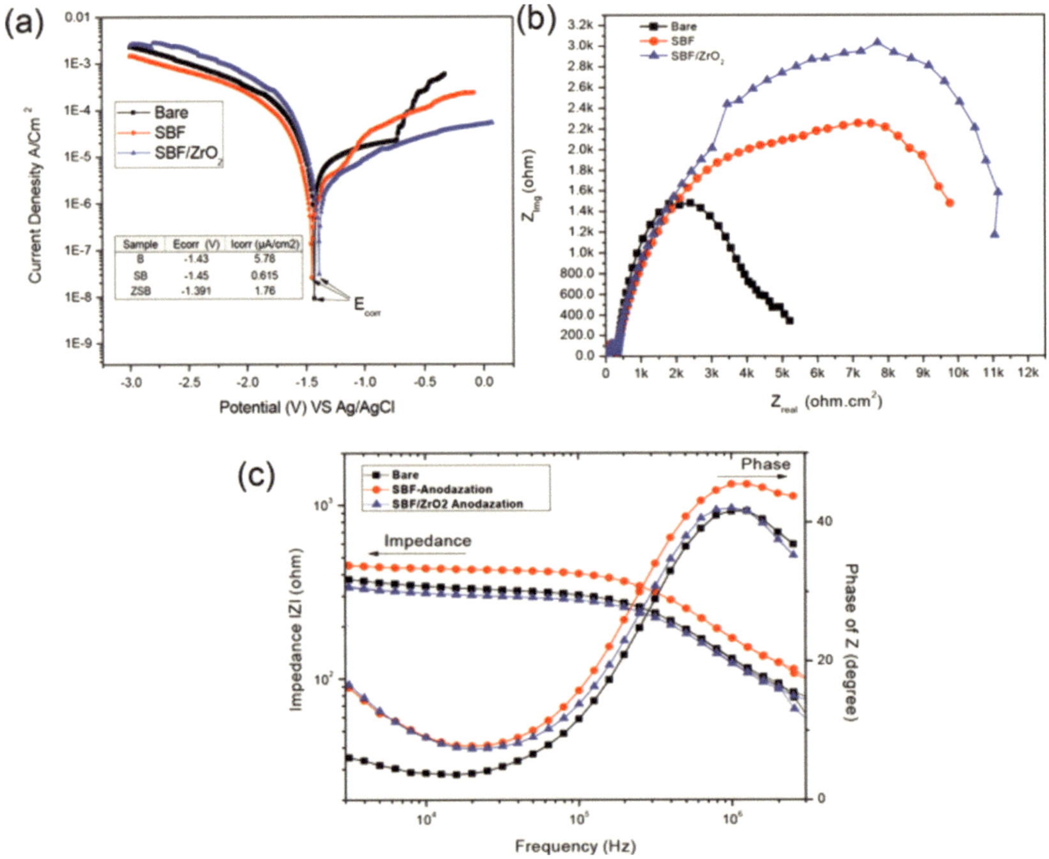

Figure 6. Corrosion evaluation of AZ31 Mg and anodizing samples in the SBF and SBF/ZrO_2 NP electrolyte. Test was performed in SBF solution at 37°C under a three-electrode system where Mg samples act as an electrode, platinum as a counter electrode and Ag/AgCl as a reference electrode. (a) Potentidynamic polarization curves, (b) Nyquist plot and (c) Bode plot diagrams.

5.2. *In vitro* immersion test

In this technique, an *in vitro* degradation rate in terms of mass loss is evaluated, before sample sterilization the initial weight of the samples was recorded. Then samples were immersed in a ratio of 30:1 volume to a weight ratio of SBF solution or Dulbecco's modified eagle's medium (DMEM) supplemented with 10% fetal bovine serum (FBS) for 168 h (1 week) under cell culture conditions (37°C, 20% O_2, 5% CO_2 and 95% humidity). The immersion medium recommended to be changed every 2–3 days to mimic the semistatic immersion test and to avoid saturation effects according to the standard ASTM-G31-72. After immersion time, the formed corrosion products were removed by treating the corroded disc with chromic acid (200 g/L Cr_2O_3 + 10 g/L $AgNO_3$) at least for 20 min at room temperature [52].

Finally, the degradation rate (DR) was calculated in mm/year using the equation [53]:

$$\mathrm{DR} = 8.7610^4 \frac{\Delta g}{A.t.\rho} \tag{7}$$

where Δg is the change in weight by grams, A is the exposed surface area of the sample in cm^2, t is the immersion time in hours and ρ is the density in g/cm^3.

While exposure of the Mg substrate to aqueous solution generates H_2 and OH^- ions along the process of its degradation reaction with the medium, because of that the fluid pH value tends to be increase around the Mg surface. However, the instability of Mg occurs at pH values less than 11, a soluble compound formation with most inorganic ions would inhibits the formation of passive films of magnesium hydroxide in the biological environment. Moreover, the released Mg ions are another factor to indicate the dissolution of Mg in the aqueous solution process according to Eq. (8).

$$\mathrm{Mg} + 2\mathrm{H_2O} \rightarrow \mathrm{Mg}^{+2} + 2\mathrm{OH} + \mathrm{H_2}\uparrow \tag{8}$$

There are various corrosion types during the Mg degradation process, including uniform corrosion [54, 55], localized corrosion [54, 55], flow-induced corrosion [55], erosion corrosion [56], galvanic corrosion [57], stress corrosion [58], atmospheric corrosion, hydrogen cracking [59] and intergranular corrosion [60]. It is worth noting that localized corrosion is always a source of stent fracture. In order to evaluate the biodegradability of Mg implant, a comparison between the anodized and the bare samples under *in vitro* conditions using SBF solution at 37° C up to 30 days usually is carried out. The corrosion rate based on mass loss and Mg^{2+} ions release rate and pH value is shown in **Figure 7a–c**. Both two factors can indicate the biodegradability behavior of Mg and its alloys. The tendency of bare samples to corrosion is significantly different from that of the anodized samples which has more corrosion resistance. The formation of $Mg(OH)_2$ as a corrosion product in bare samples generated once exposed to aqueous solution. Similarly, more Mg ions release from bare samples resulted due to the high degradation and high sensitive surface of Mg. When Mg exposed to a corrosive medium similar to the human plasma corrosion product such as $Mg(OH)_2$ and hydrogen release effect on the pH value of the surrounded solution. In short immersion time, Mg interface exhibits different features especially when treated with the CaP apatite film [61]. For example, when the surface rich with labile ions of CaP it was found to form HA nanoplates, as shown in **Figure 7e**; however, a pours structure formed in the case of apatite film enriched with ZrO_2 NPs, as shown in **Figure 7f**. In contrast, the bare sample exhibits cracks and corrosion occurrence. These results can be attributed to the formation of a porous layer of nanoparticles/plates/rods of the CaP compounds with corrosion products such as magnesium hydroxide and calcium magnesium phosphate.

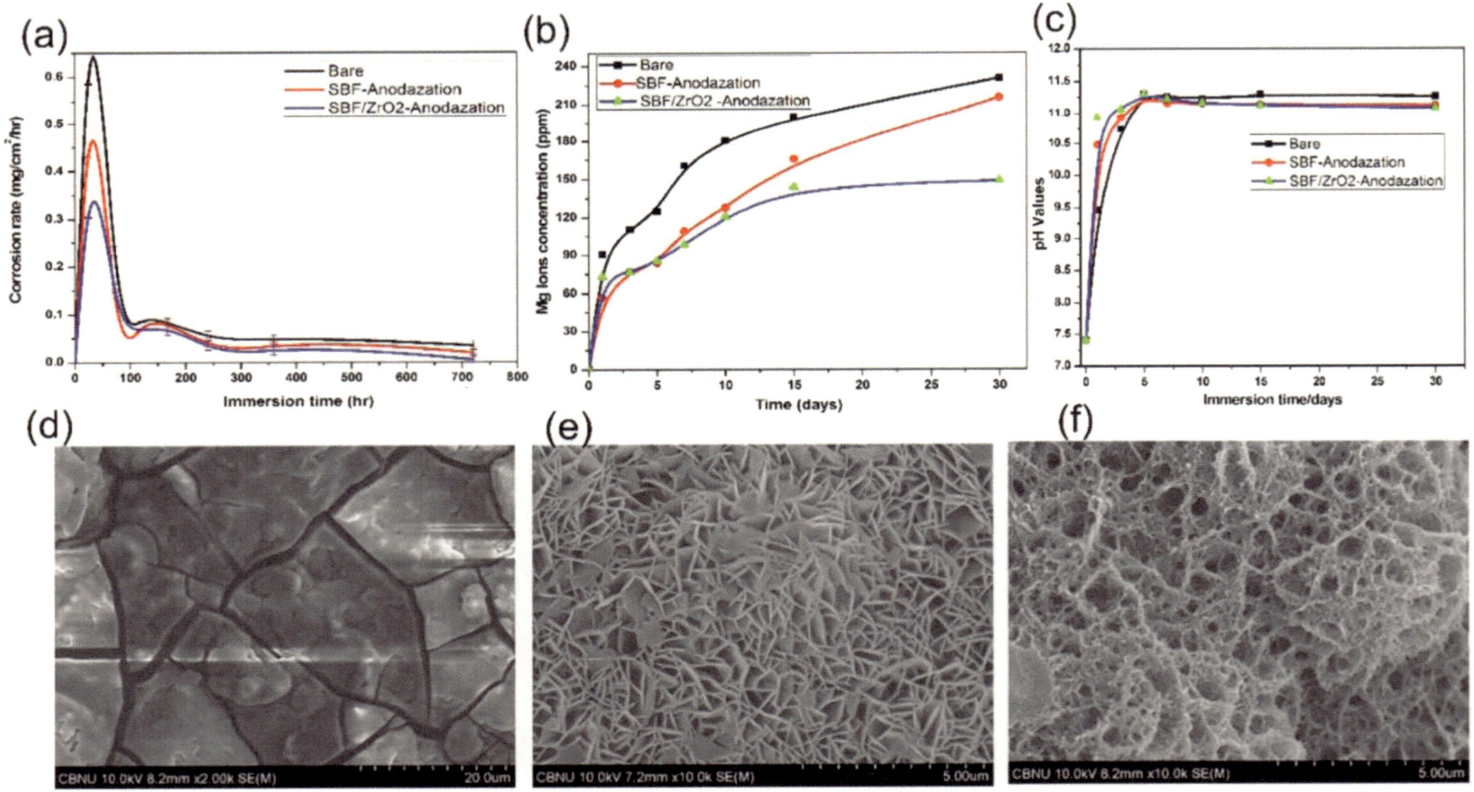

Figure 7. Immersion test results are showing, (a) corrosion rate, (b) Mg ions released and (c) pH value. FESEM images show the surface morphology of (d) bare sample (e) anodized sample in SBF (f) anodized samples in SBF/ZrO_2 NPs after 3 days of immersions in SBF solution at 37°C.

Magnesium sample employed to the anodization technique is more stable in aqueous solutions and corrosive media due to the formation of a thin ceramic layer on the Mg interface. Therefore, Mg biodegradability can be controlled and delayed. An illustrative diagram in **Figure 8b** illustrates a corrosion mechanism before and after anodization treatment in 0.9 NaCl solution. First, the corrosive solution reacts with the substrate interface and starts to corrode and induce cracks and pitting corrosion. Upon increasing the exposure time, anodized film penetrated and the solution reached substrate surface. Thereafter, both the $Mg(OH)_2$ and MgO by means of Cl^- ions penetration are converted and degraded according to the chemical equation:

$$MgO + H_2O \rightarrow Mg^{+2} + 2OH^- \quad (9)$$

Instantaneously, the corrosive solution contact substrate surface Mg^{+2} ions released and hydrogen gas evaluation occurs. As a result, Mg $(OH)_2$ will deposit and react with Cl^- ions to form $MgCl_2$ leading to corrosion occurrence according to the chemical equation:

$$Mg(OH)_2 + Cl^- \rightarrow MgCl_2 \quad (10)$$

The pitting corrosion on the metal surface is due to chloride ions; therefore, the main concept of anodization film is to block Cl^- ions and retard corrosion occurrence on the Mg surface [6].

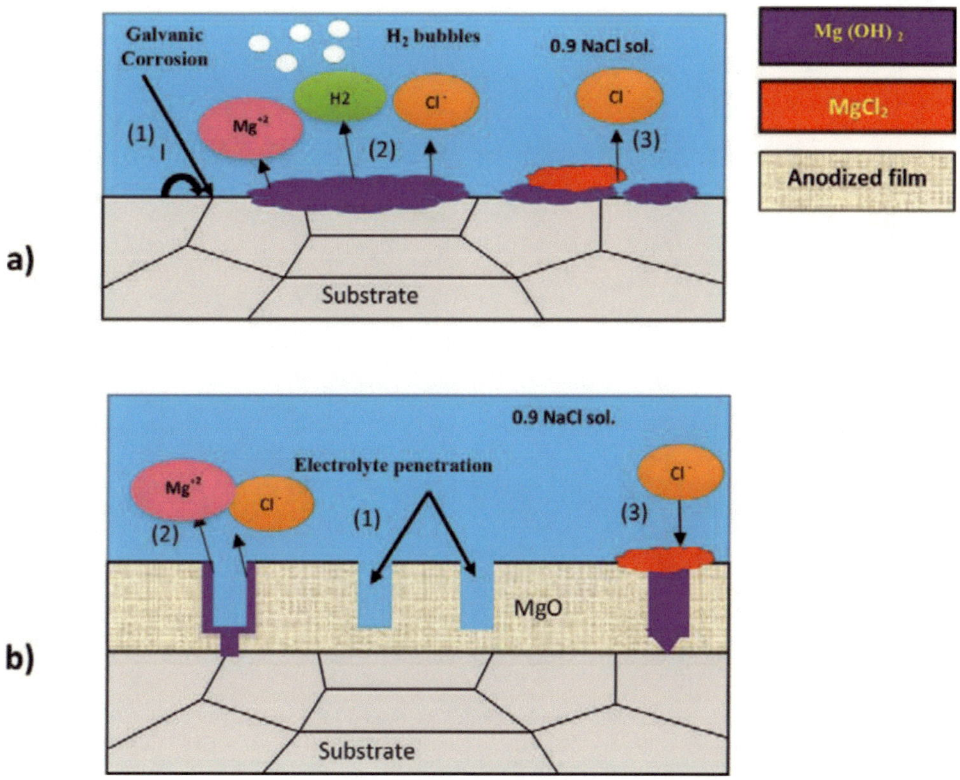

Figure 8. Schematic illustration of the degradation process: (a) bare AZ31 Mg alloy and (b) anodized samples.

6. *In vitro* biocompatibility

Biomaterials must be designed to be biocompatible; however, the majority of biomaterials community has failed to understand the biocompatibility paradigm [62].

Basically, biocompatibility is a characteristic and a complex characteristic at a system and not a material. There are different effects of materials in biological systems as, tissue processing involved in wound healing, the endothelium in contact with intravascular implant devices and the stem cells in bioreactors, the target cells in gene therapy, emphasize that there is no material with complete biocompatibility characteristics [63]. In biodegradable implant such as Mg, bare substrates without any surface modifications show few round shapes of cells on its surface. These attributed to many factors which mainly show corrosion behavior with combined hydrogen gas and induce toxicity to surround tissues. Moreover, surface tribology has additional effect, for example, a rough surface has more cell attachment comparing to smooth one in nanoscale, which behaves as accommodation for cells [64]. In addition, biomimetic nanostructure on the implant surface can enhance biocompatibility and cell proliferation. The Mg substrate that employed to surface modifications using the anodization/hydrothermal process with nanoplate structure shows flat and well-spread features among the nanoplates, as shown in **Figure 9**. Cell proliferation of the extraction of HA nanoplates on the Mg alloy surface finds higher cell proliferation. This can conclude that cells can modify their morphology to match the surface topography as shown in the inset images in **Figure 9**. These findings indicate that how surface modification can influence

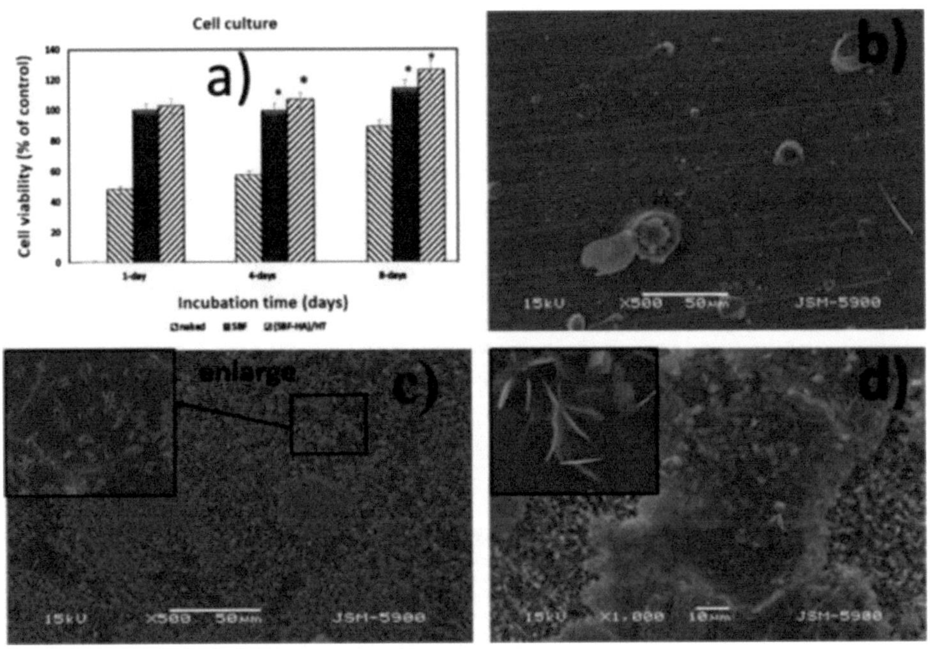

Figure 9. Figure shows the cell viability of anodized/hydrothermal treated Mg samples using cck-8 (a). FESEM images show cell attachment after 5 days (b); naked (c); (SBF-HA) (d); and (SBF-HA)/HT samples. Cell proliferation is presented in means ± STD ($n = 4$) based on ANOVA one-way test (*indicates $p < 0.05$).

surface bioactivity and cell adhesion to the implant interface. Implant surface adheres with the cells and eliminates the mismatch between the surface of the biomaterials and the connected tissue [65]. Extraction of anodized layers shows more cell viability and proliferation as shows in **Figure 10** using confocal microscopic images comparing to the bare substrates extraction.

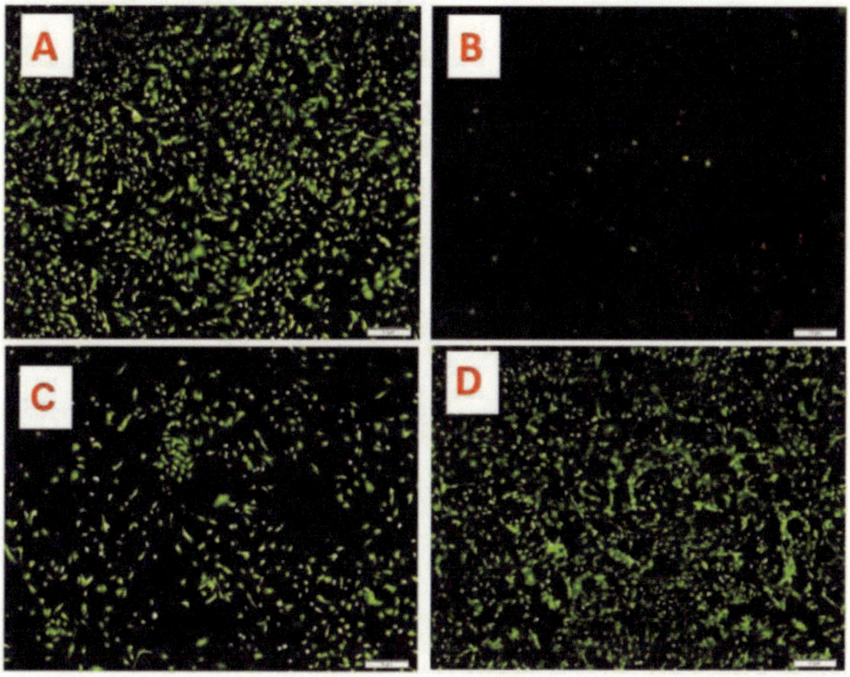

Figure 10. Microscopic florescent images for the live/dead cells of the (a) negative control, (b) positive control, (c) bare sample, (d) AZ31 Mg alloy anodized in SBF at 37°C.

7. Conclusions

Magnesium and its alloys are exhibit biodegradable in physiological media as well as its stiffness close to bone. In addition characteristics of Mg such as low weight, high specific strength and good biocompatibility bring a revolution in medical field toward new generation of biomaterials. However, the high degradation is accompanied by the hydrogen gas effect on the healing of the surrounded tissues. During its healing period, Mg implants lose their mechanical integrity before the bone heals due to the high degradation process. To overcome these limitations, different methods and techniques have been proposed to control the degradation rate of Mg to acceptable levels. Anodization as one of the surface modification techniques finds to increase the surface bioactivity and control degradation rate. In bone substitute Mg acts as a mechanical support during the healing process; moreover, the presence of apatite film on the surface of implant materials can enhance osseointegration of the defected bone. Furthermore, more research studies are devoted to Mg to be used in the future as implant materials in clinical application.

Acknowledgements

This chapter was supported by grant from the Basic Science Research Program through the National Research Foundation of Korea (NRF) by Ministry of Education, Science and Technology (Project no. 2016R1A2A2A07005160) and (Project no NRF-2015R1C1A1A02036404).

Author details

Hamouda M. Mousa[1,2], Chan Hee Park[1]* and Cheol Sang Kim[1]*

*Address all correspondence to: biochan@jbnu.ac.kr and chskim@jbnu.ac.kr

1 Department of Bionanosystem Engineering, Division of Mechanical Design Engineering, Chonbuk National University, Jeonju, Jeonbuk, Republic of Korea

2 Department of Engineering Materials and Mechanical Design, Faculty of Engineering, South Valley University, Qena, Egypt

References

[1] Zhang S, Zhang X, Zhao C, Li J, Song Y, Xie C, et al. Research on an Mg–Zn alloy as a degradable biomaterial. Acta Biomaterialia. 2010;6:626–40.

[2] Heublein B, Rohde R, Kaese V, Niemeyer M, Hartung W, Haverich A. Biocorrosion of magnesium alloys: a new principle in cardiovascular implant technology? Heart. 2003;89:651–6.

[3] Brar HS, Wong J, Manuel MV. Investigation of the mechanical and degradation properties of Mg–Sr and Mg–Zn–Sr alloys for use as potential biodegradable implant materials. Journal of the Mechanical Behavior of Biomedical Materials. 2012;7:87–95.

[4] Zhang E, Xu L, Yu G, Pan F, Yang K. In vivo evaluation of biodegradable magnesium alloy bone implant in the first 6 months implantation. Journal of Biomedical Materials Research Part A. 2009;90A:882–93.

[5] Witte F, Ulrich H, Rudert M, Willbold E. Biodegradable magnesium scaffolds: Part 1: Appropriate inflammatory response. Journal of Biomedical Materials Research Part A. 2007;81A:748–56.

[6] Staiger MP, Pietak AM, Huadmai J, Dias G. Magnesium and its alloys as orthopedic biomaterials: A review. Biomaterials. 2006;27:1728–34.

[7] Kirkland NT, Birbilis N, Walker J, Woodfield T, Dias GJ, Staiger MP. In-vitro dissolution of magnesium–calcium binary alloys: Clarifying the unique role of calcium additions in

bioresorbable magnesium implant alloys. Journal of Biomedical Materials Research Part B: Applied Biomaterials. 2010;95B:91–100.

[8] Witte F, Hort N, Vogt C, Cohen S, Kainer KU, Willumeit R, et al. Degradable biomaterials based on magnesium corrosion. Current Opinion in Solid State and Materials Science. 2008;12:63–72.

[9] Witte F, Kaese V, Haferkamp H, Switzer E, Meyer-Lindenberg A, Wirth CJ, et al. In vivo corrosion of four magnesium alloys and the associated bone response. Biomaterials. 2005;26:3557–63.

[10] Song G, Song S. A possible biodegradable magnesium implant material. Advanced Engineering Materials. 2007;9:298–302.

[11] Chen Y, Xu Z, Smith C, Sankar J. Recent advances on the development of magnesium alloys for biodegradable implants. Acta Biomaterialia. 2014;10:4561–73.

[12] Gray-Munro JE, Seguin C, Strong M. Influence of surface modification on the in vitro corrosion rate of magnesium alloy AZ31. Journal of Biomedical Materials Research Part A. 2009;91A:221–30.

[13] Zhang Y, Yan C, Wang F, Lou H, Cao C. Study on the environmentally friendly anodizing of AZ91D magnesium alloy. Surface and Coatings Technology. 2002;161:36–43.

[14] Guan R-G, Johnson I, Cui T, Zhao T, Zhao Z-Y, Li X, et al. Electrodeposition of hydroxyapatite coating on Mg-4.0Zn-1.0Ca-0.6Zr alloy and in vitro evaluation of degradation, hemolysis and cytotoxicity. Journal of Biomedical Materials Research Part A. 2012;100A:999–1015.

[15] Gao JH, Guan SK, Chen J, Wang LG, Zhu SJ, Hu JH, et al. Fabrication and characterization of rod-like nano-hydroxyapatite on MAO coating supported on Mg–Zn–Ca alloy. Applied Surface Science. 2011;257:2231–7.

[16] Chang WH, Qu B, Liao AD, Zhang SF, Zhang RF, Xiang JH. In vitro biocompatibility and antibacterial behavior of anodic coatings fabricated in an organic phosphate containing solution on Mg–1.0Ca alloys. Surface and Coatings Technology. 2016;289:75–81.

[17] Sabaghi Joni M, Fattah-alhosseini A. Effect of KOH concentration on the electrochemical behavior of coatings formed by pulsed DC micro-arc oxidation (MAO) on AZ31B Mg alloy. Journal of Alloys and Compounds. 2016;661:237–44.

[18] Dou J, Yu H, Chen C, Pan Y, Gao D, Zhang X. Formation of calcium phosphate coating on Mg-Zn-Ca alloy by micro-arc oxidation technique. Materials Letters. 2016;164:575–8.

[19] Sankara Narayanan TSN, Lee MH. Incorporation of ZrO_2 particles in the oxide layer formed on Mg by anodizing: Influence of electrolyte concentration and current modes. Journal of Colloid and Interface Science. 2016;464:36–47.

[20] Gu CD, Yan W, Zhang JL, Tu JP. Corrosion resistance of AZ31B magnesium alloy with a conversion coating produced from a choline chloride—Urea based deep eutectic solvent. Corrosion Science. 2016;106:108–16.

[21] Manoj Kumar R, Kuntal KK, Singh S, Gupta P, Bhushan B, Gopinath P, et al. Electrophoretic deposition of hydroxyapatite coating on Mg–3Zn alloy for orthopaedic application. Surface and Coatings Technology. 2016;287:82–92.

[22] Yu H, Dong Q, Dou J, Pan Y, Chen C. Preparation of Si-containing oxide coating and biomimetic apatite induction on magnesium alloy. Applied Surface Science. part A, 1 December 2016;388:148–154.

[23] Lin B, Zhong M, Zheng C, Cao L, Wang D, Wang L, et al. Preparation and characterization of dopamine-induced biomimetic hydroxyapatite coatings on the AZ31 magnesium alloy. Surface and Coatings Technology. 2015;281:82–8.

[24] Ren Y, Zhou H, Nabiyouni M, Bhaduri SB. Rapid coating of AZ31 magnesium alloy with calcium deficient hydroxyapatite using microwave energy. Materials Science and Engineering: C. 2015;49:364–72.

[25] Chen XB, Nisbet DR, Li RW, Smith PN, Abbott TB, Easton MA, et al. Controlling initial biodegradation of magnesium by a biocompatible strontium phosphate conversion coating. Acta Biomaterialia. 2014;10:1463–74.

[26] Qiu X, Wan P, Tan L, Fan X, Yang K. Preliminary research on a novel bioactive silicon doped calcium phosphate coating on AZ31 magnesium alloy via electrodeposition. Materials Science and Engineering: C. 2014;36:65–76.

[27] Zhang Y, Yan C, Wang F, Li W. Electrochemical behavior of anodized Mg alloy AZ91D in chloride containing aqueous solution. Corrosion Science. 2005;47:2816–31.

[28] Schultze JW, Lohrengel MM. Stability, reactivity and breakdown of passive films. Problems of recent and future research. Electrochimica Acta. 2000;45:2499–513.

[29] Khaselev O, Weiss D, Yahalom J. Structure and composition of anodic films formed on binary Mg–Al alloys in KOH–aluminate solutions under continuous sparking. Corrosion Science. 2001;43:1295–307.

[30] Ismail KM, Virtanen S. Electrochemical behavior of magnesium alloy AZ31 in 0.5 M KOH solution. Electrochemical and Solid-state Letters. 2007;10:C9–C11.

[31] Zhou X, Thompson Ge, Skeldon P, Wood GC, Shimizu K, Habazaki H. Film formation and detachment during anodizing of Al–Mg alloys. Corrosion Science. 1999;41:1599–613.

[32] Barchiche CE, Rocca E, Juers C, Hazan J, Steinmetz J. Corrosion resistance of plasma-anodized AZ91D magnesium alloy by electrochemical methods. Electrochimica Acta. 2007;53:417–25.

[33] Guo XW, Ding WJ, Lu C, Zhai CQ. Influence of ultrasonic power on the structure and composition of anodizing coatings formed on Mg alloys. Surface and Coatings Technology. 2004;183:359–68.

[34] Shi Z, Song G, Atrens A. Corrosion resistance of anodised single-phase Mg alloys. Surface and Coatings Technology. 2006;201:492–503.

[35] Turhan MC, Lynch RP, Jha H, Schmuki P, Virtanen S. Anodic growth of self-ordered magnesium oxy-fluoride nanoporous/tubular layers on Mg alloy (WE43). Electrochemistry Communications. 2010;12:796–9.

[36] Saji VS, Choe HC, Brantley WA. An electrochemical study on self-ordered nanoporous and nanotubular oxide on Ti–35Nb–5Ta–7Zr alloy for biomedical applications. Acta Biomaterialia. 2009;5:2303–10.

[37] Franz S, Rammelt S, Scharnweber D, Simon JC. Immune responses to implants – A review of the implications for the design of immunomodulatory biomaterials. Biomaterials. 2011;32:6692–709.

[38] Nukavarapu SP, Kumbar SG, Brown JL, Krogman NR, Weikel AL, Hindenlang MD, et al. Polyphosphazene/nano-hydroxyapatite composite microsphere scaffolds for bone tissue engineering. Biomacromolecules. 2008;9:1818–25.

[39] Guan J-J, Tian B, Tang S, Ke Q-F, Zhang C-Q, Zhu Z-A, et al. Hydroxyapatite coatings with oriented nanoplate arrays: synthesis, formation mechanism and cytocompatibility. Journal of Materials Chemistry B. 2015;3:1655–66.

[40] Yao Z, Li L, Jiang Z. Adjustment of the ratio of Ca/P in the ceramic coating on Mg alloy by plasma electrolytic oxidation. Applied Surface Science. 2009;255:6724–8.

[41] Srinivasan PB, Liang J, Blawert C, Störmer M, Dietzel W. Characterization of calcium containing plasma electrolytic oxidation coatings on AM50 magnesium alloy. Applied Surface Science. 2010;256:4017–22.

[42] Peixoto Barbosa D, Knörnschild G. Anodization of Mg-alloy AZ91 in NaOH solutions. Surface and Coatings Technology. 2009;203:1629–36.

[43] Wu H-l, Cheng Y-l, Li L-l, Chen Z-h, Wang H-m, Zhang Z. The anodization of ZK60 magnesium alloy in alkaline solution containing silicate and the corrosion properties of the anodized films. Applied Surface Science. 2007;253:9387–94.

[44] Mousa HM, Lee DH, Park CH, Kim CS. A novel simple strategy for in situ deposition of apatite layer on AZ31B magnesium alloy for bone tissue regeneration. Applied Surface Science. 2015;351:55–65.

[45] Mousa HM, Hussein KH, Woo HM, Park CH, Kim CS. One-step anodization deposition of anticorrosive bioceramic compounds on AZ31B magnesium alloy for biomedical application. Ceramics International. 2015;41:10861–70.

[46] Mousa HM, Tiwari AP, Kim J, Adhikari SP, Park CH, Kim CS. A novel in situ deposition of hydroxyapatite nanoplates using anodization/hydrothermal process onto magnesium alloy surface towards third generation biomaterials. Materials Letters. 2016; 164:144–7.

[47] Ye F, Guo H, Zhang H, He X. Polymeric micelle-templated synthesis of hydroxyapatite hollow nanoparticles for a drug delivery system. Acta Biomaterialia. 2010;6:2212–8.

[48] Paital SR, Dahotre NB. Calcium phosphate coatings for bio-implant applications: Materials, performance factors and methodologies. Materials Science and Engineering: R: Reports. 2009;66:1–70.

[49] Dorozhkin SV. Calcium orthophosphate coatings on magnesium and its biodegradable alloys. Acta Biomaterialia. 2014;10:2919–34.

[50] Kim S-M, Jo J-H, Lee S-M, Kang M-H, Kim H-E, Estrin Y, et al. Hydroxyapatite-coated magnesium implants with improved in vitro and in vivo biocorrosion, biocompatibility and bone response. Journal of Biomedical Materials Research Part A. 2014;102:429–41.

[51] Zheng YF, Gu XN, Witte F. Biodegradable metals. Materials Science and Engineering: R: Reports. 2014;77:1–34.

[52] Mousa HM, Hussein KH, Pant HR, Woo HM, Park CH, Kim CS. In vitro degradation behavior and cytocompatibility of a bioceramic anodization films on the biodegradable magnesium alloy. Colloids and Surfaces A: Physicochemical and Engineering Aspects. 2016;488:82–92.

[53] Myrissa A, Agha NA, Lu Y, Martinelli E, Eichler J, Szakács G, et al. In vitro and in vivo comparison of binary Mg alloys and pure Mg. Materials Science and Engineering: C. 2016;61:865–74.

[54] Ghali E, Dietzel W, Kainer K-U. General and localized corrosion of magnesium alloys: A critical review. Journal of Materials Engineering and Performance. 2004; 13:7–23.

[55] Ghali E, Dietzel W, Kainer K-U. Testing of general and localized corrosion of magnesium alloys: A critical review. Journal of Materials Engineering and Performance. 2004;13:517–29.

[56] Ballerini G, Bardi U, Bignucolo R, Ceraolo G. About some corrosion mechanisms of AZ91D magnesium alloy. Corrosion Science. 2005;47:2173–84.

[57] Song G. Recent progress in corrosion and protection of magnesium alloys. Advanced Engineering Materials. 2005;7:563–86.

[58] Yun Y, Dong Z, Lee N, Liu Y, Xue D, Guo X, et al. Revolutionizing biodegradable metals. Materials Today. 2009;12:22–32.

[59] Jones RH. The influence of hydrogen on the stress-corrosion cracking of low-strength Al-Mg alloys. JOM. 2003;55:42–6.

[60] Scamans GM, Holroyd NJH, Tuck CDS. The role of magnesium segregation in the intergranular stress corrosion cracking of aluminium alloys. Corrosion Science. 1987;27:329–47.

[61] Mousa HM, Hussein KH, Raslan AA, Lee J, Woo HM, Park CH, et al. Amorphous apatite thin film formation on a biodegradable Mg alloy for bone regeneration: strategy, characterization, biodegradation and in vitro cell study. RSC Advances. 2016;6:22563–74.

[62] Gallagher WM, Lynch I, Allen LT, Miller I, Penney SC, O'Connor DP, et al. Molecular basis of cell–biomaterial interaction: Insights gained from transcriptomic and proteomic studies. Biomaterials. 2006;27:5871–82.

[63] Williams DF. There is no such thing as a biocompatible material. Biomaterials. 2014; 35:10009–14.

[64] Abdal-hay A, Dewidar M, Lim JK. Biocorrosion behavior and cell viability of adhesive polymer coated magnesium based alloys for medical implants. Applied Surface Science. 2012;261:536–46.

[65] Jones FH. Teeth and bones: applications of surface science to dental materials and related biomaterials. Surface Science Reports. 2001;42:75–205.